A
FIELD GUIDE
TO LIES AND
STATISTICS

A FIELD GUIDE TO LIES AND STATISTICS

A Neuroscientist on How to
Make Sense of a Complex World

Daniel Levitin

VIKING
an imprint of
PENGUIN BOOKS

VIKING

UK | USA | Canada | Ireland | Australia
India | New Zealand | South Africa

Viking is part of the Penguin Random House group of companies
whose addresses can be found at global.penguinrandomhouse.com.

First published in the United States of America by Dutton as *A Field Guide to Lies* 2016
First published in Great Britain by Viking 2017
001

All art courtesy of the author unless otherwise noted.
Images on pp. 7, 80, 174, 185, 240 © 2016 by Dan Piraro, used by permission.
Image on p. 25 © 2016 by Alex Tabarrok, used by permission.
Image on p. 27 was drawn by the author, based on a figure under Creative Commons license
appearing on www.betterposters.blogspot.com.
Image on p. 49 © 2016 by Tyler Vigen, used by permission.
Image on p. 207 © Image redrawn by the author with permission based on a figure found at
AutismSpeaks.org.
Image on p. 247 is public domain and provided courtesy of Harrison Propser.

Set in 11.26/16.67 pt Minion Pro
Printed in Great Britain by Clays Ltd, St Ives plc

A CIP catalogue record for this book is available from the British Library

ISBN: 978–0–241–23999–5

To Shari,
whose inquisitive mind made me a better thinker

CONTENTS

PART THREE: EVALUATING THE WORLD

INTRODUCTION

THINKING, CRITICALLY

This is a book about how to spot problems with the facts you encounter, problems that may lead you to draw the wrong conclusions. Sometimes the people giving you the facts are hoping you'll draw the wrong conclusion; sometimes they don't know the difference themselves. Today, information is available nearly instantaneously, but it is becoming increasingly hard to tell what's true and what's not, to sift through the various claims we hear and to recognize when they contain misinformation, pseudo-facts, distortions, and outright lies.

There are many ways that we can be led astray by fast-talking, loose-writing purveyors of information. Here, I've grouped them into two categories, and they make the first two parts of this book: numerical and verbal. The first category includes mishandled statistics and graphs; the second includes faulty arguments, and the steps we can take to better evaluate news, statements, and reports. The last part of the book addresses what underlies our ability to determine if something is true or false: the scientific method. It grapples with the limits of what we can and cannot know, including what we know right now and don't know just yet, and includes some applications of logical thinking.

It is easy to lie with statistics and graphs because few people take the time to look under the hood and see how they work. I aim to fix that. Recognizing faulty arguments can help you to evaluate whether a chain of reasoning leads to a valid conclusion or not. Related to this is infoliteracy—recognizing that there are hierarchies in source quality, that pseudo-facts can easily masquerade as facts, and biases can distort the information we are being asked to consider, leading us to faulty conclusions.

You might object and say, "But it's not my job to evaluate statistics critically. Newspapers, bloggers, the government, Wikipedia, etc., should be doing that for us." Yes, they should, but they don't always. We—each of us—need to think critically and carefully about the numbers and words we encounter if we want to be successful at work, at play, and in making the most of our lives. This means checking the numbers, the reasoning, and the sources for plausibility and rigor. It means examining them as best as we can before we repeat them or use them to form an opinion. We want to avoid the extremes of gullibly accepting every claim we encounter or cynically rejecting every one. Critical thinking doesn't mean we disparage everything, it means that we try to distinguish between claims with evidence and those without.

Sometimes the evidence consists of numbers and we have to ask, "Where did those numbers come from? How were they collected?" Sometimes the numbers are ridiculous, but it takes some reflection to see it. Sometimes claims seem reasonable, but come from a source that lacks credibility, like a person who reports having witnessed a crime but wasn't actually there. This book can help you to avoid learning a whole lot of things that aren't so. And catch some lying weasels in their tracks.

We've created more human-made information in the last five

years than in all of human history before them. Unfortunately, found alongside things that are true is an enormous number of things that are not, in websites, videos, books, and on social media. This is not just a new problem. Misinformation has been a fixture of human life for thousands of years, and was documented in biblical times and classical Greece. The unique problem we face today is that misinformation has proliferated; it is devilishly entwined on the Internet with real information, making the two difficult to separate. And misinformation is promiscuous—it consorts with people of all social and educational classes, and turns up in places you don't expect it to. It propagates as one person passes it on to another and another, as Twitter, Facebook, Snapchat, and other social media grab hold of it and spread it around the world; the misinformation can take hold and become well known, and suddenly a whole lot of people are believing things that aren't so.

PART ONE

EVALUATING NUMBERS

It ain't what you don't know that gets you into trouble.
It's what you know for sure that just ain't so.

—MARK TWAIN

Plausibility

Statistics, because they are numbers, appear to us to be cold, hard facts. It seems that they represent facts given to us by nature and it's just a matter of finding them. But it's important to remember that *people* gather statistics. People choose what to count, how to go about counting, which of the resulting numbers they will share with us, and which words they will use to describe and interpret those numbers. Statistics are not facts. They are interpretations. And your interpretation may be just as good as, or better than, that of the person reporting them to you.

Sometimes, the numbers are simply wrong, and it's often easiest to start out by conducting some quick plausibility checks. After that, even if the numbers pass plausibility, three kinds of errors can lead you to believe things that aren't so: how the numbers were collected, how they were interpreted, and how they were presented graphically.

In your head or on the back of an envelope you can quickly determine whether a claim is plausible (most of the time). Don't just accept a claim at face value; work through it a bit.

When conducting plausibility checks, we don't care about the exact numbers. That might seem counterintuitive, but precision isn't important here. We can use common sense to reckon a lot of these: If Bert tells you that a crystal wineglass fell off a table and hit a thick carpet without breaking, that seems plausible. If Ernie says it fell off the top of a forty-story building and hit the pavement without breaking, that's not plausible. Your real-world knowledge, observations acquired over a lifetime, tells you so. Similarly, if someone says they are two hundred years old, or that they can consistently beat the roulette wheel in Vegas, or that they can run forty miles an hour, these are not plausible claims.

What would you do with this claim?

> In the thirty-five years since marijuana laws stopped being enforced in California, the number of marijuana smokers has doubled every year.

Plausible? Where do we start? Let's assume there was only one marijuana smoker in California thirty-five years ago, a very conservative estimate (there were half a million marijuana arrests nationwide in 1982). Doubling that number every year for thirty-five years would yield more than 17 billion—larger than the population of the entire world. (Try it yourself and you'll see that doubling every year for twenty-one years gets you to over a million: 1; 2; 4; 8; 16; 32; 64; 128; 256; 512; 1024; 2048; 4096; 8192; 16,384; 32,768; 65,536; 131,072; 262,144; 524,288; 1,048,576.) This claim isn't just implausible, then, it's impossible. Unfortunately, many people have trouble thinking clearly about numbers because they're intimidated by them. But as you see, nothing here requires more than elementary school arithmetic and some reasonable assumptions.

Here's another. You've just taken on a position as a telemarketer, where agents telephone unsuspecting (and no doubt irritated) prospects. Your boss, trying to motivate you, claims:

Our best salesperson made 1,000 sales a day.

Is this plausible? Try dialing a phone number yourself—the fastest you can probably do it is five seconds. Allow another five seconds for the phone to ring. Now let's assume that every call ends in a sale—clearly this isn't realistic, but let's give every advantage to this claim to see if it works out. Figure a minimum of ten seconds to make a pitch and have it accepted, then forty seconds to get the buyer's credit card number and address. That's one call per minute (5 + 5 + 10 + 40 = 60 seconds), or 60 sales in an hour, or 480 sales in a very hectic eight-hour workday with no breaks. The 1,000 just isn't plausible, allowing even the most optimistic estimates.

Some claims are more difficult to evaluate. Here's a headline from *Time* magazine in 2013:

More people have cell phones than toilets.

What to do with this? We can consider the number of people in the developing world who lack plumbing and the observation that many people in prosperous countries have more than one cell phone. The claim seems *plausible*—that doesn't mean we should accept it, just that we can't reject it out of hand as being ridiculous; we'll have to use other techniques to evaluate the claim, but it passes the plausibility test.

Sometimes you can't easily evaluate a claim without doing a bit of research on your own. Yes, newspapers and websites really ought to be doing this for you, but they don't always, and that's how runaway statistics take hold. A widely reported statistic some years ago was this:

> In the U.S., 150,000 girls and young women die of anorexia each year.

Okay—let's check its plausibility. We have to do some digging. According to the U.S. Centers for Disease Control, the annual number of deaths *from all causes* for girls and women between the ages of fifteen and twenty-four is about 8,500. Add in women from twenty-five to forty-four and you still only get 55,000. The anorexia deaths in one year cannot be three times the number of *all* deaths.

In an article in *Science*, Louis Pollack and Hans Weiss reported that since the formation of the Communication Satellite Corp.,

> The cost of a telephone call has decreased by 12,000 percent.

If a cost decreases by 100 percent, it drops to zero (no matter what the initial cost was). If a cost decreases by 200 percent, someone is paying *you* the same amount you used to pay *them* for you to take the product. A decrease of 100 percent is very rare; one of 12,000 percent seems wildly unlikely. An article in the peer-reviewed *Journal of Management Development* claimed a 200 percent reduction in customer complaints following a new customer care strategy.

Author Dan Keppel even titled his book *Get What You Pay For: Save 200% on Stocks, Mutual Funds, Every Financial Need*. He has an MBA. He should know better.

Of course, you have to apply percentages to the same baseline in order for them to be equivalent. A 50 percent reduction in salary cannot be restored by increasing your new, lower salary by 50 percent, because the baselines have shifted. If you were getting $1,000/week and took a 50 percent reduction in pay, to $500, a 50 percent increase in that pay only brings you to $750.

Percentages seem so simple and incorruptible, but they are often confusing. If interest rates rise from 3 percent to 4 percent, that is an increase of 1 percentage point, or 33 percent (because the 1 percent rise is taken against the baseline of 3, so 1/3 = .33). If interest rates fall from 4 percent to 3 percent, that is a decrease of 1 percentage point, but not a decrease of 33 percent—it's a decrease of 25 percent (because the 1 percentage point drop is now taken against the baseline of 4). Researchers and journalists are not always scrupulous about making this distinction between percentage point and percentages clear, but you should be.

The *New York Times* reported on the closing of a Connecticut textile mill and its move to Virginia due to high employment costs. The *Times* reported that employment costs, "wages, worker's compensation and unemployment insurance—are 20 times higher in Connecticut than in Virginia." Is this plausible? If it were true, you'd think that there would be a mass migration of companies out of Connecticut and into Virginia—not just this one mill—and that you would have heard of it by now. In fact, this was not true and the *Times* had to issue a correction. How did this happen? The reporter simply misread a company report. One cost, unemployment insurance, was in fact twenty times higher in Connecticut than in Virginia, but when factored in with other costs, total employment costs were really only 1.3 times higher in Connecticut, not 20 times higher. The reporter did not have training in business administration and we shouldn't expect her to. To catch these kinds of errors requires taking a step back and thinking for ourselves—which anyone can do (and she and her editors should have done).

New Jersey adopted legislation that denied additional benefits to mothers who have children while already on welfare. Some legislators believed that women were having babies in New Jersey simply to increase the amount of their monthly welfare checks. Within two months, legislators were declaring the "family cap" law a great success because births had already fallen by 16 percent. According to the *New York Times:*

> After only two months, the state released numbers suggesting that births to welfare mothers had already fallen by 16

percent, and officials began congratulating themselves on their overnight success.

Note that they're not counting pregnancies, but births. What's wrong here? Because it takes nine months for a pregnancy to come to term, any effect in the first two months cannot be attributed to the law itself but is probably due to normal fluctuations in the birth rate (birth rates are known to be seasonal).

Even so, there were other problems with this report that can't be caught with plausibility checks:

> . . . over time, that 16 percent drop dwindled to about 10 percent as the state belatedly became aware of births that had not been reported earlier. It appeared that many mothers saw no reason to report the new births since their welfare benefits were not being increased.

This is an example of a problem in the way statistics were collected—we're not actually surveying all the people that we think we are. Some errors in reasoning are sometimes harder to see coming than others, but we get better with practice. To start, let's look at a basic, often misused tool.

The pie chart is an easy way to visualize percentages—how the different parts of a whole are allocated. You might want to know what percentage of a school district's budget is spent on things like salaries, instructional materials, and maintenance. Or you might want to know what percentage of the money spent on instructional materials goes toward math, science, language arts, athletics, music, and so on. The cardinal rule of a pie chart is that the percentages have to add up to 100. Think about an actual pie—if there are nine people who each want an equal-sized piece, you can't cut it into eight. After you've reached the end of the pie, that's all there is. Still, this didn't stop Fox News from publishing this pie chart:

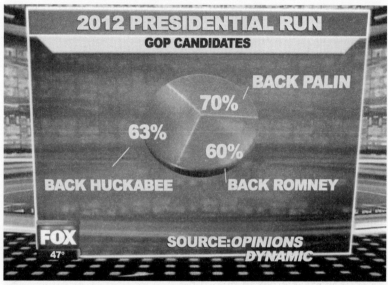

First rule of pie charts: The percentages have to add up to 100. (*Fox News, 2010*)

You can imagine how something like this could happen. Voters are given the option to report that they support more than one candidate. But then, the results shouldn't be presented as a pie chart.

FUN WITH AVERAGES

An average can be a helpful summary statistic, even easier to digest than a pie chart, allowing us to characterize a very large amount of information with a single number. We might want to know the average wealth of the people in a room to know whether our fund-raisers or sales managers will benefit from meeting with them. Or we might want to know the average price of gas to estimate how much it will cost to drive from Vancouver to Banff. But averages can be deceptively complex.

There are three ways of calculating an average, and they often yield different numbers, so people with statistical acumen usually avoid the word *average* in favor of the more precise terms *mean, median,* and *mode.* We don't say "mean average" or "median average" or simply just "average"—we say *mean, median,* or *mode.* In some cases, these will be identical, but in many they are not. If you see the word *average* all by itself, it's usually indicating the mean, but you can't be certain.

The mean is the most commonly used of the three and is calculated by adding up all the observations or reports you have and dividing by the number of observations or reports. For example, the average wealth of the people in a room is simply the total wealth divided by the number of people. If the room has ten people whose net worth is $100,000 each, the room has a total net worth of $1 million, and you can figure the mean without having to pull out a calculator: It is $100,000. If a different room has ten people whose net worth varies from $50,000 to $150,000 each, but totals $1 million, the mean is still $100,000 (because we simply take the total $1 million and divide by the ten people, regardless of what any individual makes).

The median is the middle number in a set of numbers (statisticians call this set a "distribution"): Half the observations are above it and half are below. Remember, the point of an average is to be able to represent a whole lot of data with a single number. The median does a better job of this when some of your observations are very, very different from the majority of them, what statisticians call *outliers*.

If we visit a room with nine people, suppose eight of them have a net worth of near $100,000 and one person is on the verge of bankruptcy with a net worth of negative $500,000, owing to his debts. Here's the makeup of the room:

Person 1: −$500,000
Person 2: $96,000
Person 3: $97,000
Person 4: $99,000
Person 5: $100,000
Person 6: $101,000
Person 7: $101,000
Person 8: $101,000
Person 9: $104,000

Now we take the sum and obtain a total of $299,000. Divide by the total number of observations, nine, and the mean is $33,222 per person. But the mean doesn't seem to do a very good job of characterizing the room. It suggests that your fund-raiser might not want to visit these people, when it's really only one odd person, one outlier, bringing down the average. This is the problem with the mean: It is sensitive to outliers.

The median here would be $100,000: Four people make less than that amount, and four people make more. The mode is $101,000, the number that appears more often than the others. Both the median and the mode are more helpful in this particular example.

There are many ways that averages can be used to manipulate what you want others to see in your data.

Let's suppose that you and two friends founded a small start-up company with five employees. It's the end of the year and you want to report your finances to your employees, so that they can feel good about all the long hours and cold pizzas they've eaten, and so that you can attract investors. Let's say that four employees—programmers—each earned $70,000 per year, and one employee—a receptionist/office manager—earned $50,000 per year. That's an average (mean) employee salary of $66,000 per year (4 × $70,000) + (1 × $50,000), divided by 5. You and your two friends each took home $100,000 per year in salary. Your payroll costs were therefore (4 × $70,000) + (1 × $50,000) + (3 × $100,000) = $630,000. Now, let's say your company brought in $210,000 in profits and you divided it equally among you and your co-founders as bonuses, giving you $100,000 + $70,000 each. How are you going to report this?

You could say:

Average salary of employees: $66,000
Average salary + profits of owners: $170,000

This is true but probably doesn't look good to anyone except you and your mom. If your employees get wind of this, they may feel undercompensated. Potential investors may feel that the founders are overcompensated. So instead, you could report this:

Average salary of employees: $66,000
Average salary of owners: $100,000
Profits: $210,000

That looks better to potential investors. And you can just leave out the fact that you divided the profits among the owners, and leave out that last line—that part about the profits—when reporting things to your employees. The four programmers are each going to think they're very highly valued, because they're making more than the average. Your poor receptionist won't be so happy, but she no doubt knew already that the programmers make more than she does.

Now suppose you are feeling overworked and want to persuade your two partners, who don't know much about critical thinking, that you need to hire more employees. You could do what many companies do, and report the "profits per employee" by dividing the $210,000 profit among the five employees:

Average salary of employees: $66,000
Average salary of owners: $100,000
Annual profits per employee: $42,000

Now you can claim that 64 percent of the salaries you pay to employees (42,000/66,000) comes back to you in profits, meaning you end up only having to pay 36 percent of their salaries after all those profits roll in. Of course, there is nothing in these figures to suggest that adding an employee will increase the profits—your profits may not be at all a function of how many employees there are—but for someone who is not thinking critically, this sounds like a compelling reason to hire more employees.

Finally, what if you want to claim that you are an unusually just and fair employer and that the difference between what you take in profits and what your employees earn is actually quite reasonable? Take the $210,000 in profits and distribute $150,000 of it as salary bonuses to you and your partners, saving the other $60,000 to report as "profits." This time, compute the average salary but include you and your partners in it with the salary bonuses.

> Average salary: $97,500
> Average profit of owners: $20,000

Now for some real fun:

> Total salary costs plus bonuses: $840,000
> Salaries: $780,000
> Profits: $60,000

That looks quite reasonable now, doesn't it? Of the $840,000 available for salaries and profits, only $60,000 or 7 percent went into owners' profits. Your employees will think you above reproach—who would begrudge a company owner from taking 7 percent? And it's actually not even that high—the 7 percent is divided among the three company owners to 2.3 percent each. Hardly worth complaining about!

You can do even better than this. Suppose in your first year of operation, you had only part-time employees, earning $40,000 per year. By year two, you had only full-time employees, earning the $66,000 mentioned above. You can honestly claim that average employee earnings went up 65 percent. What a great employer you are! But here you are glossing over the fact that you are comparing part-time with full-time. You would not be the first: U.S. Steel did it back in the 1940s.

In criminal trials, the way the information is presented—the framing—profoundly affects jurors' conclusions about guilt. Although they are mathematically equivalent, testifying that "the probability the suspect would match the blood drops if he were not their source is only 0.1 percent" (one in a thousand) turns out to be far more persuasive than saying "one in a thousand people in Houston would also match the blood drops."

Averages are often used to express outcomes, such as "one in X marriages ends in divorce." But that doesn't mean that statistic will apply on your street, in your bridge club, or to anyone you know. It might or might not—it's a nationwide average, and there might be certain *vulnerability factors* that help to predict who will and who will not divorce.

Similarly, you may read that one out of every five children born is Chinese. You note that the Swedish family down the street already has four children and the mother is expecting another child. This does not mean she's about to give birth to a Chinese baby—the one out of five children is on average, across all births in the world, not the births restricted to a particular house or particular neighborhood or even particular country.

Be careful of averages and how they're applied. One way that they can fool you is if the average combines samples from disparate populations. This can lead to absurd observations such as:

On average, humans have one testicle.

This example illustrates the difference between mean, median, and mode. Because there are slightly more women than men in the world, the median and mode are both zero, while the mean is close to one (perhaps 0.98 or so).

Also be careful to remember that the average doesn't tell you anything about the range. The average annual temperature in Death Valley, California, is a comfortable 77 degrees F (25 degrees C). But the range can kill you, with temperatures ranging from 15 degrees to 134 degrees on record.

Or . . . I could tell you that the *average* wealth of a hundred people in a room is a whopping $350 million. You might think this is the place to unleash a hundred of your best salespeople. But the room could have Mark Zuckerberg (net worth $35 billion) and ninety-nine people who are indigent. The average can smear across differences that are important.

Another thing to watch out for in averages is the *bimodal distribution*. Remember, the *mode* is the value that occurs most often. In many biological, physical, and social datasets, the distribution has two or more peaks—that is, two or more values that appear more than the others.

Bimodal Distribution

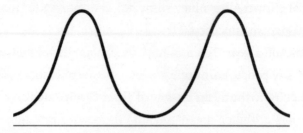

For example, a graph like this might show the amount of money spent on lunches in a week (x-axis) and how many people spent that amount (y-axis). Imagine that you've got two different groups of people in your survey, children (left hump—they're buying school lunches) and business executives (right hump—they're going to fancy restaurants). The mean and median here could be a number somewhere right between the two, and would not tell us very much about what's really going on—in fact, the mean and median in many cases are amounts that nobody spends. A graph like this is often a clue that there is heterogeneity in your sample, or that you are comparing apples and oranges. Better here is to report that it's a bimodal distribution and report the two modes. Better yet, subdivide the group into two groups and provide statistics for each.

But be careful drawing conclusions about individuals and groups based on averages. The pitfalls here are so common that they have names: the ecological fallacy and the exception fallacy. The ecological fallacy occurs when we make inferences about an individual based on aggregate data (such as a group mean), and the exception fallacy occurs when we make inferences about a group based on knowledge of a few exceptional individuals.

For example, imagine two small towns, each with only one hundred people. Town A has ninety-nine people earning $80,000 a year, and one super-wealthy person who struck oil on her property, earning $5,000,000 a year. Town B has fifty people earning $100,000 a year and fifty people earning $140,000. The mean income of Town A is $129,200 and the mean income of Town B is $120,000. Although Town A has a higher mean income, in ninety-nine out of one hundred cases, any individual you select randomly from Town B will have a higher income than an individual selected randomly from Town A. The ecological fallacy is thinking that if you select someone at random from the group with the higher mean, that individual is likely to have a higher income. The neat thing is, in the examples above, that it's not just the *mean* that is higher in Town B but also the *median* and the *mode*. (It doesn't always work out that way.)

As another example, it has been suggested that wealthy individuals are more likely to vote Republican, but evidence shows that the wealthier states tend to vote Democratic. The wealth of those wealthier states may be skewed by a small percentage of super-wealthy individuals. During the 2004 U.S. presidential election, the Republican candidate, George W. Bush, won the fifteen poorest states, and the Democratic candidate, John Kerry, won nine of the eleven wealthiest states. However, 62 percent of those with annual incomes over $200,000 voted for Bush, whereas only 36 percent of voters with annual incomes of $15,000 or less voted for Bush.

As an example of the exception fallacy, you may have read that Volvos are among the most reliable automobiles and so you decide to buy one. On your way to the dealership, you pass a Volvo mechanic and find a parking lot full of Volvos in need of repair. If you change your mind about buying a Volvo based on seeing this, you're using a relatively small number of exceptional cases to form an inference about the entire group. No one was claiming that Volvos never need repair, only that they're less likely to in the aggregate. (Hence the ubiquitous cautionary note in advertising that "individual performance may vary.") Note also that you're being unduly influenced by this in another way: The one place that Volvos needing repair will be is at a Volvo mechanic. Your "base rate" has shifted, and you cannot consider this a random sample.

Now that you're an expert on averages, you shouldn't fall for the famous misunderstanding that people tended not to live as long a hundred years ago as they do today. You've probably read that life expectancy has steadily increased in modern times. For those born in 1850, the average life expectancy for males and females was thirty-eight and forty years respectively, and for those born in 1990 it is seventy-two and seventy-nine. There's a tendency to think, then, that in the 1800s there just weren't that many fifty- and sixty-year-olds walking around because people didn't live that long. But in fact, people did live that long—it's just that infant and childhood mortality was so high that it skewed the average. If you could make it past twenty, you could live a long life back then. Indeed, in 1850 a fifty-year-old white female could expect to live to be 73.5, and a sixty-year-old could expect to live to be seventy-seven. Life expectancy has certainly increased for fifty- and sixty-year-olds today, by about ten years compared to 1850, largely due to better health care.

But as with the examples above of a room full of people with wildly different incomes, the changing averages for life expectancy at birth over the last 175 years reflect significant differences in the two samples: There were many more infant deaths back then pulling down the average.

Here is a brain-twister: The average child usually doesn't come from the average family. Why? Because of shifting baselines. (I'm using "average" in this discussion instead of "mean" out of respect for a wonderful paper on this topic by James Jenkins and Terrell Tuten, who used it in their title.)

Now, suppose you read that the average number of children per family in a suburban community is three. You might conclude then that the average child must have two siblings. But this would be wrong. This same logical problem applies if we ask whether the average college student attends the average-sized college, if the average employee earns the average salary, or if the average tree comes from the average forest. What?

All these cases involve a shift of the baseline, or sample group we're studying. When we calculate the average number of children per family, we're sampling families. A very large family and a small family each count as one family, of course. When we calculate the average (mean) number of siblings, we're sampling children. Each child in the large family gets counted once, so that the number of siblings each of them has weighs heavily on the average for sibling number. In other words, a family with ten children counts only one time in the average *family* statistic, but counts ten times in the average *number of siblings* statistic.

Suppose in one neighborhood of this hypothetical community

there are thirty families. Four families have no children, six families have one child, nine families have two children, and eleven families have six children. The average number of children per family is three, because ninety (the total number of children) gets divided by thirty (the total number of families).

But let's look at the average number of siblings. The mistake people make is thinking that if the average family has three children, then each child must have two siblings on average. But in the one-child families, each of the six children has zero siblings. In the two-child families, each of the eighteen children has one sibling. In the six-child families each of the sixty-six children has five siblings. Among the 90 children, there are 348 siblings. So although the average *child* comes from a family with three children, there are 348 siblings divided among 90 children, or an average of nearly four siblings per child.

	Families	# Children/ Family	Total # Children	Siblings
	4	0	0	0
	6	1	6	0
	9	2	18	18
	11	6	66	330
Totals	30		90	348

Average children per family: 3.0
Average siblings per child: 3.9

4 Families with 0 children

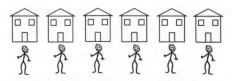

6 Families with 1 child — 6 children with 0 siblings

9 Families with 2 children — 18 children with 1 sibling

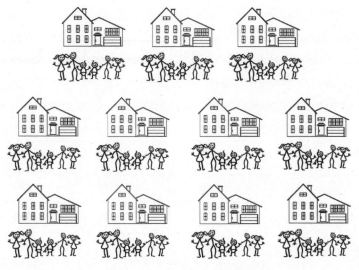

11 Families with 6 children — 66 children with 5 siblings

Consider now college size. There are many very large colleges in the United States (such as Ohio State and Arizona State) with student enrollment of more than 50,000. There are also many small colleges, with student enrollment under 3,000 (such as Kenyon College and Williams College). If we count up *schools*, we might find that the average-sized college has 10,000 students. But if we count up students, we'll find that the average student goes to a college with greater than 30,000 students. This is because, when counting students, we'll get many more data points from the large schools. Similarly, the average person doesn't live in the average city, and the average golfer doesn't shoot the average round (the total strokes over eighteen holes).

These examples involve a shift of baseline, or denominator. Consider another involving the kind of skewed distribution we looked at earlier with child mortality: The average investor does not earn the average return. In one study, the average return on a $100 investment held for thirty years was $760, or 7 percent per year. But 9 percent of the investors lost money, and a whopping 69 percent failed to reach the average return. This is because the average was skewed by a few people who made much greater than the average— in the figure below, the *mean* is pulled to the right by those lucky investors who made a fortune.

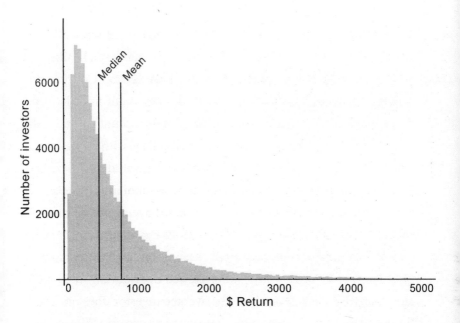

Payoff outcomes for return on a $100 investment over thirty years. Note that most people make less than the mean return, and a lucky few make more than five times the mean return.

Axis Shenanigans

The human brain did not evolve to process large amounts of numerical data presented as text; instead, our eyes look for patterns in data that are visually displayed. The most accurate but least interpretable form of data presentation is to make a table, showing every single value. But it is difficult or impossible for most people to detect patterns and trends in such data, and so we rely on graphs and charts. Graphs come in two broad types: Either they represent every data point visually (as in a scatter plot) or they implement a form of data reduction in which we summarize the data, looking, for example, only at means or medians.

There are many ways that graphs can be used to manipulate, distort, and misrepresent data. The careful consumer of information will avoid being drawn in by them.

Unlabeled Axes

The most fundamental way to lie with a statistical graph is to not label the axes. If your axes aren't labeled, you can draw or plot anything you want! Here is an example from a poster presented at a conference by a student researcher, which looked like this (I've redrawn it here):

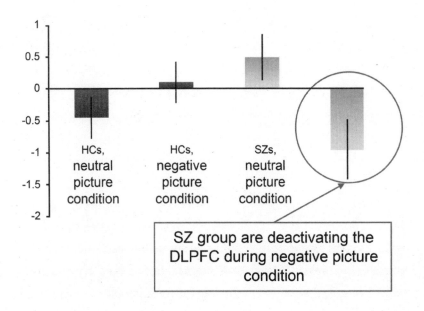

SZ group are deactivating the DLPFC during negative picture condition

What does all that mean? From the text on the poster itself (though not on this graph), we know that the researchers are studying brain activations in patients with schizophrenia (SZ). What are HCs? We aren't told, but from the context—they're being compared with SZ—we might assume that it means "healthy controls." Now, there do appear to be differences between the HCs and the SZs, but, hmmm . . . the y-axis has numbers, but . . . the units could be anything! What are we looking at? Scores on a test, levels of brain activations, number of brain regions activated? Number of Jell-O brand pudding cups they've eaten, or number of Johnny Depp movies they've seen in the last six weeks? (To be fair, the researchers subsequently published their findings in a peer-reviewed journal, and corrected this error after a website pointed out the oversight.)

In the next example, gross sales of a publishing company are plotted, excluding data from Kickstarter campaigns.

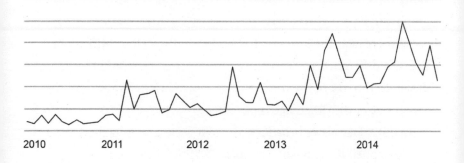

Gross Sales Excluding Kickstarter

2010   2011   2012   2013   2014

As in the previous example, but this time with the x-axis, we have numbers but we're not told what they are. In this case, it's probably self-evident: We assume that the 2010, 2011, etc., refer to calendar or fiscal years of operation, and the fact that the lines are jagged between the years suggests that the data are being tracked monthly (but without proper labeling we can only assume). The y-axis is completely missing, so we don't know what is being measured (is it units sold or dollars?), and we don't know what each horizontal line represents. The graph could be depicting an increase of sales from 50 cents a year to $5 a year, or from 50 million to 500 million units. Not to worry—a helpful narrative accompanied this graph: "It's been another great year." I guess we'll have to take their word for it.

Truncated Vertical Axis

A well-designed graph clearly shows you the relevant end points of a continuum. This is especially important if you're documenting some actual or projected change in a quantity, and you want your readers to draw the right conclusions. If you're representing crime rate, deaths, births, income, or any quantity that could take on a value of zero, then zero should be the minimum point on your graph. But if your aim is to create panic or outrage, start your y-axis somewhere near the lowest value you're plotting—this will emphasize the difference you're trying to highlight, because the eye is drawn to the size of the difference as shown on the graph, and the actual size of the difference is obscured.

In 2012, Fox News broadcast the following graph to show what would happen if the Bush tax cuts were allowed to expire:

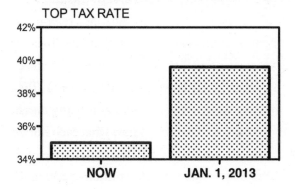

The graph gives the visual impression that taxes would increase by a large amount: The right-hand bar is six times the height of the left-hand bar. Who wants their taxes to go up by a factor of six? Viewers who are number-phobic, or in a hurry, may not take the time to examine the axis to see that the actual difference is between a tax rate of 35 percent and one of 39.6 percent. That is, if the cuts expire, taxes will only increase 13 percent, not the 600 percent that is pictured (the 4.6 percentage point increase is 13 percent of 35 percent).

If the y-axis started at zero, the 13 percent would be apparent visually:

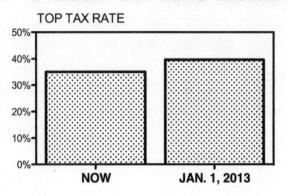

IF BUSH TAX CUTS EXPIRE

TOP TAX RATE

Discontinuity in Vertical or Horizontal Axis

Imagine a city where crime has been growing at a rate of 5 percent per year for the last ten years. You might graph it this way:

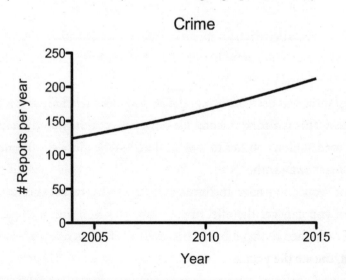

Nothing wrong with that. But suppose that you're selling home security systems and so you want to scare people into buying your product. Using all the same data, just create a discontinuity in your x-axis. This will distort the truth and deceive the eye marvelously:

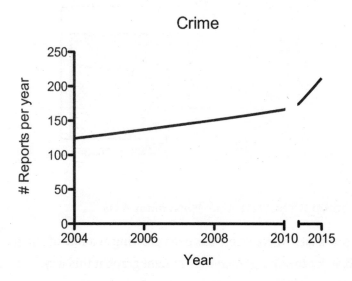

Crime

Here, the visual gives the impression that crime has increased dramatically. But you know better. The discontinuity in the x-axis crams five years' worth of numbers into the same amount of graphic real estate as was used for two years. No wonder there's an apparent increase. This is a fundamental flaw in graph making, but because most readers don't bother to look at the axes too closely, this one's easy to get away with.

And you don't have to limit your creativity to breaking the x-axis; you can get the effect by creating a discontinuity in the y-axis, and then hiding it by not breaking the line. While we're at it, we'll truncate the y-axis:

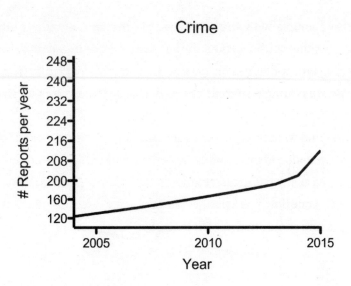

This is a bit mean. Most readers just look at that curve within the plot frame and won't notice that the tick marks on the vertical axis start out being forty reports between each, and then suddenly, at two hundred, indicate only eight reports between each. Are we having fun yet?

The honorable move is to use the first crime graph presented with the proper continuous axis. Now, to critically evaluate the statistics, you might ask if there are factors in the way the data were collected or presented that could be hiding an underlying truth.

One possibility is that the increases occur in only one particularly bad neighborhood and that, in fact, crime is *decreasing* everywhere else in the city. Maybe the police and the community have simply decided that a particular neighborhood had become unmanageable and so they stopped enforcing laws there. The city as a whole is safe—perhaps even safer than before—and one bad neighborhood is responsible for the increase.

Another possibility is that by amalgamating all the different sorts of complaints into the catchall bin of *crime*, we are overlooking a serious consideration. Perhaps *violent crime* has dropped to almost zero, and in its place, with so much time on their hands, the police are issuing hundreds more jaywalking tickets.

Perhaps the most obvious question to ask next, in your effort to understand what this statistic really means, is "What happened to the *total population* in this city during that time period?" If the population increased at any rate greater than 5 percent per year, the crime rate has actually gone down on a per-person basis. We could show this by plotting crimes committed per ten thousand people in the city:

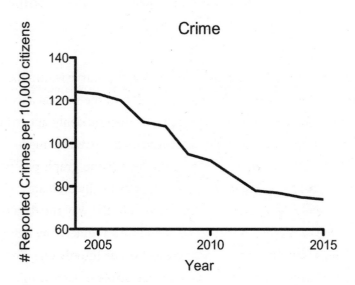

Choosing the Proper Scale and Axis

You've been hired by your local Realtor to graph the change in home prices in your community over the last decade. The prices have been steadily growing at a rate of 15 percent per year.

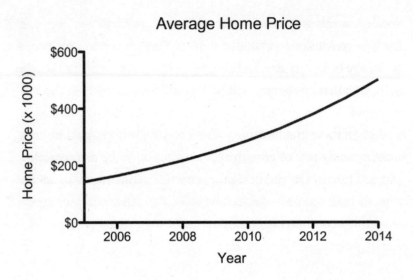

If you want to really alarm people, why not change the x-axis to include dates that you don't have data for? Adding extra dates to the x-axis artificially like this will increase the slope of the curve by compressing the viewable portion like this:

Notice how this graph tricks your eye (well, your brain) into drawing two false conclusions—first, that sometime around 1990 home prices must have been very low, and second, that by 2030 home prices will be so high that few people will be able to afford a home. Better buy one now!

Both of these graphs distort what's really going on, because they make a steady rate of growth appear, visually, to be an increasing rate of growth. On the first graph, the 15 percent growth seems twice as high on the y-axis in 2014 as it does in 2006. Many things change at a constant rate: salaries, prices, inflation, population of a species, and victims of diseases. When you have a situation of steady growth (or decline), the most accurate way to represent the data is on a logarithmic scale. The logarithmic scale allows equal percentage changes to be represented by equal distances on the y-axis. A constant annual rate of change then shows up as a straight line, as this:

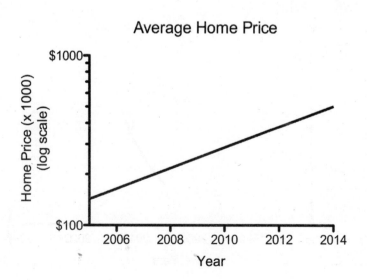

The Dreaded Double Y-Axis

The graph maker can get away with all kinds of lies simply armed with the knowledge that most readers will not look at the graph very closely. This can move a great many people to believe all kinds of things that aren't so. Consider the following graph, showing the life expectancy of smokers versus nonsmokers at age twenty-five.

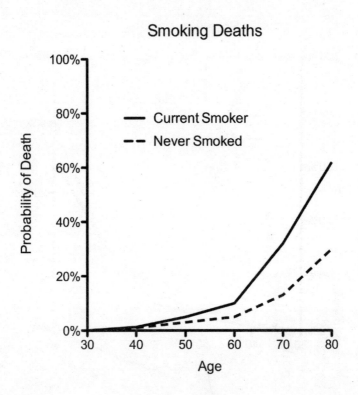

This makes clear two things: The dangers of smoking accumulate over time, and smokers are likely to die earlier than nonsmokers. The difference isn't big at age forty, but by age eighty the risk more than doubles, from under 30 percent to over 60 percent. This is a clean and accurate way to present the data. But suppose you're a young fourteen-year-old smoker who wants to convince your parents that you should be allowed to smoke. This graph is clearly not going to help you. So you dig deep into your bag of tricks and use the double y-axis, adding a y-axis to the right-hand side of the graph frame, with a different scaling factor that applies only to the non-smokers. Once you do that, your graph looks like this:

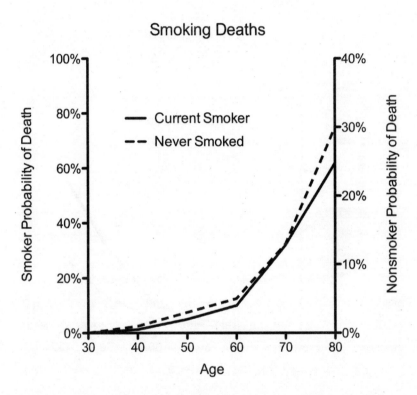

From this, it looks like you're just as likely to die from smoking as from not smoking. Smoking won't harm you—old age will! The trouble with double y-axis graphs is that you can always scale the second axis any way that you choose.

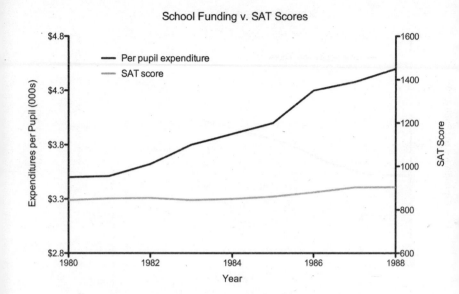

School Funding v. SAT Scores

Forbes magazine, a venerable and typically reliable news source, ran a graph very much like this one to show the relation between expenditures per public school student and those students' scores on the SAT, a widely used standardized test for college admission in the United States.

From the graph, it looks as though increasing the money spent per student (black line) doesn't do anything to increase their SAT scores (gray line). The story that some anti–government spending politicos could tell about this is one of wasted taxpayer funds. But you now understand that the choice of scale for the second (right-hand) y-axis is arbitrary. If you were a school administrator, you might simply take the exact same data, change the scale of the right-hand axis, and voilà—increasing spending delivers a better education, as evidenced by the increase in SAT scores!

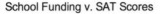

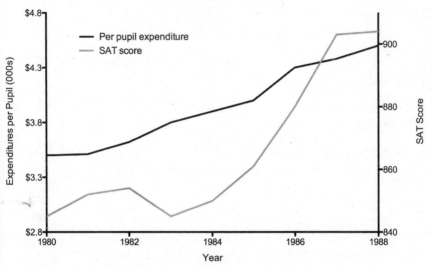

School Funding v. SAT Scores

This graph obviously tells a very different story. Which one is true? You'd need to have a measure of how the one variable changes as a function of the other, a statistic known as a correlation. Correlations range from –1 to 1. A correlation of 0 means that one variable is not related to the other at all. A correlation of -1 means that as one variable goes up, the other goes down, in precise synchrony. A correlation of 1 means that as one variable goes up, the other does too, also in precise synchrony. The first graph appears to be illustrating a correlation of 0, the second graph appears to be representing one that is close to 1. The actual correlation for this dataset is .91, a very strong correlation. Spending more on students is, at least in this dataset, associated with better SAT scores.

The correlation also provides a good estimate of how much of the result can be explained by the variables you're looking at. The correlation of .91 tells us we can explain 91 percent of students' SAT scores by looking at the amount of school expenditures per student. That is, it tells us to what extent expenditures explain the diversity in SAT scores.

A controversy about the double y-axis graph erupted in the fall of 2015 during a U.S. congressional committee meeting. Rep. Jason Chaffetz presented a graph that plotted two services provided by the organization Planned Parenthood: abortions, and cancer screening and prevention:

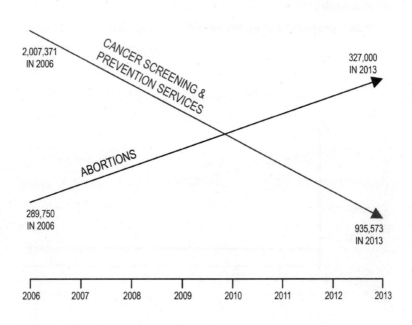

PLANNED PARENTHOOD FEDERATION OF AMERICA:
ABORTIONS UP — LIFE-SAVING PROCEDURES DOWN

2,007,371 IN 2006

CANCER SCREENING & PREVENTION SERVICES

327,000 IN 2013

ABORTIONS

289,750 IN 2006

935,573 IN 2013

2006   2007   2008   2009   2010   2011   2012   2013

The congressman was attempting to make a political point, that over a seven-year period, Planned Parenthood has increased the number of abortions it performed (something he opposes) and decreased the number of cancer screening and prevention procedures. Planned Parenthood doesn't deny this, but this distorted graph makes it seem that the number of abortion procedures exceeded those for cancer. Maybe the graph maker was feeling a bit guilty and so included the actual numbers next to the data points. Let's accept her bread crumbs and look closely. The number of abortions in 2013, the most recent year given, is 327,000. The number of cancer services was nearly three times that, at 935,573. (By the way, it's a bit suspicious that the abortion numbers are such tidy, round numbers while the cancer numbers are so precise.) This is a particularly sinister example: an implied double y-axis graph with no axes on either side!

Drawn properly, the graph would look like this:

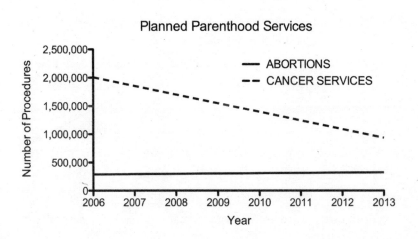

Here, we see that abortions increased modestly, compared to the reduction in cancer services.

There is another thing suspicious about the original graph: Such smooth lines are rarely found in data. It seems more likely that the graph maker simply took numbers for two particular years, 2006 and 2013, and compared them, drawing a smooth connecting line between them. Perhaps these particular years were chosen intentionally to emphasize differences. Perhaps there were great fluctuations in the intervening years of 2007–2012; we don't know. The smooth lines give the impression of a perfectly linear (straight line) function, which is very unlikely.

Graphs such as this do not always tell the story that people think they do. Is there something that could account for these data, apart from a narrative that Planned Parenthood is on a mission to perform as many abortions as it can (and to let people die of cancer at the same time)? Look at the second graph. In 2006, Planned Parenthood performed 2,007,371 cancer services, and 289,750 abortions, nearly seven times as many cancer services as abortions. By 2013, this gap had narrowed, but the number of cancer services was still nearly three times the number of abortions.

Cecile Richards, the president of Planned Parenthood, had an explanation for this narrowing gap. Changing medical guidelines for some anti-cancer services, like Pap smears, reduced the number of people for whom screening was recommended. Other changes, such as social attitudes about abortion, changing ages of the population, and increased access to health care alternatives, all influence these numbers, and so the data presented do not prove that Planned Parenthood has a pro-abortion agenda. It might—these data are just not the proof.

HIJINKS WITH HOW NUMBERS ARE REPORTED

You're trying to decide whether to buy stock in a new soft drink and you come across this graph of the company's sales figures in their annual report:

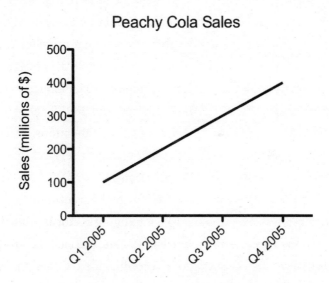

This looks promising—Peachy Cola is steadily increasing its sales. So far, so good. But a little bit of world knowledge can be applied here to good effect. The soft-drink market is very competitive. Peachy Cola's sales are increasing, but maybe not as quickly as a competitor's. As a potential investor, what you really want to see is how Peachy's sales compare to those of other companies, or to see their sales as a function of market share—Peachy's sales could go up only slightly while the market is growing enormously, and competitors are benefiting more than Peachy is. And, as this example of a useful double y-axis graph demonstrates, this may not bode well for their future:

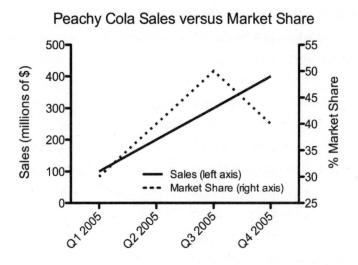

Peachy Cola Sales versus Market Share

Although unscrupulous graph makers can monkey with the scaling of the right-hand axis to make the graph appear to show anything they want, this kind of double-y-axis graph isn't scandalous because the two y-axes are representing different things, quantities that *couldn't* share an axis. This was not the case with the Planned Parenthood graph on page 40, which was reporting the same quantity on the two different axes, the number of performed procedures. That graph was distorted by ensuring that the two axes, although they measure the same thing, were scaled differently in order to manipulate perception.

It would also be useful to see Peachy's profits: Through manufacturing and distribution efficiencies, it may well be that they're making more money on a lower sales volume. Just because someone quotes you a statistic or shows you a graph, it doesn't mean it's relevant to the point they're trying to make. It's the job of all of us to make sure we get the information that matters, and to ignore the information that doesn't.

Let's say that you work in the public-affairs office for a company that manufactures some kind of device—frabezoids. For the last several years, the public's appetite for frabezoids has been high, and sales have increased. The company expanded by building new facilities, hiring new employees, and giving everyone a raise. Your boss comes into your cubicle with a somber-looking expression and explains that the newest sales results are in, and frabezoid sales have dropped 12 percent from the previous quarter. Your company's president is about to hold a big press conference to talk about the future of the company. As is his custom, he'll display a large graph on the stage behind him showing how frabezoids are doing. If word gets out about the lower sales figures, the public may think that frabezoids are no longer desirable things to have, which could then lead to an even further decline in sales.

What do you do? If you graph the sales figures honestly for the past four years, your graph would look like this:

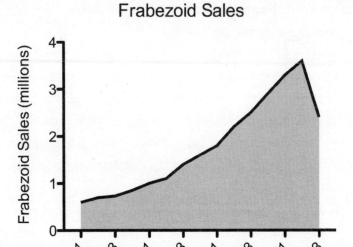

That downward trend in the curve is the problem. If only there were a way to make that curve go up.

Well, there is! The cumulative sales graph. Instead of graphing sales per quarter, graph the cumulative sales per quarter—that is, the total sales to date.

As long as you sold only one frabezoid, your cumulative graph will increase, like this one here:

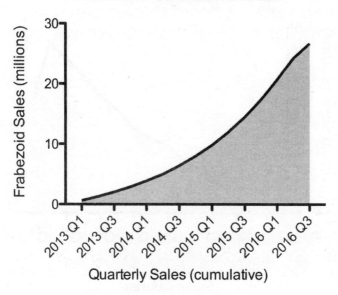

If you look carefully, you can still see a vestige of the poor sales for last quarter: Although the line is still going up for the most recent quarter, it's going up less steeply. That's your clue that sales have dropped. But our brains aren't very good at detecting rates of change such as these (what's called the first derivative in calculus, a fancy name for the slope of the line). So on casual examination, it seems the company continues to do fabulously well, and you've made a whole lot of consumers believe that frabezoids are still the hottest thing to have.

This is exactly what Tim Cook, CEO of Apple, did recently in a presentation on iPhone sales.

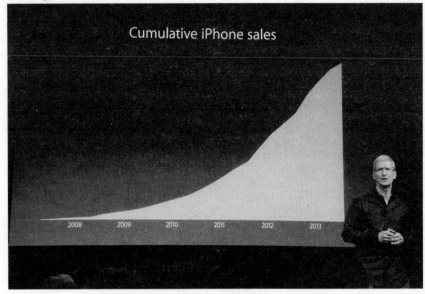

© 2013 The Verge, Vox Media Inc. (live.theverge.com/apple-iphone-5s-liveblog/)

Plotting Things That Are Unrelated

There are so many things going on in the world that some coincidences are bound to happen. The number of green trucks on the road may be increasing at the same time as your salary; when you were a kid, the number of shows on television may have increased with your height. But that doesn't mean that one is causing the other. When two things are related, whether or not one causes the other, statisticians call it a correlation.

The famous adage is that "correlation does not imply causation." In formal logic there are two formulations of this rule:

1) *Post hoc, ergo propter hoc* (after this, therefore because of this). This is a logical fallacy that arises from thinking that just because

one thing (Y) occurs after another (X), that X *caused* Y. People typically brush their teeth before going off to work in the morning. But brushing their teeth doesn't *cause* them to go to work. In this case, it is even possibly the reverse.

2) *Cum hoc, ergo propter hoc* (with this, therefore because of this). This is a logical fallacy that arises from thinking that just because two things co-occur, one must have caused the other. To drive home the point, Harvard Law student Tyler Vigen has written a book and a website that feature spurious co-occurrences—correlations—such as this one:

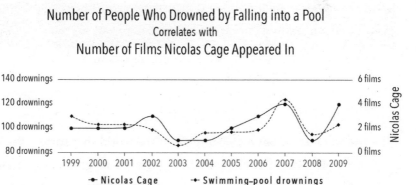

Number of People Who Drowned by Falling into a Pool
Correlates with
Number of Films Nicolas Cage Appeared In

There are four ways to interpret this: (1) drownings cause the release of new Nicolas Cage films; (2) the release of Nicolas Cage films causes drownings; (3) a third (as yet unidentified) factor causes both; or (4) they are simply unrelated and the correlation is a coincidence. If we don't separate correlation from causation, we can claim that Vigen's graph "proves" that Nic Cage was helping to prevent pool drownings, and furthermore, our best bet is to encourage him to make fewer movies so that he can ply his lifesaving skills as he apparently did so effectively in 2003 and 2008.

In some cases, there is no actual connection between items that are correlated—their correlation is simply coincidence. In other cases, one can find a causal link between correlated items, or at least spin a reasonable story that can spur the acquisition of new data.

We can rule out explanation one, because it takes time to produce and release a movie, so a spike in drownings cannot cause a spike in Nic Cage movies in the same year. What about number two? Perhaps people become so wrapped up in the drama of Cage's films that they lose focus and drown as a consequence. It may be that the same cinematic absorption also increases rates of automobile accidents and injuries from heavy machinery. We don't know until we analyze more data, because those are not reported here.

What about a third factor that caused both? We might guess that economic trends are driving both: A better economy leads to more investment in leisure activities—more films being made, more people going on vacation and swimming. If this is true, then neither of the two things depicted on the graph—Nic Cage films and drownings—caused the other. Instead, a third factor, the economy, led to changes in both. Statisticians call this the *third factor x* explanation of correlations, and there are many cases of these.

More likely, these two are simply unrelated. If we look long enough, and hard enough, we're sure to find that two unrelated things vary with each other.

Ice-cream sales increase as the number of people who wear short pants increases. Neither causes the other; the third factor *x* that causes both is the warmer temperatures of summer. The number of television shows aired in a year while you were a child may have correlated with increases in your height, but what was no doubt driving both was the passage of time during an era when (a) TV was expanding its market and (b) you were growing.

How do you know when a correlation indicates causation? One way is to conduct a controlled experiment. Another is to apply logic. But be careful—it's easy to get bogged down in semantics. Did the rain outside *cause* people to wear raincoats or was it their desire to avoid getting wet, a consequence of the rain, that caused it?

This idea was cleverly rendered by Randall Munroe in his Internet cartoon *xkcd*. Two stick figures, apparently college students, are talking. One says that he used to think correlation implied causation. Then he took a statistics class, and now he doesn't think that anymore. The other student says, "Sounds like the class helped." The first student replies, "Well, maybe."

Deceptive Illustrations

Infographics are often used by lying weasels to shape public opinion, and they rely on the fact that most people won't study what they've done too carefully. Consider this graphic that might be used to scare you into thinking that runaway inflation is eating up your hard-earned money:

That's a frightening image. But look closely. The scissors are cutting the bill not at 4.2 percent of its size, but at about 42 percent. When your visual system is pitted against your logical system, the visual system usually wins, unless you work extra diligently to overcome this visual bias. The accurate infographic would look like this but would have much less emotional impact:

Interpreting and Framing

Often a statistic will be properly created and reported, but someone—a journalist, an advocate, any non-statistician—will misreport it, either because they've misunderstood it or because they didn't realize that a small change in wording can change the meaning.

Often those who want to use statistics do not have statisticians on their staffs, and so they seek the answers to their questions from people who lack proper training. Corporations, government offices, nonprofits, and mom-and-pop grocery stores all benefit from statistics about such items as sales, customers, trends, and supply chain. Incompetence can enter at any stage, in experimental design, data collection, analysis, or interpretation.

Sometimes the statistic being reported isn't the relevant one. If you're trying to convince stockholders that your company is doing well, you might publish statistics on your annual sales, and show steadily rising numbers. But if the market for your product is expanding, sales increases would be expected. What your investors and analysts probably want to know is whether your market share has changed. If your market share is decreasing because competitors are swooping in and taking away your customers, how can you make your report look attractive? Simply fail to report the relevant statistic of market share, and instead report the sales figures. Sales are going up! Everything is fine!

The financial profiles shown on people's mortgage applications twenty-five years ago would probably not be much help in building a model for risk today. Any model of consumer behavior on a website may become out of date very quickly. Statistics on the integrity of concrete used for overpasses may not be relevant for concrete on bridges (where humidity and other factors may have caused divergence, even if both civic projects used the same concrete to begin with).

You've probably heard some variant of the claim that "four out of five dentists recommend Colgate toothpaste." That's true. What the ad agency behind these decades-old ads wants you to think is that the dentists prefer Colgate above and beyond other brands. But that's not true. The Advertising Standards Authority in the United Kingdom investigated this claim and ruled it an unfair practice because the survey that was conducted allowed dentists to recommend more than one toothpaste. In fact, Colgate's biggest competitor was named nearly as often as Colgate (a detail you won't see in Colgate's ads).

Framing came up in the section on averages and implicitly in the discussion of graphs. Manipulating the framing of any message furnishes an endless number of ways people can make you believe something that isn't so if you don't stop to think about what they're saying. The cable network C-SPAN advertises that it is "available" in 100 million homes. That doesn't mean that 100 million people are watching C-SPAN. It doesn't mean that even one person is watching it.

Framing manipulations can influence public policy. A survey of recycling yield on various streets in metropolitan Los Angeles shows that one street in particular recycles 2.2 times as much as any other street. Before the city council gives the residents of this street an award for their green city efforts, let's ask what might give rise to such a number. One possibility is that this street has more than twice as many residents as other streets—perhaps because it is longer, perhaps because there are a lot of apartment buildings on it. Measuring recycling at the level of the street is not the relevant statistic unless all streets are otherwise identical. A better statistic would be either the living unit (where you measure the recycling output of each family) or even better, because larger families probably consume more than smaller families, the individual. That is, we want to adjust the amount of recycling materials collected to take into account the number of people on the street. That is the *true frame* for the statistic.

The *Los Angeles Times* reported in 2014 about water use in the city of Rancho Santa Fe in drought-plagued California. "On a daily per capita basis, households in this area lapped up an average of nearly five times the water used by coastal Southern California homes in September, earning them the dubious distinction of being the state's biggest residential water hogs." "Households" is not the relevant frame for this statistic, and the *LA Times* was correct to report per capita—individuals; perhaps the residents of Rancho Santa Fe have larger families, meaning more showers, dishes, and flushing commodes. Another frame would look at water use per acre. Rancho Santa Fe homes tend to have larger lots. Perhaps it is desirable for fire prevention and other reasons to keep land planted with verdant vegetation, and the large lots in Rancho Santa Fe don't use more water on a per acre basis than land anywhere else.

In fact, there's a hint of this in a *New York Times* article on the issue: "State water officials warned against comparing per capita water use between districts; they said they expected use to be highest in wealthy communities with large properties."

The problem with the newspaper articles is that they frame the data to make it look as though Rancho Santa Fe residents are using more than their share of water, but the data they provide—as in the case of the Los Angeles recycling example above—don't actually show that.

Calculating proportions rather than actual numbers often helps to provide the true frame. Suppose you are northwest regional sales manager for a company that sells flux capacitors. Your sales have improved greatly, but are still no match for your nemesis in the company, Jack from the southwest. It's hardly fair—his territory is not only geographically larger but covers a much larger population. Bonuses in your company depend on you showing the higher-ups that you have the mettle to go out and get sales.

There is a legitimate way to present your case: Report your sales as a function of the area or population of the territory you serve. In other words, instead of graphing total number of flux capacitors sold, look at total number per person in the region, or per square mile. In both, you may well come out ahead.

News reports showed that 2014 was one of the deadliest years for plane crashes: 22 accidents resulted in 992 fatalities. But flying is actually safer now than it has ever been. Because there are so many more flights today than ever before, the 992 fatalities represent a dramatic decline in the number of deaths per million passengers (or per million miles flown). On any single flight on a major airline, the chances are about 1 in 5 million that you'll be killed, making it more likely that you'll be killed doing just about anything else—walking across the street, eating food (death by choking or unintentional poisoning is about 1,000 times more likely). The baseline for comparison is very important here. These statistics are spread out over a year—a year of airline travel, a year of eating and then either choking or being poisoned. We could change the baseline and look at each hour of the activities, and this would change the statistic.

Differences That Don't Make a Difference

Statistics are often used when we seek to understand whether there is a difference between two treatments: two different fertilizers in a field, two different pain medications, two different styles of teaching, two different groups of salaries (e.g., men versus women doing the same jobs). There are many ways that two treatments can differ. There can be actual differences between them; there can be confounding factors in your sample that have nothing to do with the actual treatments; there can be errors in your measurement; or there can be random variation—little chance differences that turn up, sometimes on one side of the equation, sometimes on the other, depending on when you're looking. The researcher's goal is to find stable, replicable differences, and we try to distinguish those from experimental error.

Be wary, though, of the way news media use the word "significant," because to statisticians it doesn't mean "noteworthy." In statistics, the word "significant" means that the results passed mathematical tests such as t-tests, chi-square tests, regression, and principal components analysis (there are hundreds). Statistical significance tests quantify how easily pure chance can explain the results. With a very large number of observations, even small differences that are trivial in magnitude can be beyond what our models of change and randomness can explain. These tests don't know what's noteworthy and what's not—that's a human judgment.

The more observations you have in the two groups, the more likely that you will find a difference between them. Suppose I test the annual maintenance costs of two automobiles, a Ford and a Toyota, by looking at the repair records for ten of each car. Let's say, hypothetically, the mean cost of operating the Ford is eight cents more per year. This will probably fail to meet statistical significance, and clearly a cost difference of eight cents a year is not going to be the deciding factor in which car to buy—it's just too small an amount to be concerned about. But if I look at the repair records for 500,000 vehicles, that eight-cent difference will be statistically significant. But it's a difference that doesn't matter in any real-world, practical sense. Similarly, a new headache medication may be statistically faster at curing your headache, but if it's only 2.5 seconds faster, who cares?

Interpolation and Extrapolation

You go out in your garden and see a dandelion that's four inches high on Tuesday. You look again on Thursday and it's six inches high. How high was it on Wednesday? We don't know for sure because we didn't measure it Wednesday (Wednesday's the day you got stuck in traffic on the way home from the nursery, where you bought some weed killer). But you can guess: The dandelion was probably five inches high on Wednesday. This is interpolation. Interpolation takes two data points and estimates the value that would have occurred between them if you had taken a measurement there.

How high will the dandelion be after six months? If it's growing 1 inch per day, you might say that it will grow 180 inches more in six months (roughly 180 days), for a total of 186 inches, or fifteen and a half feet high. You're using extrapolation. But have you ever seen a dandelion that tall? Probably not. They collapse under their own weight, or die of other natural causes, or get trampled, or the weed killer might get them. Interpolation isn't a perfect technique, but if the two observations you're considering are very close together, interpolation usually provides a good estimate. Extrapolation, however, is riskier, because you're making estimates outside the range of your observations.

The amount of time it takes a cup of coffee to cool to room temperature is governed by Newton's law of cooling (and is affected by other factors such as the barometric pressure and the composition of the cup). If your coffee started out at 145 degrees Fahrenheit (F), you'd observe the temperature decreasing over time like this:

Elapsed Time (mins)	Temp °F
0	145
1	140
2	135
3	130

Your coffee loses five degrees every minute. If you interpolated between two observations—say you want to know what the temperature would have been at the halfway point between measurements—your interpolation is going to be quite accurate. But if you extrapolate from the pattern, you are likely to come up with an absurd answer, such as that the coffee will reach freezing after thirty minutes.

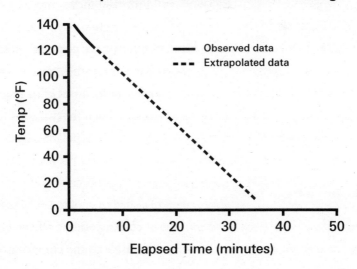

Temperature of Coffee Left Standing

The extrapolation fails to take into account a physical limit: The coffee can't get cooler than room temperature. It also fails to take into account that the rate at which the coffee cools slows down the closer it gets to room temperature. The rest of the cooling function looks like this:

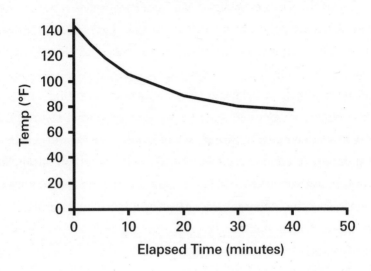

Temperature of Coffee Left Standing

Note that the steepness of the curve in the first ten minutes doesn't continue—it flattens out. This underscores the importance of two things when you're extrapolating: having a large number of observations that span a wide range, and having some knowledge of the underlying process.

Precision Versus Accuracy

When faced with the precision of numbers, we tend to believe that they are also *accurate,* but this is not the same thing. If I say "a lot of people are buying electric cars these days," you assume that I'm making a guess. If I say that "16.39 percent of new car sales are electric vehicles," you assume that I know what I'm talking about. But you'd be confusing precision for accuracy. I may have made it up. I may have sampled only a small number of people near an electric-car dealership.

Recall the *Time* magazine headline I mentioned earlier, which said that more people have cell phones than have toilets. This isn't implausible, but it is a distortion because that's *not* what the U.N. study found at all. The U.N. reported that more people had *access* to cell phones than to toilets, which is, as we know, a different thing. One cell phone might be shared among dozens of people. The lack of sanitation is still distressing, but the headline makes it sound like if you were to count, you'd find there are more cell phones in the world than there are toilets, and that is not supported by the data.

Access is one of those words that should raise red flags when you encounter them in statistics. People having access to health care might simply mean they live near a medical facility, not that the facility would admit them or that they could pay for it. As you learned above, C-SPAN is available in 100 million homes, but that doesn't mean that 100 million people are watching it. I could claim that 90 percent of the world's population has "access" to *A Field Guide to Lies* by showing that 90 percent of the population is within twenty-five miles of an Internet connection, rail line, road, landing strip, port, or dogsled route.

Comparing Apples and Oranges

One way to lie with statistics is to compare things—datasets, populations, types of products—that are different from one another, and pretend that they're not. As the old idiom says, you can't compare apples with oranges.

Using dubious methods, you could claim that it is safer to be in the military during an active conflict (such as the present war in Afghanistan) than to be stateside in the comfort of your own home. Start with the 3,482 active-duty U.S. military personnel who died in 2010. Out of a total of 1,431,000 people in the military, this gives a rate of 2.4 deaths per 1,000. Across the United States, the death rate in 2010 was 8.2 deaths per 1,000. In other words, it is more than three times safer to be in the military, in a war zone, than to live in the United States.

What's going on here? The two samples are not similar, and so shouldn't be compared directly. Active military personnel tend to be young and in good health; they are served a nutritious diet and have good health care. The general population of the United States includes the elderly, people who are sick, gang members, crackheads, motorcycle daredevils, players of mumblety-peg, and many people who have neither a nutritious diet nor good health care; their mortality rate would be high wherever they are. And active military personnel are not all stationed in a war zone—some are stationed in very safe bases in the United States, are sitting behind desks in the Pentagon, or are stationed in recruiting stations in suburban strip malls.

U.S. News & World Report published an article comparing the proportion of Democrats and Republicans in the country going back to the 1930s. The problem is that sampling methods have changed over the years. In the 1930s and '40s, sampling was typically done by in-person interviews and mail lists generated by telephone directories; by the 1970s sampling was predominantly just by telephone. Sampling in the early part of the twentieth century skewed toward those who tended to have landlines: wealthier people, who, at least at that time, tended to vote Republican. By the 2000s, cell phones were being sampled, which skewed toward the young, who tended to vote Democratic. We can't really know if the proportion of Democrats to Republicans has changed since the 1930s because the samples are incompatible. We think we're studying one thing but we're studying another.

A similar problem occurs when reporting a decline in the death rate due to motorcycle accidents now versus three decades ago. The more recent figures might include more three-wheel motorcycles compared to predominantly two-wheeled ones last century; it might compare an era when helmets were not required by law to now, where they are in most states.

Be on the lookout for changing samples before drawing conclusions! *U.S. News & World Report* (yes, them again) wrote of an increase in the number of doctors over a twelve-year period, accompanied by a significant drop in average salary. What is the takeaway message? You might conclude that now is not a good time to enter the medical profession because there is a glut of doctors, and that supply exceeding demand has lowered every doctor's salary. This might be true, but there is no evidence in the claim to support this.

An equally plausible argument is that over the twelve-year period, increased specialization and technology growth created more opportunities for doctors and so there were more available positions, accounting for the increase in the total number of doctors. What about the salary decline? Perhaps many older doctors retired, and were replaced by younger ones, who earn a smaller salary just out of medical school. There is no evidence presented either way. An important part of statistical literacy is recognizing that some statistics, as presented, simply cannot be interpreted.

Sometimes, this apples-and-oranges comparison results from

inconsistent subsamples—ignoring a detail that you didn't realize was important. For example, when sampling corn from a field that received a new fertilizer, you might not notice that some ears of corn get more sun and some get more water. Or when studying how traffic patterns affect street repaving, you might not realize that certain streets have more water runoff than others, influencing the need for asphalt repairs.

Amalgamating is putting things that are different (heterogeneous) into the same bin or category—a form of apples and oranges. If you're looking at the number of defective sprockets produced by a factory, you might combine two completely different kinds in order to make the numbers come out more favorably for your particular interests.

Take an example from public policy. You might want to survey the sexual behavior of preteens and teens. How you amalgamate (or bin) the data can have a large effect on how people perceive your data. If your agenda is to raise money for educational and counseling centers, what better way to do so than to release a statistic such as "70 percent of schoolchildren ages ten to eighteen are sexually active." We're not surprised that seventeen- and eighteen-year-olds are, but ten-year-olds! That will surely cause grandparents to reach for the smelling salts and start writing checks. But obviously, a single category of ten-year-olds to eighteen-year-olds lumps together individuals who are likely to be sexually active with those who are not. More helpful would be separate bins that put together individuals of similar age and likely similar experiences: ten to eleven, twelve to thirteen, fourteen to fifteen, sixteen to eighteen, for example.

But that's not the only problem. What do they mean by "sexually active"? What question was actually asked of the schoolchildren?

Or were the schoolchildren even asked? Perhaps it was their parents who were asked. All kinds of biases can enter into such a number. "Sexually active" is open to interpretation. Responses will vary widely depending on how it is defined. And of course respondents may not tell the truth (reporting bias).

As another example, you might want to talk about unemployment as a general problem, but this risks combining people of very different backgrounds and contributing factors. Some are disabled and can't work; some are fired with good cause because they were caught stealing or drunk on the job; some want to work but lack the training; some are in jail; some no longer want to work because they've gone back to school, joined a monastery, or are living off family money. When statistics are used to influence public policy, or to raise donations for a cause, or to make headlines, often the nuances are left out. And they can make all the difference.

These nuances often tell a story themselves about patterns in the data. People don't become unemployed for the same reasons. The likelihood that an alcoholic or a thief will become unemployed may be four times that of someone who is not. These patterns carry information that is lost in amalgamation. Allowing these factors to become part of the data can help you to see who is unemployed and why—it could lead to better training programs for people who need it, or more Alcoholics Anonymous centers in a town that is underserved by them.

If the people and agencies who track behavior use different definitions for things, or different procedures for measuring them, the data that go into the statistic can be very dissimilar, or heterogeneous. If you're trying to pin down the number of couples who live together but are not married, you might rely on data that have already been collected by various county and state agencies. But varying definitions can yield a categorization problem: What constitutes living together? Is it determined by how many nights a week they are together? By where their possessions are, where they get mail? Some jurisdictions recognize same-sex couples and some don't. If you take the data from different places using different schemes, the final statistic carries very little meaning. If the recording, collection, and measurement practices vary widely across collection points, the statistic that results may not mean what you think it means.

A recent report found that the youth unemployment rate in Spain was an astonishing 60 percent. The report amalgamated into the same category people who normally would appear in separate categories: Students who were not seeking work were counted as unemployed, alongside workers who had just been laid off and workers who were seeking jobs.

In the United States, there are *six* different indexes (numbered U1 through U6) to track unemployment (as measured by the Bureau of Labor Statistics), and they reflect different interpretations of what "unemployed" actually means. It can include people looking for a job, people who are in school but not looking, people who are seeking full-time assignments in a company where they work only part-time, and so on.

USA Today reported in July 2015 that the unemployment rate dropped to 5.3 percent, "its lowest level since April 2008." More comprehensive sources, including the AP, *Forbes*, and the *New York Times*, reported the reason for the apparent drop: Many people who were out of work gave up looking and so technically had left the workforce.

Amalgamating isn't always wrong. You might choose to combine the test scores of boys and girls in a school, especially if there is no evidence that their scores differ—in fact, it's a good idea to, in order to increase your sample size (which provides you with a more stable estimate of what you're studying). Overly broad definitions of a category (as with the sexual-activity survey mentioned earlier) or inconsistent definitions (as with the couples-living-together statistic) present problems for interpretation. When performed properly, amalgamating helps us come up with a valid analysis of data.

Suppose that you work for the state of Utah and a large national manufacturer of baby clothes is thinking about moving to your state. You're thinking that if you can show that Utah has a lot of births, you're in a better position to attract the company, so you go to the Census.gov website, and graph the results for number of births by state:

Births: United States, 2013

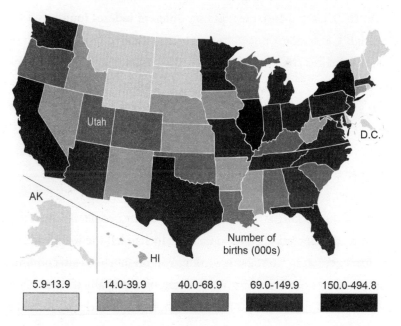

Utah looks better than Alaska, D.C., Montana, Wyoming, the Dakotas, and the small states of the Northeast. But it is hardly a booming baby state compared to California, Texas, Florida, and New York. But wait, this map you've made shows the *raw number* of births and so will be weighted heavily toward states with larger populations. Instead, you could graph the birth *rate* per thousand people in the population:

Crude Birth Rate: United States, 2013

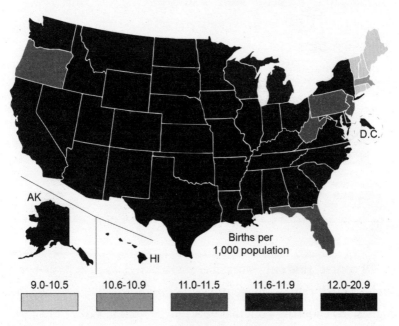

| 9.0-10.5 | 10.6-10.9 | 11.0-11.5 | 11.6-11.9 | 12.0-20.9 |

That doesn't help. Utah looks just like most of the rest of the country. What to do? Change the bins! You can play around with which range of values go into each category, those five gray-to-black bars at the bottom. By making sure that Utah's rate is in a category all by itself, you can make it stand out from the rest of the country.

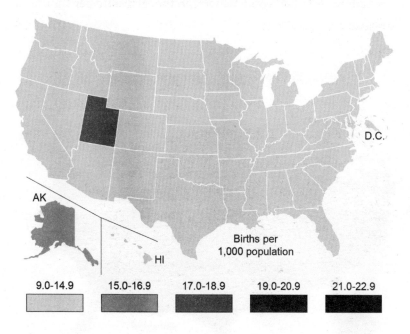

Crude Birth Rate: United States, 2013

Of course, this only works because Utah does in fact have the highest birth rate in the country—not by much, but it is still the highest. By choosing a bin that puts it all by itself in a color category, you've made it stand out. If you were trying to make a case for one of the other states, you'd have to resort to other kinds of flimflam, such as graphing the number of births per square mile, or per Walmart store, as a function of disposable income. Play around long enough and you might find a metric to make a case for any of the fifty states.

What is the *right* way, the non-lying way to present such a graph? This is a matter of judgment, but one relatively neutral way would be to bin the data so that 20 percent of the states are contained in each of the five bins, that is, an equal number of states per color category:

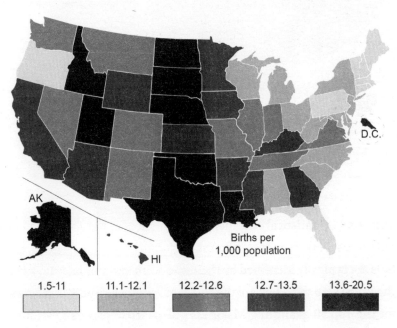

Crude Birth Rate: United States, 2013

Another would be to make the bins equal in size:

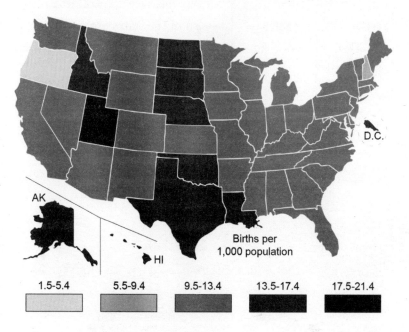

Crude Birth Rate: United States, 2013

Births per
1,000 population

| 1.5-5.4 | 5.5-9.4 | 9.5-13.4 | 13.5-17.4 | 17.5-21.4 |

This kind of statistical chicanery—using unequal bin widths in all but the last of these maps—often shows up in histograms, where the bins are typically identified by their midpoint and you have to infer the range yourself. Here are the batting averages for the 2015 season for the Top 50 qualifying Major League Baseball players (National and American Leagues):

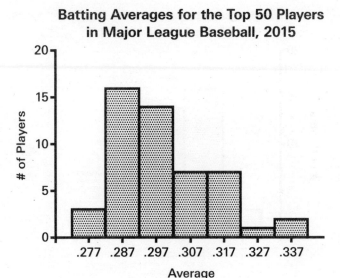

Batting Averages for the Top 50 Players in Major League Baseball, 2015

Now, suppose that you're the player whose batting average is .330, putting you in the second highest category. It's time for bonus checks and you don't want to give management any reason to deny you a bonus this year—you've already bought a Tesla. So change the bin widths, amalgamating your results with the two players who were batting .337, and now you're in with the very best players. While you're at it, close up the ensuing gap (there are no longer any batters in the .327-centered bin), creating a discontinuity in the x-axis that probably few will notice:

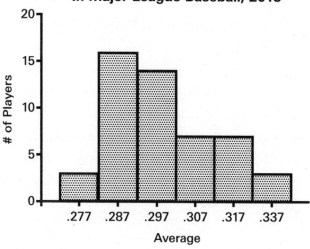

Batting Averages for the Top 50 Players in Major League Baseball, 2015

Specious Subdividing

The opposite of amalgamating is subdividing, and this can cause people to believe all kinds of things that aren't so. To claim that *x* is a leading cause of *y*, I simply need to subdivide other causes into smaller and smaller categories.

Suppose you work for a manufacturer of air purifiers, and you're on a campaign to prove that respiratory disease is the leading cause of death in the United States, overwhelming other causes like heart disease and cancer. As of today, the actual leading cause of death in the United States is heart disease. The U.S. Centers for Disease Control report that these were the top three causes of death in 2013:

Heart disease: 611,105

Cancer: 584,881

Chronic lower respiratory diseases: 149,205

Now, setting aside the pesky detail that home air purifiers may not form a significant line of defense against chronic respiratory disease, these numbers don't make a compelling case for your company. Sure, you'd like to save more than 100,000 lives annually, but to say that you're fighting the *third* largest cause of death doesn't make for a very impressive ad campaign. But wait! Heart disease isn't one thing, it's several:

Acute rheumatic fever and chronic rheumatic heart disease:
   3,260

Hypertensive heart disease: 37,144

Acute myocardial infarction: 116,793

Heart failure: 65,120

And so on. Next, break up the cancers into small subtypes. By failing to amalgamate, and creating these fine subdivisions, you've done it! Chronic lower respiratory disease becomes the number one killer. You've just earned yourself a bonus. Some food companies have used this subdivide strategy to hide the amounts of fats and sugars contained in their product.

HOW NUMBERS ARE COLLECTED

Just because there's a number on it, it doesn't mean that the number was arrived at properly. Remember, as the opening to this part of the book states, *people* gather statistics. People choose what to count, how to go about counting. There are a host of errors and biases that can enter into the collection process, and these can lead millions of people to draw the wrong conclusions. Although most of us won't ever participate in the collection process, thinking about it, critically, is easy to learn and within the reach of all of us.

Statistics are obtained in a variety of ways: by looking at records (e.g., birth and death records from a government agency, hospital, or church), by conducting surveys or polls, by observation (e.g., counting the number of electric cars that pass the corner of Main and Third Street), or by inference (if sales of diapers are going up, the birth rate is probably going up). Biases, inaccuracies, and honest mistakes can enter at any stage. Part of evaluating claims includes asking the questions "Can we really know that?" and "How do they know that?"

Sampling

Astrogeologists sample specimens of moon rock; they don't test the entire moon. Researchers don't talk to every single voter to find out which candidate is ahead, or tally every person who enters an emergency room to see how long they had to wait to be seen. To do so would be impractical or too costly. Instead, they use samples to estimate the true number. When samples are properly taken, these estimates can be very, very accurate. In public-opinion polls, an estimate of how the entire country feels about an issue (about 234 million adults over the age of twenty-one) can be obtained by interviewing only 1,067 individuals. Biopsies that sample less than one one-thousandth of an organ can be used for accurate cancer staging.

To be any good, a sample has to be representative. A sample is representative if every person or thing in the group you're studying has an equally likely chance of being chosen. If not, your sample is biased. If the cancer is only in part of an organ and you sample the wrong part, the cancer will go undiagnosed. If it's in a very small part and you take fifteen samples in that one spot, you may be led to conclude the entire organ is riddled with cancer when it's not.

We don't always know ahead of time, with biopsies or public-opinion polls, how much variability there will be. If everyone in a population were identical, we'd only need to sample one of them. If we have a bunch of genetically identical people, with identical personalities and life experience, we can find out anything we want to know about all of them by simply looking at one of them. But every group contains some heterogeneity, some differences across its members, and so we need to be careful about how we sample to ensure that we have accounted for all the differences that matter. (Not all differences matter.) For example, if we deprive a human of oxygen, we know that human will die. Humans don't differ along this dimension (although they do differ in terms of how long they can last without oxygen). But if I want to know how many pounds human beings can bench press, there is wide variation—I'd need to measure a large cross-section of different people to obtain a range and a stable average. I'd want to sample large people, small people, fat people, skinny people, men, women, children, body builders and couch potatoes, people taking anabolic steroids, and teetotalers. There are probably other factors that matter, such as how much sleep the person got the night before testing, how long it's been since they ate, whether they're angry or calm, and so on. Then there are things that we think don't matter at all: whether the air-traffic controller at the St. Hubert Airport in Quebec is male or female that day; whether a random customer at a restaurant in Aberdeen was served in a timely fashion or not. These things may make a difference to other things we're measuring (latent sexism in the air-travel industry; customer satisfaction at Northwestern dining establishments) but not to bench pressing.

The job of the statistician is to formulate an inventory of all those things that matter in order to obtain a representative sample. Researchers have to avoid the tendency to capture variables that are easy to identify or collect data on—sometimes the things that matter are not obvious or are difficult to measure. As Galileo Galilei said, the job of the scientist is to measure what is measurable and to render measurable that which is not. That is, some of the most creative acts in science involve figuring out how to measure something that makes a difference, that no one had figured out how to measure before.

But even measuring and trying to control variables that you know about poses challenges. Suppose you want to study current attitudes about climate change in the United States. You've been given a small sum of money to hire helpers and to buy a statistics program to run on your computer. You happen to live in San Francisco, and so you decide to study there. Already you're in trouble: San Francisco is not representative of the rest of the state of California, let alone the United States. Realizing this, you decide to do your polling in August, because studies show that this is peak tourist season—people from all over the country come to San Francisco then, so (you think) you'll be able to get a cross-section of Americans after all.

But wait: Are the kinds of people who visit San Francisco representative of others? You'd skew toward people who can afford to travel there, and people who want to spend their vacation in a city, as opposed to, say, a national park. (You might also skew toward liberals, because San Francisco is a famously liberal city.)

So you decide that you can't afford to study U.S. attitudes, but you can study attitudes about climate change among San Franciscans. You set up your helpers in Union Square and stop passersby with a short questionnaire. You instruct them to seek out people of different ages, ethnicities, styles of dress, with and without tattoos— in short, a cross-section with variability. But you're still in trouble. You're unlikely to encounter the bedridden, mothers with small children, shift workers who sleep during the day, and the hundreds of thousands of San Francisco residents who, for various reasons, don't come to Union Square, a part of town known for its expensive shops and restaurants. Sending half your helpers down to the Mission District helps solve the problem of different socioeconomic statuses being represented, but it doesn't solve your other problems. The test of a random sample is this: Does everyone or -thing in the whole group have an equal chance of being measured by you and your team? Here the answer is clearly no.

So you conduct a *stratified* random sample. That is, you identify the different strata or subgroups of interest and draw people from them in proportion to the whole population. You do some research about climate change and discover that attitudes do not seem to fall upon racial lines, so you don't need to create subsamples based on race. That's just as well because it can be difficult, or offensive, to make assumptions about race, and what do you do with people of mixed race? Put them in one category or another, or create a new category entirely? But then what? A category for Americans who identify as black-white, black-Hispanic, Asian-Persian, etc.? The categories then may become so specific as to create trouble for your analysis because you have too many distinct groups. Another hurdle: You want age variability, but people don't always feel comfortable telling you their age. You can choose people who are obviously under forty and obviously over forty, but then you miss people who are in their late thirties and early forties.

To even out the problem of people who aren't out and about during the day, you decide to go to people's homes door-to-door. But if you go during the day, you miss working people; if you go during the evening, you miss nightclubbers, shift workers, nighttime church-service attendees, moviegoers, and restaurant-goers. Once you've stratified, how do you get a random sample within your subgroups? All the same problems described above still persist—creating the subgroups doesn't solve the problem that within the subgroup you still have to try to find a fair representation of all the *other* factors that might make a difference to your data. It starts to feel like we will have to sample all the rocks on the moon to get a good analysis.

Don't throw in the trowel—er, towel. Stratified random sampling is better than non-stratified. If you take a poll of college students drawn at random to describe their college experience, you may end up with a sample of students only from large state colleges—a random sample is more likely to choose them because there are so many more of them. If the college experience is radically different

"After sampling every bird that frequents the sidewalk outside this building, we've concluded that what birds really love is bagels!"

at small, private liberal-arts colleges, you need to ensure that your sample includes students from them, and so your stratified sample will ensure you ask people from different-sized schools. Random sampling should be distinguished from convenience or quota sampling—the practice of just talking to people whom you know or people on the street who look like they'll be helpful. Without a random sample, your poll is very likely to be biased.

Collecting data through sampling therefore becomes a never-ending battle to avoid sources of bias. And the researcher is never entirely successful. Every time we read a result in the newspaper that 71 percent of the British are in favor of something, we should reflexively ask, "Yes, but 71 percent of *which* British?"

Add to this the fact that any questions we ask of people are only a sample of all the possible questions we could be asking them, and their answers may only be a sample of what their complex attitudes and experiences are. To make matters worse, they may or may not understand what we're asking, and they may be distracted while they're answering. And, more frequently than pollsters like to admit, people sometimes give an intentionally wrong answer. Humans are a social species; many try to avoid confrontation and want to please, and so give the answer they think the pollster wants to hear. On the other hand, there are disenfranchised members of society and nonconformists who will answer falsely just to shock the pollster, or as a way to try on a rebellious persona to see if it feels good to shock and challenge.

Achieving an unbiased sample isn't easy. When hearing a new statistic, ask, "What biases might have crept in during the sampling?"

Samples give us estimates of something, and they will almost always deviate from the true number by some amount, large or small, and that is the margin of error. Think of this as the price you pay for not hearing from everyone in the population under study, or sampling every moon rock. Of course, there can be errors even if you've interviewed or measured every single person in the population, due to flaws or biases in the measurement device. The margin of error does not address underlying flaws in the research, only the degree of error in the sampling procedure. But ignoring those deeper possible flaws for the moment, there is another measurement or statistic that accompanies any rigorously defined sample: the confidence interval.

The margin of error is how accurate the results are, and the confidence interval is how confident you are that your estimate falls within the margin of error. For example, in a standard two-way poll (in which there are two possibilities for respondents), a random sample of 1,067 American adults will produce a margin of error around 3 percent in either direction (we write ±3%). So if a poll shows that 45 percent of Americans support Candidate A, and 47 percent support Candidate B, the true number is somewhere between 42 and 48 percent for A, and between 44 and 50 percent for B. Note that these ranges overlap. What this means is that the two-percentage-point difference between Candidate A and Candidate B is within the margin of error: We can't say that one of them is truly ahead of the other, and so the election is too close to call.

How confident are we that the margin of error is 3 percent and not more? We calculate a confidence interval. In the particular case I mentioned, I reported the 95 percent confidence interval. That means that if we were to conduct the poll a hundred times, using the same sampling methods, ninety-five out of those hundred times the interval we obtain will contain the true value. Five times out of a hundred, we'd obtain a value outside that range. The confidence interval doesn't tell us how far out of the range—it could be a small difference or a large one; there are other statistics to help with that.

We can set the confidence level to any value we like, but 95 percent is typical. To achieve a narrower confidence interval you can do one of two things: *increase* the sample size for a given confidence level; or, for a given sample size, *decrease* the confidence level. For a given sample size, changing your confidence level from 95 to 99 will increase the size of the interval. In most instances the added expense or inconvenience isn't worth it, given that a variety of external circumstances can change people's minds the following day or week anyway.

Note that for very large populations—like that of the United States—we only need to sample a very small percentage, in this case, less than .0005 percent. But for smaller populations—like that of a corporation or school—we require a much larger percentage. For a company with 10,000 employees, we'd need to sample 964 (almost 10 percent) to obtain the 3 percent margin with 95 percent confidence, and for a company of 1,000 employees, we'd need to sample nearly 600 (60 percent).

Margin of error and confidence interval apply to sampling of any kind, not just samples with people: sampling the proportion of electric cars in a city, of malignant cells in the pancreas, or of mercury in fish at the supermarket. In the figure on page 84, margin of error and sample size are shown for a confidence interval of 95 percent.

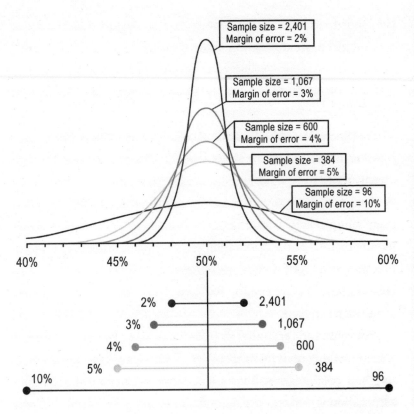

The formula for calculating the margin of error (and confidence interval) is in the notes at the end of the book, and there are many online calculators to help. If you see a statistic quoted and no margin of error is given, you can calculate the margin yourself, knowing the number of people who were surveyed. You'll find that in many cases, the reporter or polling organization doesn't provide this information. This is like a graph without axes—you can lie with statistics very easily by failing to report the margin of error or confidence interval. Here's one: My dog Shadow is the leading gubernatorial candidate in the state of Mississippi, with 76 percent of voters favoring him over other candidates (with an unreported margin of error of ±76 percent; vote for Shadow!!!).

Sampling Biases

While trying to obtain a random sample, researchers sometimes make errors in judgment about whether every person or thing is equally likely to be sampled.

An infamous error was made in the 1936 U.S. presidential election. The *Literary Digest* conducted a poll and concluded that Republican Alf Landon would win over the incumbent Democrat, President Roosevelt. The *Digest* had polled people who were magazine readers, car owners, or telephone customers, not a random sample. The conventional explanation, cited in many scholarly and popular publications, is that in 1936, this skewed heavily toward the wealthy, who were more likely to vote Republican. In fact, according to a poll conducted by George Gallup in 1937, this conventional explanation is wrong—car and telephone owners were more likely to back Roosevelt. The bias occurred in that Roosevelt backers were far less likely to participate in the poll. This sampling bias was recognized by Gallup, who conducted his own poll using a random sample, and correctly predicted the outcome. The Gallup poll was born. And it became the gold standard for political polling until it misidentified the winner in the 2012 U.S. presidential election. An investigation uncovered serious flaws in their sampling procedures, ironically involving telephone owners.

Just as polls based on telephone directories skewed toward the wealthy in the 1930s and 1940s, now landline sampling skews toward older people. All phone sampling assumes that people who own phones are representative of the population at large; they may or may not be. Many Silicon Valley workers use Internet applications for their conversations, and so phone sampling may under-represent high-tech individuals.

If you want to lie with statistics and cover your tracks, take the average height of people near the basketball court; ask about income by sampling near the unemployment office; estimate statewide incidence of lung cancer by sampling only near a smelting plant. If you don't disclose how you selected your sample, no one will know.

Participation Bias

Those who are willing to participate in a study and those who are not may differ along important dimensions such as political views, personalities, and incomes. Similarly, those who answer a recruitment notice—those who volunteer to be in your study—may show a bias toward or against the thing you're interested in. If you're trying to recruit the "average" person in your study, you may bias participation merely by telling them ahead of time what the study is about. A study about sexual attitudes will skew toward those more willing to disclose those attitudes and against the shy and prudish. A study about political attitudes will skew toward those who are willing to discuss them. For this reason, many questionnaires, surveys, and psychological studies don't indicate ahead of time what the research question is, or they disguise the true purpose of the study with a set of irrelevant questions that the researcher isn't interested in.

The people who complete a study may well be different from

those who stop before it's over. Some of the people you contact simply won't respond. This can create a bias when the types of people who respond to your survey are different from the ones who don't, forming a special kind of sampling bias called non-response error.

Let's say you work for Harvard University and you want to show that your graduates tend to earn large salaries just two years after graduation. You send out a questionnaire to everyone in the graduating class. Already you're in trouble: People who have moved without telling Harvard where they went, who are in prison, or who are homeless won't receive your survey. Then, among the ones who respond, those who have high incomes and good feelings about what Harvard did for them might be more likely to fill out the survey than those who are jobless and resentful. The people you don't hear from contribute to non-response error, sometimes in systematic ways that distort the data.

If your goal in conducting the Harvard income-after-two-years survey is to show that a Harvard education yields a high salary, this survey may help you show that to most people. But the critical thinker will realize that the kinds of people who attend Harvard are not the same as the average person. They tend to come from higher-income families, and this is correlated with a student's future earnings. Harvard students tend to be go-getters. They might have earned as high a salary if they had attended a college with a lesser reputation, or even no college at all. (Mark Zuckerberg, Matt Damon, and Bill Gates are financially successful people who dropped out of Harvard.)

If you simply can't reach some segment of the population, such as military personnel stationed overseas, or the homeless and institutionalized, this sampling bias is called *coverage error* because some members of the population from which you want to sample cannot be reached and therefore have no chance of being selected.

If you're trying to figure out what proportion of jelly beans in a jar are red, orange, and blue, you may not be able to get to the bottom of the jar. Biopsies of organs are often limited to where the surgeon can collect material, and this is not necessarily a representative sample of the organ. In psychological studies, experimental subjects are often college undergraduates, who are not representative of the general population. There is a great diversity of people in this country, with differing attitudes, opinions, politics, experiences, and lifestyles. Although it would be a mistake to say that all college students are similar, it would be equally mistaken to say that they represent the rest of the population accurately.

Reporting Bias

People sometimes lie when asked their opinions. A Harvard graduate may overstate her income in order to appear more successful than she is, or may report what she thinks she should have made if it weren't for extenuating circumstances. Of course, she may understate as well so that the Harvard Alumni Association won't hit her up for a big donation. These biases may or may not cancel each other out. The average we end up with in a survey of Harvard graduates' salaries is only the average of what they reported, not what they actually earn. The wealthy may not have a very good idea of their annual income because it is not all salary—it includes a great many other things that vary from year to year, such as income from investments, dividends, bonuses, royalties, etc.

Maybe you ask people if they've cheated on an exam or on their taxes. They may not believe that your survey is truly confidential and so may not want to report their behavior truthfully. (This is a problem with estimating how many illegal immigrants in the U.S. require health care or are crime victims; many are afraid to go to hospitals and police stations for fear of being reported to immigration authorities.)

Suppose you want to know what magazines people read. You could ask them. But they might want to make a good impression on you. Or they might want to think of themselves as more refined in their tastes than they actually are. You may find that a great many more people report reading the *New Yorker* or the *Atlantic* than sales indicate, and a great many fewer people report reading *Us Weekly* and the *National Enquirer.* People don't always tell the truth in surveys. So here, you're not actually measuring what they read, you're measuring snobbery.

So you come up with a plan: You'll go to people's houses and see what magazines they actually have in their living rooms. But this too is biased: It doesn't tell you what they actually read, it only tells you what they choose to keep after they've read it, or choose to display for impression management. Knowing what magazines people read is harder to measure than knowing what magazines people *buy* (or display). But it's an important distinction, especially for advertisers.

What factors underlie whether an individual identifies as multiracial? If they were raised in a single racial community, they may be less inclined to think of themselves as mixed race. If they experienced discrimination, they may be more inclined. *We* might define multiraciality precisely, but it doesn't mean that people will report it the way we want them to.

Lack of Standardization

Measurements must be standardized. There must be clear, replicable, and precise procedures for collecting data so that each person who collects it does it in the same way. Each person who is counting has to count in the same way. Take Gleason grading of tumors—it is only relatively standardized, meaning that you can get different Gleason scores, and hence cancer stage labels, from different pathologists. (In Gleason scoring, a sample of prostate tissue is examined under a microscope and assigned a score from 2 to 10 to indicate how likely it is that a tumor will spread.) Psychiatrists differ in their opinions about whether a certain patient has schizophrenia or not. Statisticians disagree about what constitutes a sufficient demonstration of psychic phenomena. Pathology, psychiatry, parapsychology, and other fields strive to create well-defined procedures that anyone can follow and obtain the same results, but in almost all measurements, there are ambiguities and room for differences of opinion. If you are asked to weigh yourself, do you do so with or without clothes on, with or without your wallet in your pocket? If you're asked to take the temperature of a steak on the grill, do you measure it in one spot or in several and take the average?

Measurement Error

Participants may not understand a question the way the researcher thought they would; they may fill in the wrong bubble on a survey, or in a variety of unanticipated ways, they may not give the answer that they intended. Measurement error occurs in every measurement, in every scientific field. Physicists at CERN reported that they had measured neutrinos traveling faster than the speed of light, a finding that would have been among the most important of the last hundred years. They reported later that they had made an error in measurement.

Measurement error turns up whenever we quantify anything. The 2000 U.S. presidential election came down to measurement error (and to unsuccessfully recording people's intentions): Different teams of officials, counting the same ballots, came up with different numbers. Part of this was due to disagreements over how to count a dimpled chad, a hanging chad, etc.—problems of definition—but even when strict guidelines were put in place, differences in the count still showed up.

We've all experienced this: When counting pennies in our penny jar, we get different totals if we count twice. When standing on a bathroom scale three times in a row, we get different weights. When measuring the size of a room in your house, you may get slightly different lengths each time you measure. These are explainable occurrences: The springs in your scale are imperfect mechanical devices. You hold the tape measure differently each time you use it, it slips from its resting point just slightly, you read the sixteenths of an inch incorrectly, or the tape measure isn't long enough to measure the whole room so you have to mark a spot on the floor and take the measurement in two or three pieces, adding to the possibility of error. The measurement tool itself could have variability (indeed, measurement devices have accuracy specifications attached to them, and the higher-priced the device, the more accurate it tends to be). Your bathroom scale may only be accurate to within half a pound, a postal scale within half an ounce (one thirty-second of a pound).

A 1960 U.S. Census study recorded sixty-two women aged fifteen to nineteen with twelve or more children, and a large number of fourteen-year-old widows. Common sense tells us that there can't be many fifteen- to nineteen-year-olds with twelve children, and fourteen-year-old widows are very uncommon. Someone made an error here. Some census-takers might have filled in the wrong box on a form, accidentally or on purpose to avoid having to conduct time-consuming interviews. Or maybe an impatient (or impish) group of responders to the survey made up outlandish stories and the census-takers didn't notice.

In 2015 the New England Patriots were accused of tampering with their footballs, deflating them to make them easier to catch. They claimed measurement error as part of their defense. Inflation pressures for the footballs of both teams that day, the Pats and the Indianapolis Colts, were taken after halftime. The Pats' balls were tested first, followed by the Colts'. The Colts' balls would have been in a warm locker room or office longer, giving them more time to warm up and thus increase pressure. A federal district court accepted this, and other testimony, and ruled there was insufficient evidence of tampering.

Measurement error also occurs when the instrument you're using to measure—the scale, ruler, questionnaire, or test—doesn't actually measure what you intended it to measure. Using a yardstick to measure the width of a human hair, or using a questionnaire about depression when what you're really studying is motivation (they may be related but are not identical), can create this sort of error. Tallying which candidates people support financially is not the same as knowing how they'll vote; many people give contributions to several candidates in the same race.

Much ink has been spilled over tests or surveys that purport to show one thing but show another. The IQ test is among the most misinterpreted tests around. It is used to assess people's intelligence, as if intelligence were a single quantity, which it is not—it manifests itself in different forms, such as spatial intelligence, artistic intelligence, mathematical intelligence, and so forth. And IQ tests are known to be biased toward middle-class white people. What we usually want to know when we look at IQ test results is how suitable a person is for a particular school program or job. IQ tests can predict performance in these situations, but probably not because the person with a high IQ score is necessarily more intelligent, but because that person has a history of other advantages (economic, social) that show up in an IQ test.

If the statistic you encounter is based on a survey, try to find out what questions were asked and if these seem reasonable and unbiased to you. For any statistic, try to find out how the subject under study was measured, and if the people who collected the data were skilled in such measurements.

Definitions

How something is defined or categorized can make a big difference in the statistic you end up with. This problem arises in the natural sciences, such as in trying to grade cancer cells or describe rainfall, and in the social sciences, such as when asking people about their opinions or experiences.

Did it rain today in the greater St. Louis area? That depends on how you define rain. If only one drop fell to the ground in the 8,846 square miles that comprise "greater St. Louis" (according to the U.S. Office of Management and Budget), do we say it rained? How many drops have to fall over how large an area and over how long a period of time before we categorize the day as one with rainfall?

The U.S. Bureau of Labor Statistics has two different ways of measuring inflation based on two different definitions. The Personal Consumption Expenditures (PCE) and the Consumer Price Index (CPI) can yield different numbers. If you're comparing two years or two regions of the country, of course you need to ensure that you're using the same index each time. If you simply want to make a case about how inflation rose or fell recently, the unscrupulous statistic user would pick whichever of the two made the most impact, rather than choosing the one that is most appropriate, based on an understanding of their differences.

Or what does it mean to be homeless? Is it someone who is sleeping on the sidewalk or in a car? They may have a home and are not able or choose not to go there. What about a woman living on a friend's couch because she lost her apartment? Or a family who has sold their house and is staying in a hotel for a couple of weeks while they wait for their new house to be ready? A man happily and comfortably living as a squatter in an abandoned warehouse? If we compare homelessness across different cities and states, the various jurisdictions may use different definitions. Even if the definition becomes standardized across jurisdictions, a statistic you encounter may not have defined homelessness the way that you would. One of the barriers to solving "the homelessness problem" in our large cities is that we don't have an agreed-upon definition of what it is or who meets the criteria.

Whenever we encounter a news story based on new research, we need to be alert to how the elements of that research have been defined. We need to judge whether they are acceptable and reasonable. This is particularly critical in topics that are highly politicized, such as abortion, marriage, war, climate change, the minimum wage, or housing policy.

And nothing is more politicized than, well, politics. A definition can be wrangled and twisted to anyone's advantage in public-opinion polling by asking a question just-so. Imagine that you've been hired by a political candidate to collect information on his opponent, Alicia Florrick. Unless Florrick has somehow managed to appeal to everyone on every issue, voters are going to have gripes. So here's what you do: Ask the question "Is there anything at all that you disagree with or disapprove of, in anything the candidate has said, even if you support her?" Now almost everyone will have some gripe, so you can report back to your boss that "81 percent of people disapprove of Florrick." What you've done is collected data on one thing (even a single minor disagreement) and swept it into a pile of similar complaints, rebranding them as "disapproval." It almost sounds fair.

Things That Are Unknowable or Unverifiable

GIGO is a famous saying coined by early computer scientists: garbage in, garbage out. At the time, people would blindly put their trust into anything a computer output indicated because the output had the illusion of precision and certainty. If a statistic is composed of a series of poorly defined measures, guesses, misunderstandings, oversimplifications, mismeasurements, or flawed estimates, the resulting conclusion will be flawed.

Much of what we read should raise our suspicions. Ask yourself: Is it possible that someone can know this? A newspaper reports the proportion of suicides committed by gay and lesbian teenagers. Any such statistic has to be meaningless, given the difficulties in knowing which deaths are suicides and which corpses belong to gay versus straight individuals. Similarly, the number of deaths from starvation in a remote area, or the number of people killed in a genocide during a civil war, should be suspect. This was borne out by the wildly divergent casualty estimates provided by observers during the Iraq-Afghanistan-U.S. conflict.

A magazine publisher boasts that the magazine has 2 million readers. How do they know? They don't. They assume some proportion of every magazine sold is shared with others—what they call the "pass along" rate. They assume that every magazine bought by a library is read by a certain number of people. The same applies to books and e-books. Of course, this varies widely by title. Lots of people *bought* Stephen Hawking's *A Brief History of Time*. Indeed, it's said to be the most purchased and least finished book of the last thirty years. Few probably passed it along, because it looks impressive to have it sitting there in the living room. How many readers does a magazine or book have? How many listeners does a podcast have? We don't know. We know how many were sold or downloaded, that is all (although recent developments with e-books will probably be changing that long-standing status quo).

The next time that you read that the average New Zealander flosses 4.36 times a week (a figure I just made up, but it may be as accurate as any estimate), ask yourself: How could anyone know such a thing? What data are they relying on? If there were hidden cameras in bathrooms, that would be one thing, but more likely, it's people reporting to a survey taker, and only reporting what they remember—or want to believe is true, because we are always up against that.

PROBABILITIES

Did you believe me when I said few people *probably* passed along *A Brief History of Time*? I was using the term loosely, as many of us do, but the topic of mathematical probability confronts the very limits of what we can and cannot know about the world, stretching from the behavior of subatomic particles like quarks and bosons to the likelihood that the world will end in our lifetimes, from people playing a state lottery to trying to predict the weather (two endeavors that may have similar rates of success).

Probabilities allow us to quantify future events and are an important aid to rational decision making. Without them, we can become seduced by anecdotes and stories. You may have heard someone say something like "I'm not going to wear my seat belt because I heard about a guy who died in a car crash *because* he was wearing one. He got trapped in the car and couldn't get out. If he hadn't been wearing the seat belt, he would have been okay."

Well, yes, but we can't look at just one or two stories. What are the relative risks? Although there are a few odd cases where the seat belt *cost* someone's life, you're far more likely to die when *not* wearing one. Probability helps us look at this quantitatively.

We use the word *probability* in different ways to mean different things. It's easy to get swept away thinking that a person means one thing when they mean another, and that confusion can cause us to draw the wrong conclusion.

One kind of probability—*classic probability*—is based on the idea of symmetry and equal likelihood: A die has six sides, a coin has two sides, a roulette wheel has thirty-eight slots (in the United States; thirty-seven slots in Europe). If there is no manufacturing defect or tampering that favors one outcome over another, each outcome is equally likely. So the probability of rolling any particular number on a die is one out of six, of getting heads on a coin toss is one out of two, of getting any particular slot on the roulette wheel is one out of thirty-seven or thirty-eight.

Classic probability is restricted to these kinds of well-defined objects. In the classic case, we know the parameters of the system and thus can calculate the probabilities for the events each system will generate. A second kind of probability arises because in daily life we often want to know something about the likelihood of other events occurring, such as the probability that a drug will work on a patient, or that consumers will prefer one beer to another. In this second case, we need to estimate the parameters of the system because we don't know what those parameters are.

To determine this second kind of probability, we make observations or conduct experiments and count the number of times we get the outcome we want. These are called *frequentist* probabilities. We administer a drug to a group of patients and count how many people get better—that's an experiment, and the probability of the drug working is simply the proportion of people for whom it worked (based on the *frequency* of the desired outcome). If we run the experiment on a large number of people, the results will be close to the true probability, just like public-opinion polling.

Both classic and frequentist probabilities deal with recurring, replicable events and the proportion of the time that you can expect to obtain a particular outcome under substantially the same conditions. (Some hard-liner probabilists contend they have to be *identical* conditions, but I think this takes it too far because in the limit, the universe is never *exactly* the same, due to chance variations.) When you conduct a public-opinion poll by interviewing people at random, you're in effect asking them under identical conditions, even if you ask some today and some tomorrow—provided that some big event that might change their minds didn't occur in between. When a court witness testifies about the probability of a suspect's DNA matching the DNA found on a revolver, she is using *frequentist* probability, because she's essentially counting the number of DNA fragments that match versus the number that don't. Drawing a card from a deck, finding a defective widget on an assembly line, asking people if they like their brand of coffee are all examples of classic or frequentist probabilities that are recurring, replicable events (the card is classic, the widget and coffee are frequentist).

A third kind of probability differs from these first two because it's not obtained from an experiment or a replicable event—rather, it expresses an opinion or degree of belief about how likely a particular event is to occur. This is called *subjective probability* (one type of this is Bayesian probability, after the eighteenth-century statistician Thomas Bayes). When a friend tells you that there's a 50 percent chance that she's going to attend Michael and Julie's party this weekend, she's using Bayesian probability, expressing a strength of belief that she'll go. What will the unemployment rate be next year? We can't use the frequentist method because we can't consider next year's unemployment as a set of observations taken under identical or even similar conditions.

Let's think through an example. When a TV weather reporter says that there is a 30 percent chance of rain tomorrow, she didn't conduct experiments on a bunch of identical days with identical conditions (if such a thing even exists) and then count the outcomes. The 30 percent number expresses her degree of belief (on a scale of one to a hundred) that it will rain, and is meant to inform you about whether you want to go to the trouble of grabbing your galoshes and umbrella.

If the weather reporter is well calibrated, it will rain on exactly 30 percent of the days for which she says there is a 30 percent chance of rain. If it rains on 60 percent of those days, she's underestimated by a large amount. The issue of calibration is relevant only with subjective probabilities.

By the way, getting back to your friend who said there is a 50 percent chance she'll attend a party, a mistake that many non–critical thinkers make is in assuming that if there are two possibilities, they must be equally likely. Cognitive psychologists Amos Tversky and Daniel Kahneman described parties and other scenarios to people in an experiment. At a particular party, for example, people might be told that 70 percent of the guests are writers and 30 percent are engineers. If you bump into someone with a tattoo of Shakespeare, you might correctly assume that person to be one of the writers; if you bump into someone wearing a Maxwell's equations T-shirt, you might correctly assume that they are one of the engineers. But what if you bump into someone at random in the party and you've got nothing to go on—no Shakespeare tattoo, no math T-shirt—what's the probability that this person is an engineer? In Tversky and Kahneman's experiments, people tended to say, "Fifty-fifty," apparently confusing the two possible outcomes with two equally likely outcomes.

Subjective probability is the only kind of probability that we have at our disposal in practical situations in which there is no experiment, no symmetry equation. When a judge instructs the jury to return a verdict if the "preponderance of evidence" points toward the defendant's guilt, this is a subjective probability—each juror needs to decide for themselves whether a preponderance has been reached, weighing the evidence according to their own (and possibly idiosyncratic, not objective) internal standards and beliefs.

When a bookmaker lays odds for a horse race, he is using subjective probability—while it might be informed by data on the horses' track records, health, and the jockeys' history, there is no natural symmetry (meaning it's not a classic probability) and there is no experiment being conducted (meaning it's not a frequentist probability). The same is true for baseball and other sporting events. A bookie might say that the Royals have an 80 percent chance of winning their next game, but he's not using probability in a mathematical sense; this is just a way he—and we—use language to give the patina of numerical precision. The bookie can't turn back the hands of time and watch the Royals play the same game again and again, counting how many times they win it. He might well have crunched numbers or used a computer to inform his estimate, but at the end of the day, the number is just a guess, an indication of his degree of confidence in his prediction. A telltale piece of evidence that this is subjective is that different pundits come up with different answers.

Subjective probabilities are all around us and most of us don't even realize it—we encounter them in newspapers, in the boardroom, and in sports bars. The probability that a rogue nation will set off an atomic bomb in the next twelve months, that interest rates will go up next year, that Italy will win the World Cup, or that soldiers will take a particular hill are all subjective, not frequentist:

They are one-time, nonreplicable events. And the reputations of pundits and forecasters depend on their accuracy.

Combining Probabilities

One of the most important rules in probability is the multiplication rule. If two events are independent—that is, if the outcome of one does not influence the outcome of the other—you obtain the probability of *both* of them happening by multiplying the two probabilities together. The probability of getting heads on a coin toss is one half (because there are only two equally likely possibilities: heads and tails). The probability of getting a heart when drawing from a deck of cards is one-quarter (because there are only four equally likely possibilities: hearts, diamonds, clubs, and spades). If you toss a coin and draw a card, the probability of getting both heads and a heart is calculated by multiplying the two individual probabilities together: $\frac{1}{2} \times \frac{1}{4} = \frac{1}{8}$. This is called a joint probability.

You can satisfy yourself that this is true by listing all possible cases and then counting how many times you get the desired outcome:

Head	Heart	Tail	Heart
Head	Diamond	Tail	Diamond
Head	Club	Tail	Club
Head	Spade	Tail	Spade

I'm ignoring the very rare occasions on which you toss the coin and it lands exactly on its side, or it gets carried off by a seagull while it's in midair, or you have a trick deck of cards with all clubs.

We can similarly ask about the joint probability of three events: getting heads on a coin toss, drawing a heart from a deck of cards, and the next person you meet having the same birthday as you (the probability of that is roughly 1 out of 365.24—although births cluster a bit and some birthdates are more common than others, this is a reasonable approximation).

You may have visited websites where you are asked a series of multiple-choice questions, such as "Which of the following five streets have you lived on?" and "Which of the following five credit cards do you have?" These sites are trying to authenticate you, to be sure that you are who they think you are. They're using the multiplication rule. If you answer six of these questions in a row, each with a probability of only one in five (.2) that you'll get it right, the chances of you getting them right by simply guessing are only .2 × .2 × .2 × .2 × .2 × .2, or .000064—that's about 6 chances in 100,000. Not as strict as what you find in DNA courtroom testimony, but not bad. (If you're wondering why they don't just ask you a bunch of short-answer, fill-in questions, where you have to provide the entire answer yourself, instead of using multiple choice, it's because there are too many variants of correct answers. Do you refer to your credit card as being with Chase, Chase Bank, or JPMorgan Chase? Did you live on North Sycamore Street, N. Sycamore Street, or N. Sycamore St.? You get the idea.)

When the Probability of Events Is Informed by Other Events

The multiplication rule only applies if the events are independent of one another. What events are not independent? The weather, for example. The probability of it freezing tonight *and* freezing tomorrow night are not independent events—weather patterns tend to remain for more than one day, and although freak freezes are known to occur, your best bet about tomorrow's overnight temperatures is to look at today's. You *could* calculate the number of nights in the year in which temperatures drop below freezing—let's say it's thirty-six where you live—and then state that the probability of a freeze tonight is 36 out of 365, or roughly 10 percent, but that doesn't take the dependencies into account. If you say that the probability of it freezing two nights in a row during winter is 10% × 10% = 1% (following the multiplication rule), you'd be underestimating the probability because the two nights' events are not independent; tomorrow's weather forecast is informed by today's.

The probability of an event can also be informed by the particular sample that you're looking at. The probability of it freezing tonight is obviously affected by the area of the world you're talking about. That probability is higher at the forty-fourth parallel than the tenth. The probability of finding someone over six foot six is greater if you're looking at a basketball practice than at a tavern frequented by jockeys. The subgroup of people or things you're looking at is relevant to your probability estimate.

Conditional Probabilities

Often when looking at statistical claims, we're led astray by examining an entire group of random people when we really should be looking at a subgroup. What is the probability that you have pneumonia? Not very high. But if we know more about you and your particular case, the probability may be higher or lower. This is known as a *conditional probability*.

We frame two different questions:

1. What is the probability that a person drawn at random from the population has pneumonia?
2. What is the probability that a person *not* drawn at random, but one who is exhibiting three symptoms (fever, muscle pain, chest congestion) has pneumonia?

The second question involves a conditional probability. It's called that because we're not looking at every possible condition, only those people who match the condition specified. Without running through the numbers, we can guess that the probability of pneumonia is greater in the second case. Of course, we can frame the question so that the probability of having pneumonia is lower than for a person drawn at random:

1. What is the probability that a person *not* drawn at random, but one who has just tested negative for pneumonia three times in a row, and who has an especially robust immune system, and has just minutes ago finished first place in the New York City Marathon, has pneumonia?

Along the same lines, the probability of you developing lung cancer is not independent of your family history. The probability of a waiter bringing ketchup to your table is not independent of what you ordered. You can calculate the probability of any person selected at random developing lung cancer in the next ten years, or the probability of a waiter bringing ketchup to a table calculated over all tables. But we're in the lucky position of knowing that these events are dependent on other behaviors. This allows us to narrow the population we're studying in order to obtain a more accurate estimate. For example, if your father and mother both had lung cancer, you want to calculate the probability of you contracting lung cancer by looking at other people in this select group, people whose parents had lung cancer. If your parents didn't have lung cancer, you want to look at the relevant subgroup of people who lack a family history of it (and you'll likely come up with a different figure). If you want to know the probability that your waiter will bring you ketchup, you might look at only the tables of those patrons who ordered hamburgers or fries, not those who ordered tuna tartare or apple pie.

Ignoring the dependence of events (assuming independence) can have serious consequences in the legal world. One was the case of Sally Clark, a woman from Essex, U.K., who stood trial for murdering her second child. Her first child had died in infancy, and his death had been attributed to SIDS (sudden infant death syndrome, or crib death). The prosecutors argued that the odds of having two children die of SIDS were so low that she must have murdered the second child. The prosecution's witness, a pediatrician, cited a study that said SIDS occurred in 1 out of 8,543 infant deaths. (Dr. Meadow's expertise in pediatrics does not make him an expert statistician or epidemiologist—this sort of confusion is the basis for many faulty judgments and is discussed in Part 3 of this book; an expert in one domain is not automatically an expert in another, seemingly related, domain.)

Digging deeper, we might question the figure of 8,543 deaths. How do they know that? SIDS is a diagnosis of exclusion—that is, there is no test that medical personnel can perform to conclude a death was by

SIDS. Rather, if doctors are not able to find the cause, and they've ruled out everything else, they label it SIDS. Not being able to find something is not proof that it didn't occur, so it is plausible that some of the deaths attributed to SIDS were actually the result of less mysterious causes, such as poisoning, suffocation, heart defect, etc.

For the sake of argument, however, let's assume that SIDS is the cause of 1 out of 8,543 infant deaths as the expert witness testified. He further testified that the odds of two SIDS deaths occurring in the same family were $\frac{1}{8543}$ x $\frac{1}{8543}$, or 1 in 73 million. ("Coincidence? I think *not!*" the prosecutor might have shouted during his summation.) This calculation—this application of the multiplication rule—assumes the deaths are independent, but they might not be. Whatever caused Mrs. Clark's first child to die suddenly might be present for both children by virtue of them being in the same household: Two environmental factors associated with SIDS are secondhand smoke and putting a baby to sleep on its stomach. Or perhaps the first child suffered from a congenital defect of some sort; this would have a relatively high probability of appearing in the second child's genome (siblings share 50 percent of their DNA). By this way of thinking, there was a 50 percent chance that the second child would die due to a factor such as this, and so now Mrs. Clark looks a lot less like a child murderer. Eventually, her husband found evidence in the hospital archives that the second child's death had a microbiological cause. Mrs. Clark was acquitted, but only after serving three years in prison for a crime she didn't commit.

There's a special notation for conditional probabilities. The probability of a waiter bringing you ketchup, given that you just ordered a hamburger, is written:

P(ketchup | hamburger)

where the vertical bar | is read as *given*. Note that this notation leaves out a lot of the words from the English-language description, so that the mathematical expression is succinct.

The probability of a waiter bringing you ketchup, given that you just ordered a hamburger *and* you asked for the ketchup, is noted:

P(ketchup | hamburger ∧ asked)

where the ∧ is read as *and*.

Visualizing Conditional Probabilities

The relative incidence of pneumonia in the United States in one year is around 2 percent—six million people out of the 324 million in the country are diagnosed each year (of course there are no doubt many undiagnosed cases, as well as individuals who may have more than one case in a year, but let's ignore these details for now). Therefore the probability of any person drawn at random having pneumonia is approximately 2 percent. But we can home in on a better estimate if we know something about that particular person. If you show up at the doctor's office with coughing, congestion, and a fever, you're no longer a person drawn at random—you're someone in a doctor's office showing these symptoms. You can methodically update your belief that something is true (that you have pneumonia) in light of new evidence. We do this by applying Bayes's rule to calculate a conditional probability: What is the probability that I have pneumonia *given* that I show symptom x? This kind of updating can become increasingly refined the more information you have. What is the probability that I have pneumonia *given* that I have these symptoms, and *given* that I have a family history of it, and *given* that I just spent three days with someone who has it? The probabilities climb higher and higher.

You can calculate the probabilities using the formula for Bayes's rule (found in the Appendix), but an easy way to visualize and compute conditional probabilities is with the fourfold table, describing all possible scenarios: You did or didn't order a hamburger, and you did or didn't receive ketchup:

Ordered Hamburger

		YES	NO
Received Ketchup	**YES**		
	NO		

Then, based on experiments and observation, you fill in the various values, that is, the frequencies of each event. Out of sixteen customers you observed at a restaurant, there was one instance of someone ordering a hamburger with which they received ketchup, and two instances with which they didn't. These become entries in the left-hand column of the table:

Ordered Hamburger

		YES	NO
Received Ketchup	**YES**	I	5
	NO	2	8

Similarly, you found that five people who didn't order a hamburger received ketchup, and eight people did not. These are the entries in the right-hand column.

Next, you sum the rows and columns:

Ordered Hamburger

		YES	NO	
Received	**YES**	1	5	6
Ketchup	**NO**	2	8	10
		3	13	16

Now, calculating the probabilities is easy. If you want to know the probability that you received ketchup *given* that you ordered a hamburger, you start with the given. That's the left-hand vertical column.

Ordered Hamburger

		YES	NO	
Received	**YES**	1	5	6
Ketchup	**NO**	2	8	10
		3	13	16

Three people ordered hamburgers altogether—that's the total at the bottom of the column. Now what is the probability of receiving ketchup *given* you ordered a hamburger? We look now at the "YES received ketchup" square in the "YES ordered hamburger" column, and that number is 1. The conditional probability, P(ketchup|hamburger) is then just one out of three. And you can visualize the logic: three people ordered a hamburger; one of them got ketchup and two didn't. We ignore the right-hand column for this calculation.

We can use this to calculate any conditional probability, including the probability of receiving ketchup if you *didn't* order a hamburger: Thirteen people didn't order a hamburger, five of them got ketchup, so the probability is five out of thirteen, or about 38 percent. In this particular restaurant, you're more likely to get ketchup if you didn't order a hamburger than if you did. (Now fire up your critical thinking. How could this be? Maybe the data are driven by people who ordered fries. Maybe all the hamburgers served already have ketchup on them.)

Medical Decision Making

This way of visualizing conditional probabilities is useful for medical decision making. If you take a medical test, and it says you have some disease, what is the probability you actually have the disease? It's not 100 percent, because the tests are not perfect—they produce false positives (reporting that you have the disease when you don't) and false negatives (reporting that you don't have the disease when you do).

The probability that a woman has breast cancer is 0.8 percent. If she has breast cancer, the probability that a mammogram will indicate it is only 90 percent because the test isn't perfect and it misses some cases. If the woman does not have breast cancer, the probability of a positive result is 7 percent. Now, suppose a woman, drawn at random, has a positive result—what is the probability that she actually has breast cancer?

We start by drawing a fourfold table and filling in the possibilities: The woman actually has breast cancer or doesn't, and the test can report that she does or that she doesn't. To make the numbers work out easily—to make sure we're dealing with whole numbers—let's assume we're talking about 10,000 women.

That's the total population, and so that number goes in the lower right-hand corner of the figure, outside the boxes.

Test Result

		YES	NO	
Actually Has Breast Cancer	YES			————
	NO			————
				10,000

Unlike the hamburger-ketchup example, we fill in the margins first, because that's the information we were given. The probability of breast cancer is 0.8 percent, or 80 out of 10,000 people. That number goes in the margin of the top row. (We don't yet know how to fill in the boxes, but we will in a second.) And because the row has to add up to 10,000, we know that the margin for the bottom row has to equal

$$10,000 - 80 = 9,920.$$

Test Result

		YES	NO	
Actually Has Breast Cancer	**YES**			80
	NO			9,920
				10,000

We were told that the probability that the test will show a positive *if* breast cancer exists is 90 percent. Because probabilities have to add up to 100 percent, the probability that the test will *not* show a positive result if breast cancer exists has to be 100 percent – 90 percent, or 10 percent. For the eighty women who actually have breast cancer (the margin for the top row), we now know that 90 percent of them will have a positive test result (90% of 80 = 72) and 10 percent will have a negative result (10% of 80 = 8). This is all we need to know how to fill in the boxes on the top row.

Test Result

		YES	NO	
Actually Has Breast Cancer	**YES**	72	8	80
	NO			9,920
				10,000

We're not yet ready to calculate the answer to questions such as "What is the probability that I have breast cancer given that I had a positive test result?" because we need to know how many people will have a positive test result. The missing piece of the puzzle is in the original description: 7 percent of women who don't have breast cancer will still show a positive result. The margin for the lower row tells us 9,920 women don't have breast cancer; 7 percent of them = 694.4. (We'll round to 694.) That means that 9,920 − 694 = 9,226 goes in the lower right square.

Test Result

		YES	NO	
Actually Has Breast Cancer	**YES**	72	8	80
	NO	694	9,226	9,920
		766	9,234	10,000

Finally, we add up the columns.

If you're among the millions of people who think that having a positive test result means you definitely have the disease, you're wrong. The conditional probability of having breast cancer given a positive test result is the upper left square divided by the left column's margin total, or $72/766$. The good news is that, *even with a positive mammogram, the probability of actually having breast cancer is* 9.4 percent. This is because the disease is relatively rare (less than 1 in 1,000) and the test for it is imperfect.

Test Result

		YES	NO	
Actually Has Breast Cancer	**YES**	72	8	80
	NO	694	9,226	9,920
		766	9,234	10,000

Conditional Probabilities Do Not Work Backward

We're used to certain symmetries in math from grade school: If x = y then y = x. 5 + 7 = 7 + 5. But some concepts don't work that way, as we saw in the discussion above on probability values (if the probability of a false alarm is 10 percent, that doesn't mean that the probability of a hit is 90 percent).

Consider the statistic:

> Ten times as many apples are sold in supermarkets as in roadside stands.

A little reflection should make it apparent that this does not mean you're more likely to find an apple on the day you want one by going to the supermarket: The supermarket may have more than ten times the number of customers as roadside stands have, but even with its greater inventory, it may not keep up with demand. If you see a random person walking down the street with an apple, and you have no information about where they bought it, the probability is higher that they bought it at a supermarket than at a roadside stand.

We can ask, as a conditional probability, what is the probability that this person bought it at a supermarket given that they have an apple?

P(was in a supermarket | found an apple to buy)

It is not the same as what you might want to know if you're craving a Honeycrisp:

P(found an apple to buy | was in a supermarket)

This same asymmetry pops up in various disguises, in all manner of statistics. If you read that more automobile accidents occur at seven p.m. than at seven a.m., what does that mean? Here, the language of the statement itself is ambiguous. It could either mean you're looking at the probability that it was seven p.m. given that an accident occurred, or the probability that an accident occurred given that it was seven p.m. In the first case, you're looking at all accidents and seeing how many were at seven p.m. In the second case, you're looking at how many cars are on the road at seven p.m., and seeing what proportion of them are involved in accidents. What?

Perhaps there are far more cars on the road at seven p.m. than any other time of day, and far fewer accidents per thousand cars. That would yield more accidents at seven p.m. than any other time, simply due to the larger number of vehicles on the road. It is the accident *rate* that helps you determine the safest time to drive.

Similarly, you may have heard that most accidents occur within three miles of home. This isn't because that area is more dangerous per se, it's because the majority of trips people take are short ones, and so the three miles around the home is much more traveled. In most cases, these two different interpretations of the statement will not be equivalent:

P(7 p.m. | accident) ≠ P(accident | 7 p.m.)

The consequences of such confusion are hardly just theoretical: Many court cases have hinged on a misapplication of conditional probabilities, confusing the direction of what is known. A forensics expert may compute, correctly, that the probability of the blood found at the crime scene matching the defendant's blood type by chance is only 1 percent. This is *not* at all the same as saying that there is only a 1 percent chance the defendant is innocent. What? Intuition tricks us again. The forensics expert is telling us the probability of a blood match *given* that the defendant is innocent:

P(blood match | innocence)

Or, in plain language, "the probability that we would find a match if the defendant were actually innocent." That is not the same as the number you really want to know: What is the probability that the defendant is innocent *given* that the blood matched:

P(blood match | innocence) ≠ P(innocence | blood match)

Many innocent citizens have been sent to prison because of this misunderstanding. And many patients have made poor decisions about medical care because they thought, mistakenly, that

P(positive test result | cancer) = P(cancer | positive test result)

And it's not just patients—doctors make this error all the time (in one study 90 percent of doctors treated the two different probabilities the same). The results can be horrible. One surgeon persuaded ninety women to have their healthy breasts removed if they were in a high-risk group. He had noted that 93 percent of breast cancers occurred in women who were in this high-risk group. Given that a woman had breast cancer, there was a 93 percent chance she was in this group: P(high-risk group | breast cancer) = .93. Using a fourfold table for a sample of 1,000 typical women, and adding the additional information that 57 percent of women fall into this high-risk group, and that the probability of a woman having breast cancer is 0.8 percent (as mentioned earlier), we can calculate P(breast cancer | high-risk group), which is the statistic a woman needs to know before consenting to the surgery (numbers are rounded to the nearest integer):

High-Risk Group

		YES	NO	
Actually Has Breast Cancer	**YES**	7	1	8
	NO	563	429	992
		570	430	1,000

The probability that a woman has cancer, given that she is in this high-risk group is not 93 percent, as the surgeon erroneously thought, but only 7/570, or 1 percent. The surgeon overestimated the cancer risk by nearly one hundred times the actual risk. And the consequences were devastating.

The fourfold tables might feel like a strange little exercise, but actually what you're doing here is scientific and critical thinking, laying out the numbers visually in order to make the computation easier. And the results of those computations allow you to quantify the different parts of the problem, to help you make more rational, evidence-based decisions. They are so powerful, it's surprising that they're not taught to all of us in high school.

Thinking About Statistics and Graphs

Most of us have difficulty figuring probabilities and statistics in our heads and detecting subtle patterns in complex tables of numbers. We prefer vivid pictures, images, and stories. When making decisions, we tend to overweight such images and stories, compared to statistical information. We also tend to misunderstand or misinterpret graphics.

Many of us feel intimidated by numbers and so we blindly accept the numbers we're handed. This can lead to bad decisions and faulty conclusions. We also have a tendency to apply critical thinking only to things we disagree with. In the current information age, pseudo-facts masquerade as facts, misinformation can be indistinguishable from true information, and numbers are often at the heart of any important claim or decision. Bad statistics are everywhere. As sociologist Joel Best says, it's not just because the other guys are all lying weasels. Bad statistics are produced by people—often sincere, well-meaning people—who aren't thinking critically about what they're saying.

The same fear of numbers that prevents many people from analyzing statistics prevents them from looking carefully at the numbers in a graph, the axis labels, and the story that they tell. The world is full of coincidences and bizarre things are very likely to happen—but just because two things change together doesn't mean that one caused the other or that they are even related by a hidden *third factor x*. People who are taken in by such associations or co-incidences usually have a poor understanding of probability, cause and effect, and the role of randomness in the unfolding of events. Yes, you could spin a story about how the drop in the number of pirates over the last three hundred years and the coinciding rise in global temperatures must surely indicate that pirates were essential to keeping global warming under control. But that's just sloppy thinking, and is a misinterpretation of the evidence. Sometimes the purveyors of this sort of faulty logic know better and hope that you won't notice; sometimes they have been taken in themselves. But now you know better.

PART TWO

EVALUATING WORDS

A lie which is half a truth is ever the blackest of lies.

—ALFRED, LORD TENNYSON

How Do We Know?

We are a storytelling species, and a social species, easily swayed by the opinions of others. We have three ways to acquire information: We can discover it ourselves, we can absorb it implicitly, or we can be told it explicitly. Much of what we know about the world falls in this last category—somewhere along the line, someone told us a fact or we read about it, and so we know it only secondhand. We rely on people with expertise to tell us.

I've never seen an atom of oxygen or a molecule of water, but there is a body of literature describing meticulously conducted experiments that lead me to believe these exist. Similarly, I haven't verified firsthand that Americans landed on the moon, that the speed of light is 186,000 miles per second, that pasteurization really kills bacteria, or that humans normally have twenty-three chromosomes. I don't know firsthand that the elevator in my building has been properly designed and maintained, or that my doctor actually went to medical school. We rely on experts, certifications, licenses, encyclopedias, and textbooks.

But we also need to rely on ourselves, on our own wits and powers of reasoning. Lying weasels who want to separate us from our money, or get us to vote against our own best interests, will try to snow us with pseudo-facts, confuse us with numbers that have no basis, or distract us with information that, upon closer examination, is not actually relevant. They will masquerade as experts.

The antidote to this is to analyze claims we encounter the way we analyze statistics and graphs. The skills necessary should not be beyond the ability of most fourteen-year-olds. They are taught in law schools and schools of journalism, sometimes in business schools and graduate science programs, but rarely to the rest of us, to those who need it most.

If you like watching crime dramas, or reading investigative journalism pieces, many of the skills will be familiar—they resemble the kinds of evaluations that are made during court cases. Judges and juries evaluate competing claims and try to discover the truth within. There are codified rules about what constitutes real evidence; in the United States, documents that haven't been authenticated are generally not allowed, nor is "hearsay" testimony, although there are exceptions.

Suppose someone points you to a website that claims that listening to Mozart music for twenty minutes a day will make us smarter. Another website says it's not true. A big part of the problem here is that the human brain often makes up its mind based on emotional considerations, and then seeks to justify them. And the brain is a very powerful self-justifying machine. It would be nice to believe that all you have to do is listen to beautiful music for twenty minutes to suddenly take your place at the head of the IQ line. It takes effort to evaluate claims like this, probably more time than it would take to listen to *Eine Kleine Nachtmusik*, but it is necessary to avoid drawing incorrect conclusions. Even the smartest of us can be fooled. Steve Jobs delayed treatment for his pancreatic cancer while he followed the advice (given in books and websites) that a change in diet could provide a cure. By the time he realized the diet wasn't working, the cancer had progressed too far to be treated.

Determining the truthfulness or accuracy of a source is not always possible. Consider the epigram opening Part One:

> It ain't what you don't know that gets you into trouble. It's
> what you know for sure that just ain't so.

I saw this at the opening of the feature film *The Big Short*, which attributed it to Mark Twain, and I felt I had seen it somewhere before; Al Gore also used it in his film *An Inconvenient Truth* nine years earlier with the same attribution. But in fact-checking the *Field Guide*, I could not find any evidence that Twain ever said this. The attribution and quote itself are prime examples of what the quote is trying to warn us against. The directors, writers, and producers of both films didn't do their homework—what they thought they knew for sure turned out not to be true at all.

A little web research pulled up an article in *Forbes* that claims it is a misattribution. The author, Nigel Rees, cites *Respectfully Quoted*, a dictionary of quotations compiled by the U.S. Library of Congress. That book reports various formulations of the remark in *Everybody's Friend, or Josh Billing's Encyclopedia and Proverbial Philosophy of Wit and Humor* (1874). "There you are, you see," writes Rees. "Mark Twain is a better-known humorist than 'Josh Billings' and so the quote drifts towards him."

Rees continues:

And not only him. In a 1984 presidential debate, Walter Mondale came up with this: "I'm reminded a little bit of what Will Rogers once said of Hoover. He said 'It's not what he doesn't know that bothers me, it's what he knows for sure just ain't so.'"

Who's right? With difficult matters such as this, it is often helpful to consult an expert. I asked Gretchen Lieb, a research librarian at Vassar who works as the liaison to the English Department, and who provided this insightful analysis:

> Quotations are tricky things. They're the literary equivalent of statistics, really, in terms of lies, damn lies, etc. Older quotations are almost like translations from another language, too, in terms of being interpretations rather than verbatim, especially in the case of this circle, since these authors wrote in a sort of fantasy dialect, à la Huckleberry Finn, that is difficult to read and downright disturbing to us now in some cases.
>
> I could go check numerous other books of quotations, such as Oxford, etc., but that would be so twentieth century.
>
> Have you come across HathiTrust? It's the corpus of books from research libraries that is behind Google Books, and it's a gold mine, especially for pre-1928 printed materials.
>
> Here's the Josh Billings attribution in "Respectfully Quoted" (we have it as an e-book; I didn't need to walk away from my desk!), and it cites the Oxford Dictionary of Quotations, which I tend to use more than Bartlett's:

"The trouble with people is not that they don't know but that they know so much that ain't so." Attributed to Josh Billings (Henry Wheeler Shaw) by *The Oxford Dictionary of Quotations*, 3d ed., p. 491 (1979). Not verified in his writings, although some similar ideas are found in *Everybody's Friend, or Josh Billing's Encyclopedia and Proverbial Philosophy of Wit and Humor* (1874). Original spelling is corrected: "What little I do know I hope I am certain of." (p. 502) "Wisdom don't consist in knowing more that is new, but in knowing less that is false." (p. 430) "I honestly believe it is better to know nothing than to know what ain't so." (p. 286)

By the way, regarding the Walter Mondale attribution to Will Rogers, *Respectfully Quoted* notes that this has not been found in Rogers's work.

Here is a link to Billings's book, where you can search for the phrase "ain't so" and get the idea of what lies therein: http://hdl.handle.net/2027/njp.32101067175438.

Not verifiable. Plus, if you search for Mark Twain, you find that this compendium/encyclopedia writer cites fellow humorist and smartypants Mark Twain as his most trusted correspondent, so they're having a conversation and bouncing clever aphorisms, or as Billings would say, "affurisms," off of each other. Who knows who said what?

I usually roll my eyes when people, especially politicians, quote Mark Twain or Will Rogers, and think to myself, H. L. Mencken, we hardly know you. Critical minds like his are in short supply these days. Poor Josh Billings. Being the second most famous humorist puts you on precarious ground a hundred years later.

So here's an odd case of a quote that appears to have been utterly fabricated, both in its content and its attribution. The basic idea was contained in Billings, although it's not clear if that idea came from him, Twain, or perhaps their buddy Bret Harte. Will Rogers gets put in the mix because, well, it just sort of *sounds* like something he would say.

The quote that opens Part Two was given to me by an acquaintance who misremembered it as:

> The blackest lie is a partial truth that leads you to the wrong conclusion.

It sounded plausible. It would be just like Tennyson to give color to an abstract noun, and to mix the metaphysical with the practical. I only found out the actual quote ("A lie which is half a truth is ever the blackest of lies") when fact-checking for this book. So it goes, as Kurt Vonnegut would say.

In the presence of new or conflicting claims, we can make an informed and evidence-based choice about what is true. We examine the claims for ourselves, and make a decision, acting as our own judge and jury. And as part of the process we usually do well to seek expert opinions. How do we identify them?

Identifying Expertise

The first thing to do when evaluating a claim by some authority is to ask who or what established their authority. If the authority comes from having been a witness to some event, how credible a witness are they?

Venerable authorities can certainly be wrong. The U.S. government was mistaken about the existence of weapons of mass destruction (WMDs) in Iraq in the early 2000s, and, in a less politically fraught case, scientists thought for many years that humans had twenty-four pairs of chromosomes instead of twenty-three. Looking at what the acknowledged authorities say is not the last step in evaluating claims, but it is a good early step.

Experts talk in two different ways, and it is vital that you know how to tell these apart. In the first way, they review facts and evidence, synthesizing them and forming a conclusion based on the evidence. Along the way, they share with you what the evidence is, why it's relevant, and how it helped them to form their conclusion. This is the way science is supposed to be, the way court trials proceed, and the way the best business decisions, medical diagnoses, and military strategies are made.

The second way experts talk is to just share their opinions. They are human. Like the rest of us, they can be given to stories, to spinning loose threads of their own introspections, what-ifs, and untested ideas. There's nothing wrong with this—some good, testable ideas come from this sort of associative thinking—but it should not be confused with a logical, evidence-based argument. Books and articles for popular audiences by pundits and scientists often contain this kind of rampant speculation, and we buy them because we are impressed by the writer's expertise and rhetorical talent. But properly done, the writer should also lift the veil of authority, let you look behind the curtain, and see at least some of the evidence for yourself.

The term *expert* is normally reserved for people who have undertaken special training, devoted a large amount of time to developing their expertise (e.g., MDs, airline pilots, musicians, or athletes), and whose abilities or knowledge are considered high relative to others'. As such, expertise is a social judgment—we're comparing one person's skill to the skill level of other people in the world. Expertise is relative. Einstein was an expert on physics sixty years ago; he would probably not be considered one if he were still alive today and hadn't added to his knowledge base what Stephen Hawking and so many other physicists now know. Expertise also falls along a continuum. Although John Young is one of only twelve people to have walked on the moon, it would probably not be accurate to say that Captain Young is an *expert* on moonwalking, although he knows more about it than almost anyone else in the world.

Individuals with similar training and levels of expertise will not necessarily agree with one another, and even if they do, these experts are not always right. Many thousands of expert financial analysts make predictions about stock prices that are completely wrong, and some small number of novices turn out to be right. Every British record company famously rejected the Beatles' demo tape, and a young producer with no expertise in popular music, George Martin, signed them to EMI. Xerox PARC, the inventors of the graphical interface computer, didn't see any future for personal computers; Steve Jobs, who had no business experience at all, thought they were wrong. The success of newcomers in these domains is generally understood to be because stock prices and popular taste are highly unpredictable and chaotic. Stuff happens. So it's not that experts are never wrong, it's just that, statistically, they're more likely to be right.

Many inventors and innovators were told "it will never work" by experts, with the Wright brothers and their fellow would-be inventors of motorized flight being an example *par excellence*. The Wright brothers were high school dropouts, with no formal training in aeronautics or physics. Many experts with formal training declared that heavier-than-air flight would never be possible. The Wrights were self-taught, and their perseverance made them de facto experts themselves when they built a functional heavier-than-air airplane, and proved the other experts wrong. Michael Lewis's baseball story *Moneyball* shows how someone can beat the experts by rejecting conventional wisdom and applying logic and statistical analysis to an old problem; Oakland A's manager Billy Beane built a competitive team by using player performance metrics that other teams undervalued, bringing his team to the playoffs two years in a row, and substantially increasing the team's worth.

Experts are often licensed, or hold advanced degrees, or are recognized by other authorities. A Toyota factory-certified mechanic can be considered an expert on Toyotas. The independent or self-taught mechanic down the street may have just as much expertise, and may well be better and cheaper. It's just that the odds aren't as good, and it can be difficult to figure that out for yourself. It's just averages: The average licensed Toyota mechanic is going to know more about fixing your Toyota than the average independent. Of course, there are exceptions and you have to bring your own logic to bear on this. I knew a Mercedes mechanic who worked for a Mercedes dealership for twenty-five years and was among their most celebrated and top-rated mechanics. He wanted to shorten his commute and be his own boss so he opened up his own shop. His thirty-five years of experience (by the time I knew him) gave him more expertise than many of the dealer's younger mechanics. Or another case: An independent may specialize in certain repairs that the dealer rarely performs, such as transmission overhaul or reupholstering. You're better off having your differential rebuilt by an independent who does five of those a month than a dealer who probably only did it once in vocational school. It's like the saying about surgeons: If you need one, you want the doctor who has performed the same operation you're going to get two hundred times, not once or twice, no matter how well those couple of operations went.

In science, technology, and medicine, experts' work appears in peer-reviewed journals (more on those in a moment) or on patents. They may have been recognized with awards such as a Nobel Prize, an Order of the British Empire, or a National Medal of Science. In business, experts may have had experience such as running or starting a company, or amassing a fortune (Warren Buffett, Bill Gates). Of course, there are smaller distinctions as well—salesperson of the month, auto mechanic of the year, community "best of" awards (e.g., best Mexican restaurant, best roofing contractor).

In the arts and humanities, experts may hold university positions or their expertise may be acknowledged by those with university or governmental positions, or by expert panels. These expert panels are typically formed by soliciting advice from previous winners and well-placed scouts—this is how the Nobel and the MacArthur "genius" award nomination and selection panels are constituted.

If people in the arts and humanities have won a prize, such as the Nobel, Pulitzer, Kennedy Center Honors, Polaris Music Prize, Juno, National Book Award, Newbery, or Man Booker Prize, we conclude they are among the experts at their craft. Peer awards are especially useful in judging expertise. ASCAP, an association whose membership is limited to professional songwriters, composers, and music publishers, presents awards voted on by its members; the award is meaningful because those who bestow it constitute a panel of peer experts. The Grammys and the Academy Awards are similarly voted on by peers within the music and film industry, respectively.

You might be thinking, "Wait a minute. There are always elements of politics and personal taste in such awards. My favorite actor/singer/writer/dancer has never won an award, and I'll bet I could find thousands of people who think she's as good as this year's award winner." But that's a different matter. The award system is generally biased toward ensuring that every winner is deserving, which is not the same as saying that every deserving person is a winner. (Recall the discussion of asymmetries earlier.) Those who are recognized by bona fide, respectable awards have usually risen to a level of expertise. (Again, there are exceptions, such as the awarding of a Grammy in 1990, which was later retracted, to lip-syncers Milli Vanilli; or the awarding of a Pulitzer Prize to

Washington Post reporter Janet Cooke, which was withdrawn two days later when it was discovered that the winning story was fraudulent. Novelist Gabriel García Márquez quipped that Cooke should've been awarded the Nobel Prize for *literature*.) When an expert has been found guilty of fraud, does it negate their expertise? Perhaps. It certainly impacts their credibility—now that you know they've lied once, you should be on guard that they may lie again.

Expertise Is Typically Narrow

Dr. Roy Meadow, the pediatrician who testified in the case of the alleged baby killer Sally Clark, had no expertise in medical statistics or epidemiology. He *was* in the medical profession, and the prosecutor who put him on the stand undoubtedly hoped that jurors would assume he had this expertise. William Shockley was awarded a Nobel Prize in physics as one of three inventors of the transistor. Later in life, he promoted strongly racist views that took hold, probably because people assumed that if he was smart enough to win a Nobel, he must know things that others don't. Gordon Shaw, who "discovered" the now widely discredited Mozart effect, was a physicist who lacked training in behavioral science; people probably figured, as they did with Shockley, "He's a physicist—he must be really smart." But intelligence and experience tend to be domain-specific, contrary to the popular belief that intelligence is a single, unified quantity. The best Toyota mechanic in the world may not be able to diagnose what's wrong with your VW, and the best tax attorney may not be able to give the best advice for a breach-of-contract suit. A physicist is probably not the best person to ask about social science.

There's a special place in our hearts (but hopefully not our rational minds) for actors who use their character's image to hawk products. As believable as Sam Waterston was as the trustworthy, ethical district attorney Jack McCoy in *Law & Order*, as an actor he has no special insight into banking and investments, although his commercials for TD Ameritrade were compelling. A generation earlier, Robert Young, who was much loved on TV's *Marcus Welby, M.D.*, did commercials for Sanka. Actors Chris Robinson (*General Hospital*) and Peter Bergman (*All My Children*) hawked Vicks Formula 44; due to FTC regulations (the so-called white coat rule) the actors had to speak a disclaimer that became a widely known catchphrase: "I'm not a doctor, but I play one on TV." Apparently, gullible viewers mistook the actors' authority in a television drama for authority in the real world of medicine.

Source Hierarchy

Some publications are more likely to consult true experts than others, and there exists a hierarchy of information sources. Some sources are simply more consistently reliable than others. In academia, peer-reviewed articles are generally more accurate than books, and books by major publishers are generally more accurate than self-published books (because major publishers are more likely to review and edit the material and have a greater financial incentive to do so). Award-winning newspapers such as the *New York Times*, the *Washington Post*, and the *Wall Street Journal* earned their reputations by being consistently accurate in their coverage of news. They strive to obtain independent verifications for any news story. If one government official tells them something, they get corroboration from another. If a scientist makes a claim, they contact other scientists who don't have any stake in the finding to hear independent opinions. They do make mistakes; even *Times* reporters have been found guilty of fabrications, and the "newspaper of record" prints errata every day. Some people, including Noam Chomsky, have argued that the *Times* is a vessel of propaganda, reporting news about the U.S. government without a proper amount of skepticism. But again, like with auto mechanics, it's a matter of averages—the great majority of what you read in the *New York Times* is likelier to be true than what you read in, for example, the *New York Post*.

Reputable sources want to be certain of facts before publishing them. Many sources have emerged on the Web that do not hold to the same standards, and in some cases, they can break news stories and do so accurately before the more traditional and cautious media do. Many of us learned of Michael Jackson's death from TMZ.com before the traditional media reported it. TMZ was willing to run the story based on less evidence than were the *Los Angeles Times* or NBC. In that particular case, TMZ turned out to be right, but you can't count on this sort of reporting.

A number of celebrity death reports that circulated on Twitter were found to be false. In 2015 alone, these included Carlos Santana, James Earl Jones, Charles Manson, and Jackie Chan. A 2011 fake tweet caused a sell-off of shares for the company Audience, Inc., during which its stock lost 25 percent. Twitter itself saw its shares climb 8 percent—temporarily—after false rumors of a takeover were tweeted, based on a bogus website made to look a great deal like Bloomberg.com's. As the *Wall Street Journal* reported, "The use of false rumors and news reports to manipulate stocks is a centuries-old ruse. The difference today is that the sheer ubiquity and amount of information that courses through markets makes it difficult for traders operating at high speeds to avoid a well-crafted hoax." And it happens to the best of us. Veteran reporter (and part of a team of journalists that was awarded a 1999 Pulitzer Prize) Jonathan Capehart wrote a story for the *Washington Post* based on a tweet by a nonexistent congressman in a nonexistent district.

As with graphs and statistics, we don't want to blindly believe everything we encounter from a good source, nor do we want to automatically reject everything from a questionable source. You shouldn't trust everything you read in the *New York Times*, or reject everything you read on TMZ. Where something appears goes to the credibility of the claim. And, as in a court trial, you don't want to rely on a single witness, you want corroborating evidence.

The Website Domain

The three-digit suffix of the URL indicates the domain. It pays to familiarize yourself with the domains in your country because some of the domains have restrictions, and that can help you establish a site's credibility for a given topic. In the United States, for example, .edu is reserved for nonprofit educational institutions like Stanford.edu (Stanford University); .gov is reserved for official government agencies like CDC.gov (the Centers for Disease Control); .mil for U.S. military organizations, like army.mil. The most famous is probably .com, which is used for commercial enterprises like GeneralMotors.com. Others include .net, .nyc, and .management, which carry no restrictions (!). Caveat emptor. BestElectrical Service.nyc might actually be in New Jersey (and their employees might not even be licensed to work in New York); AlphaAnd OmegaConsulting.management may not know the first or the last thing about management.

Knowing the domain can also help to identify any potential bias. You're more likely to find a neutral report from an educational or nonprofit study (found on a .edu, .gov, or .org site) than on a commercial site, although such sites may also host student blogs and unsupported opinions. And educational and nonprofits are not without bias: They may present information in a way that maximizes donations or public support for their mission. Pfizer.com may be biased in their discussions about drugs made by competing companies, such as GlaxoSmithKline, and Glaxo of course may be biased toward their own products.

Note that you don't always want neutrality. When searching for the owner's manual for your refrigerator, you probably want to visit the (partisan) manufacturer's website (e.g., Frigidaire.com) rather than a site that could be redistributing an outdated or erroneous version of the manual. That .gov site may be biased toward government interests, but a .gov site can give you most accurate info on laws, tax codes, census figures, or how to register your car. CDC .gov and NIH.gov probably have more accurate information about most medical issues than a .com because they have no financial interest.

Who Is Behind It?

Could the website be operating under a name meant to deceive you? The Vitamin E Producers Association might create a website called

NutritionAndYou.info, just to make you think that their claims are unbiased. The president of the grocery chain Whole Foods was caught masquerading as a customer on the Web, touting the quality of his company's groceries. Many rating sites, including Yelp! and Amazon, have found their ratings ballot boxes stuffed by friends and family of the people and products being rated. People are not always who they appear to be on the Web. Just because a website is named U.S. Government Health Service, that doesn't mean it is run by the government; a site named Independent Laboratories doesn't mean that it is independent—it could well be operated by an automobile manufacturer who wants to make its cars look good in not-so-independent tests.

In the 2014 congressional race for Florida's thirteenth district, the local GOP offices created a website with the name of their Democratic opponent, Alex Sink, to trick people into thinking they were giving money to her; in reality, the money went to her opponent, David Jolly. The site, contribute.sinkforcongress2014.com, used Sink's color scheme and featured a smiling photo of her, very similar to the photo on her own site.

Working together, across the aisle to break the gridlock in Washington

Illustration of the website for Democratic Congressional candidate Alex Sink

Illustration of the GOP website used to solicit money for Alex Sink's Republican opponent, David Jolly

The GOP's site does say that the money will be used to defeat Sink, so it's not outright fraud, but let's face it—most people don't take the time to read such things carefully. The most eye-catching parts of the trick site are the large photo of Sink, and the headline Alex Sink | Congress, which strongly implies that the site is *for* Alex Sink, not against her. Not to be outdone, Democrats responded with the same trick, creating the site www.JollyForCongress.com to collect money meant for Sink's rival.

Dentec Safety Specialists and Degil Safety Products are competing companies with similar services and products. Dentec has a website, DentecSafety.com, to market their products, and Degil has a website, DegilSafety.com. However, Degil also registered Dentec Safety.ca to redirect Canadian customers to their own site in order to steal customers. A court case ruled that Degil had to pay Dentec $10,000 and to abandon DentecSafety.ca.

An online vendor operated the website GetCanadaDrugs.com. A court found the site name to be "deceptively misdescriptive." Major points included that the pharmaceutical products did not all originate in Canada, and that only around 5 percent of the website's customers were Canadian. The domain name has now ceased to exist.

Knowing the domain name is helpful but hardly a foolproof verification system. MartinLutherKing.org sounds like a site that would provide information about the great orator and civil rights leader. Because it is a .org site, you might conclude that there is no ulterior motive of profit. The site proclaims that it offers "a true historical examination" of Martin Luther King. Wait a minute. Most people don't begin an utterance by saying, "What I am about to tell you is true." The BBC doesn't begin every news item saying, "This is true." Truth is the default position and we assume others are being truthful with us. An old joke goes, "How do you know that someone is lying to you? Because they begin with the phrase *to be perfectly honest*." Honest people don't need to preface their remarks this way.

What MartinLutherKing.org contains is a shameful assortment of distortions, anti-Semitic rants, and out-of-context quotes. Who runs the site? Stormfront, a white-supremacy, neo-Nazi hate group. What better way to hide a racist agenda than by promising "the truth" about a great civil rights leader?

Institutional Bias

Are there biases that could affect the way a person or organization structures and presents the information? Does this person or organization have a conflict of interest? A claim about the health value of almonds made by the Almond Growers' Association is not as credible as one made by an independent testing laboratory.

When judging an expert, keep in mind that experts can be biased without even realizing it. For the same tumor, a surgical oncologist may advise surgery, while a radiation oncologist advises radiation and a medical oncologist advises chemotherapy. A psychiatrist may recommend drugs for depression while a psychologist recommends talk therapy. As the old saying goes, if you have a hammer, everything looks like a nail. Who's right? You might have to look at the statistics yourself. Or find a neutral party who has assessed the various possibilities. This is what meta-analyses accomplish in science and medicine. (Or at least they're supposed to.) A meta-analysis is a research technique whereby the results of dozens or hundreds of studies from different labs are analyzed together to determine the weight of evidence supporting a particular claim. It's the reason companies bring in an auditor to look at their accounting records or a financial analyst to decide what a company they seek to buy is really worth. Insiders at the company to be acquired certainly are expert in their own company's financial situation, but they are clearly biased. And not always in the direction you'd think. They may inflate the value of the company if they want to sell, or deflate it if they are worried about a hostile takeover.

Who Links to the Web Page?

A special Google search allows you to see who else links to a web page you land on. Type "link:" followed by the website URL, and Google will return all the sites that link to it. (For example, link:breastcancer .org shows you the two hundred sites that have links to it.) Why might you want to do this? If a consumer protection agency, Better Business Bureau, or other watchdog organization links to a site, you might want to know whether they're praising or condemning it. The page could be the exhibit in a lawsuit. Or it could be linked by an authoritative source, such as the American Cancer Society, as a valuable resource.

Alexa.com tells you about the demographics of site visitors—what country they are from, their educational background, and what sites people visited immediately before visiting the site in question. This information can give you a better picture of who is using the site and a sense of their motivations. A site with drug information that is visited by doctors is probably a more trusted source than one that isn't. Reviews about a local business from people who are from your town are probably more relevant to you than reviews by people who are out of state.

Peer-Reviewed Journals

In peer-reviewed publications, scholars who are at arm's length from one another evaluate a new experiment, report, theory, or claim. They must be expert in the domain they're evaluating. The method is far from foolproof, and peer-reviewed findings are sometimes overturned, or papers retracted. Peer review is not the only system to rely on, but it provides a good foundation in helping us to draw our own conclusions, and like democracy, it's the best such system we have. If something appears in *Nature,* the *Lancet,* or *Cell,* for example, you can be sure it went through rigorous peer review. As when trying to decide whether to trust a tabloid or a serious news organization, the odds are better that a paper published in a peer-reviewed journal is correct.

In a scientific or scholarly article, the report should include footnotes or other citations to peer-reviewed academic literature. Claims should be justified, facts should be documented through citations to respected sources. Ten years ago, it was relatively easy to know whether a journal was reputable, but the lines have become blurred with the proliferation of open-access journals that will print anything for a fee, in a parallel world of pseudo-academia. Reference librarians can help you distinguish the two. Journals that appear on indexes such as PubMed (maintained by the U.S. National Library of Medicine) are selected for their quality; articles you return from a regular search are not. Scholar.Google.com is more restrictive than Google or other search engines, limiting search results to scholarly and academic papers, although it does not vet the journals and many pseudo-academic papers are included. It does do a good job of weeding out things that don't even *resemble* scholarly research, but that's a double-edged sword: That can make it more difficult to know what to believe because so many of the results appear to be valid. Jeffrey Beall, a research librarian at the University of Colorado, Denver, has developed a blacklist of what he calls predatory open-access journals (which often charge high fees to authors). His list has grown from twenty publishers four years ago to more than three hundred today. Other sites exist that help you to vet research papers, such as the Social Science Research Network (ssrn.com).

Regulated Authority

On the Web, there is no central authority to prevent people from making claims that are untrue, no way to shut down an offending site other than going through the costly procedure of obtaining a court injunction.

Off the Web, the lay of the land can be easier to see. Textbooks and encyclopedias undergo careful peer review for accuracy (although that content is sometimes changed under political pressure by school boards and legislatures). Articles at major newspapers in democratic countries are rigorously sourced compared to the untrustworthy government-controlled newspapers of Iran or North Korea, for example. If a drug manufacturer makes a claim, the FDA in the United States (Health Canada in Canada, or similar agencies in other countries) had to certify it. If an ad appears on television, the FTC will investigate claims that it is untrue or misleading (in Canada this is done by the ASC, Advertising Standards Canada; in the U.K. by the ASA, the Advertising Standards Authority; Europe uses a self-regulation organization called the EASA, European Advertising Standards Alliance; many other countries have equivalent mechanisms).

The lying weasels who make fraudulent claims can face punishment, but often the punishment is meager and doesn't serve as much of a deterrent. Energy-drink company Red Bull paid more than $13 million in 2014 to settle a class-action lawsuit for misleading consumers with promises of increased physical and mental performance. In 2015, Target agreed to pay $3.9 million to settle claims that the prices it charged in-store were higher than those it advertised, and that it misrepresented the weights of products. Grocery retailer Whole Foods was similarly charged in 2015 with misrepresenting the weight of its prepackaged food items. Kellogg's paid $4 million to settle a lawsuit over misleading ads that claimed its Frosted Mini-Wheats were "clinically shown to improve kids' attentiveness by 11 percent." While these amounts might sound like a lot to us, to Red Bull ($7.7 billion in revenue for 2014), Kellogg's ($14.6 billion), and Target ($72.6 billion) these fines are little more than a rounding error in their accounting.

Is the Information Current? Discredited?

Unlike books, newspapers, and conventional sources, Web pages seldom carry a date; graphs, charts, and tables don't always reveal the time period they apply to. You can't assume that the "Sales Earnings Year to Date" you read on a Web page today actually covers today in the "To Date," or even that it applies to this year.

Because Web pages are relatively cheap and easy to create, people often abandon them when they're done with them, move on to other projects, or just don't feel like updating them anymore. They become the online equivalent of an abandoned storefront with a lighted neon sign saying "open" when, in fact, the store is closed.

For the various reasons already mentioned—fraud, incompetence, measurement error, interpretation errors—findings and claims become discredited. Individuals who were found guilty in properly conducted trials become exonerated. Vehicle airbags that underwent multiple inspections get recalled. Pundits change their minds. Merely looking at the newness of a site is not enough to ensure that it hasn't been discredited. New sites pop up almost weekly claiming things that have been thoroughly debunked. There are many websites dedicated to exposing urban myths, such as Snopes.com, or to collating retractions, such as RetractionWatch.com.

During the fall of 2015 leading up to the 2016 U.S. presidential elections, a number of people referred to fact-checking websites to verify the claims made by politicians. Politicians have been lying at least since Quintus Cicero advised his brother Marcus to do so in 64 B.C.E. What we have that Cicero didn't is real-time verification. This doesn't mean that all the verifications are accurate or unbiased, dear reader—you still need to make sure that the verifiers don't have a bias for or against a particular candidate or party.

Politifact.com, a site operated by the *Tampa Bay Times,* won a Pulitzer Prize for their reporting, which monitors and fact-checks speeches, public appearances, and interviews by political figures, and uses a six-point meter to rate statements as True, Mostly True, Half True, Mostly False, False, and—at the extreme end of false— Pants on Fire, for statements that are not accurate and completely ridiculous (from the children's playground taunt "Liar, liar, pants on fire"). The *Washington Post* also runs a fact-checking site with ratings from one to four Pinocchios, and awards the prized Geppetto Checkmark for statements and claims that "contain the truth, the whole truth, and nothing but the truth."

As just one example, presidential candidate Donald Trump spoke at a rally on November 21, 2015, in Birmingham, Alabama. To support his position that he would create a Muslim registry in the United States to combat the threat of terrorism from within the country, he recounted watching "thousands and thousands" of Muslims in Jersey City cheering as the World Trade Center came tumbling down on 9/11/2001. ABC News reporter George Stephanopoulos confronted Trump the following day on camera, noting that the Jersey City police denied this happened. Trump responded that he saw it on television, with his own eyes, and that it was very well covered. Politifact and the *Washington Post* checked all records of television broadcasts and news reports for the three months following the attacks and found no evidence to support Trump's claim. In fact, Paterson, New Jersey, Muslims had placed a banner on the city's main street that read "The Muslim Community Does Not Support Terrorism." Politifact summarized its findings, writing that Trump's recollection "flies in the face of all evidence we could find. We rate this statement Pants on Fire." The *Washington Post* gave it their Four-Pinocchio rating.

During the same campaign, Hillary Clinton claimed "all of my grandparents" were immigrants. According to Politifact (and based on U.S. census records), only one grandparent was born abroad; three of her four grandparents were born in the United States.

Copied and Pasted, Reposted, Edited?

One way to fool people into thinking that you're really knowledgeable is to find knowledgeable-sounding things on other people's Web pages and post them to your own. While you're at it, why not add your own controversial opinions, which will now be enrobed in the scholarship of someone else, and increase hits to your site? If you've got a certain ideological ax to grind, you can do a hatchet job by editing someone else's carefully supported argument to promote the position opposite of theirs. The burden is on all of us to make sure that we're reading the original, unadulterated information, not someone's mash-up of it.

Supporting Information

Unscrupulous hucksters count on the fact that most people don't bother reading footnotes or tracking down citations. This makes it really easy to lie. Maybe you'd like your website to convince people that your skin cream has been shown to reverse the aging process by ten years. So you write an article and pepper it with footnotes that lead to Web pages that are completely irrelevant to the argument. This will fool a lot of people, because most of them won't actually follow up. Those who do may go no further than seeing that the URL you point to is a relevant site, such as a peer-reviewed journal on aging or on dermatology, even though the article cited says nothing about your product.

Even more diabolically, the citation may actually be peripherally related, but not relevant. You might claim that your skin cream contains Vitamin X and that Vitamin X has been shown to improve skin health and quality. So far, so good. But how? Are the studies of Vitamin X reporting on people who spread it on their skin or people who took it orally? And at what dosage? Does your skin product even have an adequate amount of Vitamin X?

Terminology Pitfalls

You may read on CDC.gov that the incidence of a particular disease is 1 in 10,000 people. But then you stumble on an article at NIH.gov that says the same disease has a prevalence of 1 in 1,000. Is there a misplaced comma here, a typo? Aren't incidence and prevalence the same thing? Actually, they're not. The incidence of a disease is the number of new cases (incidents) that will be reported in a given period of time, for example, in a year. The prevalence is the number of existing cases—the total number of people who have the disease. (And sometimes, people who are afraid of numbers make the at-a-glance error that 1 in 1,000 is less than 1 in 10,000, focusing on that large number with all the zeros instead of the word *in*.)

Take multiple sclerosis (MS), a demyelination disease of the brain and spinal cord. About 10,400 new cases are diagnosed each year in the United States, leading to an incidence of 10,400/322,000,000, or 3.2 cases per 100,000 people—in other words, a 0.0032 percent chance of contracting it. Compare that to the total number of people in the United States who already have it, 400,000, leading to a prevalence rate of 400,000/322,000,000, or 120 cases per 100,000, a 0.12 percent chance of contracting it at some point during your lifetime.

In addition to incidence and prevalence, a third statistic, mortality, is often quoted—the number of people who die from a disease, typically within a particular period of time. For coronary heart disease, 1.1 million new cases are diagnosed each year, 15.5 million Americans currently have it, and 375,295 die from it each year. The probability of being diagnosed with heart disease this year is 0.3 percent, about a hundred times more likely than getting MS; the probability of having it right now is nearly 5 percent, and the probability of dying from it in any given year is 0.1 percent. The probability of dying from it at some point in your life is 20 percent. Of course, as we saw in Part One, all of this applies to the aggregate of all Americans. If we know more about a particular person, such as their family history of heart disease, whether or not they smoke, their weight and age, we can make more refined estimates, using conditional probabilities.

The incidence rate for a disease can be high while the prevalence and mortality rates can be relatively low. The common cold is an example—there are many millions of people who will get a cold during the year (high incidence), but in almost every case it clears up quickly, and so the prevalence—the number of people who have it at any given time—can be low. Some diseases are relatively rare, chronic, and easily managed, so the incidence can be low (not many cases in a year) but the prevalence high (all those cases add up, and people continue to live with the disease) and the mortality is low.

When evaluating evidence, people often ignore the numbers and axis labels, as we've seen, but they also often ignore the verbal descriptors, too. Recall the maps of the United States showing "Crude Birth Rate" in Part One. Did you wonder what "crude birth rate" is? You could imagine that a birth rate might be adjusted by several factors, such as whether the birth is live or not, whether the child survives beyond some period of time, and so on. You might think that because the dictionary definition of the word "crude" is that it is something in a natural or raw state, not yet processed or refined (think crude oil) it must mean the raw, unadulterated, unadjusted number. But it doesn't. Statisticians use the term crude birth rate to count live births (thus it is an adjusted number that subtracts stillborn infants). In trying to decide whether to open a diaper business, you want the crude birth rate, not the total birth rate (because total birth rate includes babies who didn't survive birth).

By the way, a related statistic, the crude death rate, refers to the number of people who die at any age. If you subtract this from the crude birth rate, you get a statistic that public policy makers are (and Thomas Malthus was) very interested in: the RNI, rate of natural increase of a population.

Overlooked, Undervalued Alternative Explanations

When evaluating a claim or argument, ask yourself if there is another reason—other than the one offered—that could account for the facts or observations that have been reported. There are always alternative explanations; our job is to weigh them against the one(s) offered and determine whether the person drawing the conclusion has drawn the most obvious or likely one.

For example, if you pass a friend in the hall and they don't return your hello, you might conclude that they're mad at you. But alternative explanations are that they didn't see you, were late for a meeting, were preoccupied, were part of a psychology experiment, have taken a vow of silence for an hour, or were temporarily invaded by bodysnatchers. (Or maybe permanently invaded.)

Alternative explanations come up a great deal in pseudoscience and counterknowledge, and they come up often in real science too. Physics researchers at CERN reported that they had discovered neutrinos traveling faster than light. That would have upended a century of Einsteinian theory. It turns out it was just a loose cable in the linear accelerator that caused a measurement error. This underscores the point that a methodological flaw in an extremely complicated experiment is almost always the more likely explanation than something that would cause us to completely rewrite our understanding of the nature of the universe.

Similarly, if a Web page cites experiments showing that a brand-new, previously unheard-of cocktail of vitamins will boost your IQ by twenty points—and the drug companies don't want you to know!—you should wonder how likely it is that nobody else has heard of this, and if an alternative explanation for the claim is simply that someone is trying to make money.

Mentalists, fortune-tellers, and psychics make a lot of money performing seemingly impossible feats of mind reading. One explanation is that they have tapped into a secret, hidden force that goes against everything we know about cause and effect and the nature of space-time. An alternative explanation is that they are magicians, using magic tricks, and simply lying about how they do what they do. Lending credence to the latter view is that professional magicians exist, including James Randi, who, so far, has been able to use clever illusions to duplicate every single feat performed by a mentalist. And often, the magicians—in an effort to discredit the self-proclaimed psychics—will tell you how they did the tricks. In fairness, I suppose that it's possible that it is the *magicians* who are trying to deceive us—they are really psychics who are afraid to reveal their gifts to us (possibly for fear of exploitation, kidnapping, etc.) and they are only *pretending* to use clever illusions. But again, look at the two possibilities: One causes us to throw out everything we know about nature and science, and the other doesn't. Any psychologist, law enforcement officer, businessperson, divorced spouse, foreign service worker, spy, or lawyer can tell you that people lie; they do so for a variety of reasons and with sometimes alarming frequency and alacrity. But if you're facing a claim that seems unlikely, the more likely (alternative) explanation is that the person telling it to you is lying in one way or another.

People who try to predict the future without using psychic powers—military leaders, economists, business strategists—are often wildly off in their predictions because they fail to consider alternative explanations. This has led to a business practice called *scenario planning*—considering all possible outcomes, even those that seem unlikely. This can be very difficult to do, and even experts fail. In 1968, Will and Ariel Durant wrote:

> In the United States the lower birth rate of the Anglo-Saxons has lessened their economic and political power; and the higher birth rate of Roman Catholic families suggests that by the year 2000 the Roman Catholic Church will be the dominant force in national as well as in municipal or state governments.

What they failed to consider was that, during those intervening thirty-two years, many Catholics would leave the Church, and many would use birth control in spite of the Church's prohibitions. Alternative scenarios to their view in 1968 were difficult to imagine.

Social and artistic predictions get upended too: Experts said around the time of the Beatles that "guitar bands are on their way out." The reviews of Beethoven's Fifth Symphony on its debut included a number of negative pronouncements that no one would ever want to hear it again. Science also gets upended. Experts said that fast-moving trains would never work because passengers would die of asphyxiation. Experts thought that light moved through an invisible "ether." Science and life are not static. All we can do is evaluate the weight of evidence and judge for ourselves, using the best tools we have at our disposal. One of those tools that is under-used is employing creative thinking to imagine alternatives to the way we've been thinking all along.

Alternative explanations are often critical to legal arguments in criminal trials. The framing effects we saw in Part One, and the failure to understand that conditional probabilities don't work backward, have led to many false convictions.

Proper scientific reasoning entails setting up two (or more) hypotheses and presenting the probabilities for both. In a court-room, attorneys shouldn't be focusing on the probability of a match, but the probability of two possible scenarios: What is the probability that the blood samples came from the same source, versus the prob-ability that they did not? More to the point, we need to compare the probability of a match given that the subject is guilty with the prob-ability of a match given that the subject is innocent. Or we could compare the probability that the subject is innocent given the data, versus the probability that the subject is guilty given the data. We also need to know the accuracy of the measures. The FBI announced in 2015 that microscopic hair analyses were incorrect 90 percent of the time. Without these pieces of information, it is impossible to decide the case fairly or accurately. That is, if we talk only in terms of a match, we're considering only one-sided evidence, the probabil-ity of a match given the hypothesis that the criminal was at the scene of the crime. What we don't know is the probability of a match given alternative hypotheses. And the two need to be compared.

This comes up all the time. In one case in the U.K., the suspect, Dennis Adams, was accused based solely on DNA evidence. The victim failed to pick him out of a lineup, and in court said that Adams did not look like her assailant. The victim added that Adams appeared two decades older than the assailant. In addition, Adams had an alibi for the night in question, which was corroborated by testimony from a third party. The only evidence the prosecution presented at trial was the DNA match. Now, Adams had a brother, whom the DNA would also have matched, but there was no additional evidence that the brother had committed the crime, and so investigators didn't consider the brother. But they also lacked additional evidence against Dennis—the *only* evidence they had was the DNA match. No one in the trial considered the alternative hypothesis that it might have been Dennis's brother. . . . Dennis was convicted both in the original trial and on appeal.

Built by the Ancients to Be Seen from Space

You may have heard the speculation that human life didn't really evolve on Earth, that a race of space aliens came down and seeded the first human life. This by itself is not implausible, it's just that there is no real evidence supporting it. That doesn't mean it's not true, and it doesn't mean we shouldn't look for evidence, but the fact that something *could* be true has limited utility—except perhaps for science fiction.

A 2015 story in the *New York Times* described a mysterious formation on the ground in Kazakhstan that could be seen only from space.

Satellite pictures of a remote and treeless northern steppe reveal colossal earthworks—geometric figures of squares, crosses, lines and rings the size of several football fields, recognizable only from the air and the oldest estimated at 8,000 years old.

The largest, near a Neolithic settlement, is a giant square of 101 raised mounds, its opposite corners connected by a diagonal cross, covering more terrain than the Great Pyramid of Cheops. Another is a kind of three-limbed swastika, its arms ending in zigzags bent counterclockwise.

It's easy to get carried away and imagine that these great designs were a way for ancient humans to signal space aliens, perhaps following strict extraterrestrial instructions. Perhaps it was an ancient spaceship landing pad, or a coded message, something like "Send more food." We humans are built that way—we like to imagine things that are out of the ordinary. We are the storytelling species.

Setting aside the rather obvious fact than any civilization capable of interstellar flight must have had a more efficient communication technology at their disposal than arranging large mounds of dirt on the ground, an alternative explanation exists. Fortunately, the *New York Times* (although not every other outlet that reported the story) provides it, in a quote from Dimitriy Dey, the discoverer of the mysterious stones:

> "I don't think they were meant to be seen from the air," Mr. Dey, 44, said in an interview from his hometown, Kostanay, dismissing outlandish speculations involving aliens and Nazis.

(Long before Hitler, the swastika was an ancient and near-universal design element.) He theorizes that the figures built along straight lines on elevations were "horizontal observatories to track the movements of the rising sun."

An ancient sundial explanation seems more likely than space aliens. It doesn't mean it's true, but part of information literacy and evaluating claims is uncovering plausible alternatives, such as this.

The Missing Control Group

The so-called Mozart effect was discredited because the experiments, showing that listening to Mozart for twenty minutes a day temporarily increased IQ, lacked a control group. That is, one group of people was given Mozart to listen to, and one group of people was given nothing to do. Doing nothing is not an adequate control for doing something, and it turns out if you give people something to do—almost anything—the effect disappears. The Mozart effect wasn't driven by Mozart's music increasing IQ, it was driven by the boredom of doing nothing temporarily decreasing effective IQ.

If you bring twenty people with headaches into a laboratory and give them your new miracle headache drug and ten of them get better, you haven't learned anything. Some headaches are going to get better on their own. How many? We don't know. You'd need to have a control group of people with similar ages and backgrounds, and reporting similar pain. And because just the belief that you might get better can lead to health improvements, you have to give the control group something that enables that belief as much as the medicine under study. Hence the well-known placebo, a pill that is made to look exactly like the miracle headache drug so that no one knows who is receiving what until after the experiment is over.

Malcolm Gladwell spread an invalid conclusion in his book *David and Goliath* by suggesting that people with dyslexia might actually have an advantage in life, leading many parents to believe that their dyslexic children should not receive the educational remedies they need. Gladwell fell for the missing control condition. We don't know how much *more* successful his chosen dyslexics might have been if they had been able to improve their condition.

The missing control group shows up in everyday conversation, where it's harder to spot than in scientific claims, simply because we're not looking for it there. You read—and validate—a new study showing that going to bed every night and waking up every morning at the same time increases productivity and creativity. An artist friend of yours, successful by any measure, counters that she's always just slept whenever she wanted, frequently pulling all-nighters and sometimes sleeping for twenty hours at a time, and she's done just fine. But there's a missing control group. How much *more* productive and creative might she have been with a regular sleep schedule? We don't know.

Two twins were separated at birth and reared apart—one in Nazi Germany and the other in Trinidad and Venezuela. One was raised as a Roman Catholic who joined the Hitler Youth, the other as a Jew. They were reunited twenty-one years later and discovered a bizarre list of similar behaviors that many fascinated people could only attribute to genetics: Both twins scratched their heads with their ring finger, both thought it was funny to sneak up on strangers and sneeze loudly. Both men wore short, neatly trimmed mustaches and rectangular wire-rimmed glasses, rounded at the corner. Both wore blue shirts with epaulets and military-style pockets. Both had the same gait when walking, and the same way of sitting in chairs. Both loved butter and spicy food, flushed the toilet before and after using it, and read the endings of books first. Both wrapped tape around pens and pencils to get a better grip.

Stories like this may cause you to wonder about how our behaviors are influenced by our genes. Or if we're all just automatons, and our actions are predetermined. How else to explain such coincidences?

Well, there are two ways, and they both boil down to a missing control group. A social psychologist might say that the world tends to treat people who look alike in similar ways. The attractive are treated differently from the unattractive, the tall differently from the short. If there's something about your face that just looks honest and free of self-interest, people will treat you differently from how they would if your face suggests otherwise. The brothers' behaviors were shaped by the social world in which they live. We'd need a control group of people who are not related, but who still look astonishingly alike, and were raised separately, in order to draw any firm conclusions about this "natural experiment" of the twins separated at birth.

A statistician or behavioral geneticist would say that of the thousands upon thousands of things that we do, it is likely that any two strangers will share some striking similarities in dress, grooming, penchant for practical jokes, or odd proclivities if you just look long enough and hard enough. Without this control group—bringing strangers together and taking an inventory of their habits—we don't know whether the fascinating story about the twins is driven by genetics or pure chance. It may be that genetics plays a role here, but probably not as large a role as we might think.

Cherry-picking

Our brains are built to make stories as they take in the vastness of the world with billions of events happening every second. There are apt to be some coincidences that don't really mean anything. If a long-lost friend calls just as you're thinking of her, that doesn't mean either of you has psychic powers. If you win at roulette three times in a row, that doesn't mean you're on a streak and should bet your last dollar on the next spin. If your non-certified mechanic fixes your car this time, it doesn't mean he'll be able to do it next time—he may just have gotten lucky.

Say you have a pet hypothesis, for example, that too much Vitamin D causes malaise; you may well find evidence to support that view. But if you're looking only for supporting evidence, you're not doing proper research, because you're ignoring the contradictory evidence—there might be a little of this or a lot, but you don't know because you haven't looked. Colloquially, scientists call this "cherry-picking" the data that suit your hypothesis. Proper research demands that you keep an open mind about any issue, and try to valiantly consider the evidence for and against, and then form an evidence-based (not a "gee, I wish this were so"–based) conclusion.

A companion to the cherry-picking bias is selective windowing. This occurs when the information you have access to is unrepresentative of the whole. If you're looking at a city through the window of a train, you're only seeing a part of that city, and not necessarily a representative part—you have visual access only to the part of the city with train tracks running through it, and whatever biases may attach to that. Trains make noise. Wealthier people usually occupy houses away from the noise, so the people who are left living near the tracks tend to have lower income. If all you know of a city is who lives near the tracks, you are not seeing the entire city.

This is of course related to the discussion in Part One about data gathering (how data are collected), and the importance of obtaining representative samples. We're trying to understand the nature of the world—or at least a new city that the train's passing through—and we want to consider alternative explanations for what we're seeing or being told. A good alternative explanation with broad applicability is that you're only seeing part of the whole picture, and the part you're not seeing may be very different.

Maybe your sister is proudly displaying her five-year-old daughter's painting. It may be magnificent! If you love the painting, frame it! But if you're trying to figure out whether to invest in the child's future as the world's next great painter, you'll want to ask some questions: Who cropped it? Who selected it? How big was the original? How many drawings did the little Picasso make before this one? What came before and what came after? Through selective windowing, you may be seeing part of a series of brilliant drawings or a lovely little piece of a much larger (and unimpressive) work that was identified and cropped by the teacher.

We see selective windowing in headlines too. A headline might announce that "three times more Americans support this new legislation than oppose it." Even if you satisfy yourself, based on the steps in Part One of the *Field Guide*, that the survey was conducted on a representative and sufficiently large sample of Americans, you can't conclude that the majority of Americans support the legislation. It could well be that 1 percent oppose it, 3 percent support it, and 94 percent remain undecided. Translate this same kind of monkey-shines to an election headline stating that five times as many

Republicans support Candidate A than Candidate B for the presidential primaries. That may be true, but the headline might leave out that Candidate C is polling with 80 percent of the vote.

Try tossing a coin ten times. You "know" that it should come up heads half the time. But it probably won't. Even if you toss it 1,000 times, you probably won't get exactly 500 heads. Theoretical probabilities are achieved only with an infinite number of trials. The more coin tosses, the closer you'll get to fifty-fifty heads/tails. It's counterintuitive, but there's a probability very close to 100 percent that somewhere in that sequence you'll get five heads in a row. Why is this so counterintuitive? We didn't evolve brains with a sufficient understanding of what randomness looks like. It's not usually heads-tails-heads-tails, but there are going to be runs (also called streaks) even in a random sequence. This makes it easy to fool someone. Just make a cell phone video recording of yourself tossing a coin 1,000 times in a row. Before each toss, say, "This is going to be the first of five heads in a row." Then, if you get a head, before the next toss, say, "This is going to be the second of five heads in a row." If the next one is a tail, start over. If it's not, before you make the next toss, say, "This is going to be the third of five heads in a row." Then just edit your video so that it only includes those five in a row. No one will be any the wiser! If you want to really impress people, go for ten in a row! (There's roughly a 38 percent chance of that happening in 1,000 tosses. Looking at this another way, if you ask a hundred people in a room to toss a coin five times, there is a 96 percent chance that one of them will get five heads in a row.)

The kinds of experiences that a seventy-five-year-old socialite has with the New York City police department are likely to be very different from those of a sixteen-year-old boy of color; their experiences are selectively windowed by what they see. The sixteen-year-old may report being stopped repeatedly without cause, being racially profiled and treated like a criminal. The seventy-five-year-old may fail to understand how this could be. "All *my* experiences with those officers have been so *nice.*"

Paul McCartney and Dick Clark bought up all the celluloid film of their television appearances in the 1960s, ostensibly so that they could control the way their histories are told. If you're a scholar doing research, or a documentarian looking for archival footage, you're limited to what they choose to release to you. When looking at data or evidence to support a claim, ask yourself if what you're being shown is likely to be representative of the whole picture.

Selective Small Samples

Small samples are usually not representative.

Suppose you're responsible for marketing a new hybrid car. You want to make claims about its fuel efficiency. You send a driver out in the vehicle and find that the car gets eighty miles to the gallon. That looks great—you're done! But maybe you just got lucky. Your competitor does a larger test, sending out five drivers in five vehicles and gets a figure closer to sixty miles per gallon. Who's right? You both are! Suppose that your competitor reported the results like this:

Test I: 58 mpg
Test 2: 38 mpg
Test 3: 69 mpg
Test 4: 54 mpg
Test 5: 80 mpg

Road conditions, ambient temperature, and driving styles create a great deal of variability. If you were lucky (and your competitor unlucky) your one driver might produce an extreme result that you then report with glee. (And of course, if you want to cherry-pick, you just ignore tests one through four). But if the researcher is pursuing the truth, a larger sample is necessary. An independent lab that tested fifty different excursions might find that the average is something completely different. In general, anomalies are more likely to show up in small samples. *Larger samples more accurately reflect the state of the world.* Statisticians call this *the law of large numbers.*

If you look at births in a small rural hospital over a month and see that 70 percent of the babies born are boys, compared to 51 percent in a large urban hospital, you might think there is something funny going on in the rural hospital. There might be, but that isn't enough evidence to be sure. The small sample is at work again. The large hospital might have reported fifty-one out of a hundred births were boys, and the small might have reported seven out of ten. As with the coin toss mentioned above, the statistical average of fifty-fifty is most recognizable in large samples.

How many is enough? This is a job for a professional statistician, but there are rough-and-ready rules you can use when trying to make sense of what you're reading. For population surveys (e.g., voting preferences, toothpaste preferences, and such), sample-size calculators can readily be found on the Web. For determining the local incidence of something (rates such as how many births are boys, how many times a day the average person reports being hungry) you need to know something about the base rate (or incidence rate) of the thing you're looking for. If a researcher wanted to know how many cases of albinism are occurring in a particular community, and then examined the first 1,000 births and found none, it would be foolish to draw any conclusions: Albinism occurs in only 1 in 17,000 births. One thousand births is too small a sample—"small" relative to the scarcity of the thing you're looking for. On the other hand, if the study was on the incidence of preterm births, 1,000 should be more than enough because they occur in one in nine births.

Statistical Literacy

Consider a street game in which a hat or basket contains three cards, each with two sides: One card is red on both sides, one white on both sides, and one is red on one side and white on the other. The con man draws one card from the hat and shows you one side of it and it is red. He bets you $5 that the other side is also red. He wants you to think that there is a fifty-fifty chance that this is so, so you're willing to bet against him, that is, that the other side is just as likely to be white. You might reason something like this:

He's showing me a red side. So he has pulled either the red-red card or the red-white card. That means that the other side is either red or white with equal probability. I can afford to take this bet because even if I don't win this time, I will win soon after.

Setting aside the gambler's fallacy—many people have lost money by doubling down on roulette only to find out that chance is not a self-correcting process—the con man is relying on you (counting on you?) to make this erroneous assignment of probability, and usually talking fast in order to fractionate your attention. It's helpful to work it out pictorially.

Here are the three cards:

Red	Red	White
White	Red	White

If he is showing you a red side, it could be any one of *three* sides that he's showing you. In two of those cases, the other side is red and in only one case the other side is white. So there is a two in three chance that if he showed you red the other side will be red, not a one in two chance. This is because most of us fail to account for the fact that on the double-red card, he could be showing you *either* side. If you had trouble with this, don't feel bad—similar mistakes were made by mathematical philosopher Gottfried Wilhelm Leibniz and many more recent textbook authors. When evaluating claims based on probabilities, try to understand the underlying model. This can be difficult to do, but if you recognize that probabilities are tricky, and recognize the limitations most of us have in evaluating them, you'll be less likely to be conned. But what if everyone around you is agreeing with something that is, well, wrong? The exquisite new clothes the emperor is wearing, perhaps?

COUNTERKNOWLEDGE

Counterknowledge, a term coined by the U.K. journalist Damian Thompson, is misinformation packaged to look like fact and that some critical mass of people have begun to believe. Examples come from science, current affairs, celebrity gossip, and pseudo-history. It includes claims that lack supporting evidence, and claims for which evidence exists that clearly contradicts them. Take the pseudo-historical claims that the Holocaust, moon landings, or the attacks of September 11, 2001, in the United States never happened, but were part of massive conspiracies. (Counterknowledge doesn't always involve conspiracies—only sometimes.)

Part of what helps counterknowledge spread is the intrigue of imagining *what if it were true?* Again, humans are a storytelling species, and we love a good tale. Counterknowledge initially attracts us with the patina of knowledge by using numbers or statistics, but further examination shows that these have no basis in fact—the purveyors of counterknowledge are hoping you'll be sufficiently impressed (or intimidated) by the presence of numbers that you'll blindly accept them. Or they cite "facts" that are simply untrue.

Damian Thompson tells the story of how these claims can take hold, get under our skin, and cause us to doubt what we know . . .

that is, until we apply a rational analysis. Thompson recalls the time a friend, speaking of the 9/11 attacks in the United States, "grabbed our attention with a plausible-sounding observation: 'Look at the way the towers collapsed vertically, instead of toppling over. Jet fuel wouldn't generate enough heat to melt steel. Only controlled explosions can do that.'"

The anatomy of this counterknowledge goes something like this:

The towers collapsed vertically: This is true. We've seen footage.

If the attack had been carried out the way they told us, you'd expect the building to topple over: This is an unstated, hidden premise. We don't know if this is true. Just because the speaker is asserting it doesn't make it true. This is a claim that requires verification.

Jet fuel wouldn't generate enough heat to melt steel: We don't know if this is true either. And it ignores the fact that other flammables—cleaning products, paint, industrial chemicals—may have existed in the building so that once a fire got going, they added to it.

If you're not a professional structural engineer, you might find these premises plausible. But a little bit of checking reveals that professional structural engineers have found nothing mysterious about the collapse of the towers.

It's important to accept that in complex events, not everything is explainable, because not everything was observed or reported. In the assassination of President John F. Kennedy, the Zapruder film is the only photographic evidence of the sequence of events, and it is incomplete. Shot on a consumer-grade camera, the frame rate is only 18.3 frames per second and it is low-resolution. There are many unanswered questions about the assassination, and indications that evidence was mishandled, many eyewitnesses were never questioned, and many unexplained deaths of people who claimed or were presumed to know what really happened. There may well have been a conspiracy, but the mere fact that there are unanswered questions and inconsistencies is not proof of one. An unexplained headache with blurred vision is not evidence of a rare brain tumor—it is more likely something less dramatic.

Scientists and other rational thinkers distinguish between things that we know are almost certainly true—such as photosynthesis or that the Earth revolves around the sun—and things that are *probably* true, such as that the 9/11 attacks were the result of hijacked airplanes, not a U.S. government plot. There are different amounts of evidence, and different kinds of evidence, weighing in on each of these topics. And a few holes in an account or a theory does not discredit it. A *handful* of unexplained anomalies does not discredit or undermine a well-established theory that is based on *thousands* of pieces of evidence. Yet these anomalies are typically at the heart of all conspiratorial thinking, Holocaust revisionism, anti-evolutionism, and 9/11 conspiracy theories. The difference between a false theory and a true theory is one of probability. Thompson dubs something counterknowledge when it runs contrary to real knowledge and has some social currency.

When Reporters Lead Us Astray

News reporters gather information about important events in two different ways. These two ways are often incompatible with each other, resulting in stories that can mislead the public if the journalists aren't careful.

In *scientific investigation* mode, reporters are in a partnership with scientists—they report on scientific developments and help to translate them into a language that the public can understand, something that most scientists are not good at. The reporter reads about a study in a peer-reviewed journal or press release. By the time a study reaches peer review, usually three to five unbiased and established scientists have reviewed the study and accepted its accuracy and its conclusions. It is not usually the reporter's job to establish the weight of scientific evidence supporting every hypothesis, auxiliary hypothesis, and conclusion; that has already been done by the scientists writing the paper.

Now the job splits off into two kinds of reporters. The serious investigative reporter, such as for the *Washington Post*, or the *Wall Street Journal*, will typically contact a handful of scientists *not associated* with the research to get their opinions. She will seek out opinions that go against the published report. But the vast majority of reporters consider that their work is done if they simply report on the story as it was published, translating it into simpler language.

In *breaking news* mode, reporters try to figure out something that's going on in the world by gathering information from sources—witnesses to events. This can be someone who witnessed a holdup in Detroit or a bombing in Gaza or a buildup of troops in

Crimea. The reporter may have a single eyewitness, or try to corroborate with a second or third. Part of the reporter's job in these cases is to ascertain the veracity and trustworthiness of the witness. Questions such as "Did you see this yourself?" or "Where were you when this happened?" help to do so. You'd be surprised at how often the answer is no, or how often people lie, and it is only through the careful verifications of reporters that inconsistencies come to light.

So in Mode One, journalists report on scientific findings, which themselves are probably based on thousands of observations and a great amount of data. In Mode Two, journalists report on events, which are often based on the accounts of only a few eyewitnesses.

Because reporters have to work in both these modes, they sometimes confuse one for the other. They sometimes forget that the plural of anecdote is not data; that is, a bunch of stories or casual observations do not make science. Tangled in this is our expectation that newspapers should entertain us as we learn, tell us stories. And most good stories show us a chain of actions that can be related in terms of cause and effect. Risky mortgages were repackaged into AAA-rated investment products, and that led to the housing collapse of 2007. Regulators ignored the buildup of debris above the Chinese city of Shenzhen, and in 2015 it collapsed and created an avalanche that toppled thirty-three buildings. These are not scientific experiments, they are events that we try to make sense of, to make stories out of. The burden of proof for news articles and scientific articles is different, but without an explanation, even a tentative one, we don't have much of a story. And newspapers, magazines, books—people—need stories.

This is the core reason why rumors, counterknowledge, and pseudo-facts can be so easily propagated by the media, as when Geraldo Rivera contributed to a national panic about Satanists taking over America in 1987. There have been similar media scares about alien abduction and repressed memories. As Damian Thompson notes, "For a hard-pressed news editor, anguished testimony trumps dry and possibly inconclusive statistics every time."

Perception of Risk

We assume that newspaper space given to crime reporting is a measure of crime rate. Or that the amount of newspaper coverage given over to different causes of death correlates to risk. But assumptions like this are unwise. About five times more people die each year of stomach cancer than of unintentional drowning. But to take just one newspaper, the *Sacramento Bee* reported no stories about stomach cancer in 2014, but three on unintentional drownings. Based on news coverage, you'd think that drowning deaths were far more common than stomach-cancer deaths. Cognitive psychologist Paul Slovic showed that people dramatically overweight the relative risks of things that receive media attention. And part of the calculus for whether something receives media attention is whether or not it makes a good story. A death by drowning is more dramatic, more sudden, and perhaps more preventable than death by stomach cancer—all elements that make for a good, though tragic, tale. So drowning deaths are reported more, leading us to believe, erroneously, that they're more common. Misunderstandings of risk can lead us to ignore or discount evidence we could use to protect ourselves.

*"You're awfully cavalier considering how much tsunamis
have been in the headlines lately."*

Using this principle of misunderstood risk, unscrupulous or simply
uninformed amateur statisticians with a media platform can easily
bamboozle us into believing many things that are not so.

A front-page headline in the *Times* (U.K.) in 2015 announced
that 50 percent of Britons would contract cancer in their lifetimes,
up from 33 percent. This could rise to two-thirds of today's children,
posing a risk that the National Health Service will be overwhelmed
by the number of cancer patients. What does that make you think?
That there is a cancer epidemic on the rise? Perhaps something
about our modern lifestyle with healthless junk food, radiation-
emitting cell phones, carcinogenic cleaning products, and radiation coming through a hole in the ozone layer is suspect. Indeed,

this headline could be used to promote an agenda by any number of profit-seeking stakeholders—health food companies, sunblock manufacturers, holistic medicine practitioners, and yoga instructors.

Before you panic, recognize that this figure represents all kinds of cancer, including slow-moving ones like prostate cancer, melanomas that are easily removed, etc. It doesn't mean that everyone who contracts cancer will die. Cancer Research UK (CRUK) reports that the percentage of people beating cancer has doubled since the 1970s, thanks to early detection and improved treatment.

What the headline ignores is that, thanks to advances in medicine, people are living longer. Heart disease is better controlled than ever and deaths from respiratory diseases have decreased dramatically in the last twenty-five years. The main reason why so many people are dying of cancer is that they're not dying of other things first. You have to die of *something*. This idea was contained in the same story in the *Times,* if you read that far (which many of us don't; we just stop at the headline and then fret and worry). Part of what the headline statistic reflects is that cancer is an old-person's disease, and many of us now will live long enough to get it. It is not necessarily a cause for panic. This would be analogous to saying, "Half of all cars in Argentina will suffer complete engine failure during the life the car." Yes, of course—the car has to be put out of service for some reason. It could be a broken axle, a bad collision, a faulty transmission, or an engine failure, but it has to be something.

Persuasion by Association

If you want to snow people with counterknowledge, one effective technique is to get a whole bunch of verifiable facts right and then add only one or two that are untrue. The ones you get right will have the ring of truth to them, and those intrepid Web explorers who seek to verify them will be successful. So you just add one or two untruths to make your point and many people will haplessly go along with you. You persuade by associating bogus facts or counterknowledge with actual facts and actual knowledge.

Consider the following argument:

1. Water is made up of hydrogen and oxygen.
2. The molecular symbol for water is H_2O.
3. Our bodies are made up of more than 60 percent water.
4. Human blood is 92 percent water.
5. The brain is 75 percent water.
6. Many locations in the world have contaminated water.
7. Less than 1 percent of the world's accessible water is drinkable.
8. You can only be sure that the quality of your drinking water is high if you buy bottled water.
9. Leading health researchers recommend drinking bottled water, and the majority drink bottled water themselves.

Assertions one through seven are all true. Assertion eight doesn't follow logically, and assertion nine, well . . . who are the leading health researchers? And what does it mean that they drink bottled water themselves? It could be that at a party, restaurant, or on an airplane, when it is served and there are no alternatives, they'll drink it. Or does it mean that they scrupulously avoid all other forms of water? There is a wide chasm between these two possibilities.

The fact is that bottled water is at best no safer or healthier than most tap water in developed countries, and in some cases less safe because of laxer regulations. This is based on reports by a variety of reputable sources, including the Natural Resources Defense Council, the Mayo Clinic, *Consumer Reports*, and a number of reports in peer-reviewed journals.

Of course, there are exceptions. In New York City; Montreal; Flint, Michigan; and many other older cities, the municipal water supply is carried by lead pipes and the lead can leech into the tap water and cause lead poisoning. Periodic treatment-plant problems lead city governments to impose a temporary advisory on tap water. And when traveling in Third World countries, where regulation and sanitation standards are lower, bottled water may be the best bet. But tap-water standards in industrialized nations are among the most stringent standards in any industry—save your money and skip the plastic bottle. The argument of pseudo-scientific health advocates as typified by the above does not, er, hold water.

PART THREE

EVALUATING
THE WORLD

Nature permits us to calculate only probabilities. Yet science has not collapsed.

—RICHARD P. FEYNMAN

How Science Works

The development of critical thinking over many centuries led to a paradigm shift in human thought and history: the scientific revolution. Without its development and practice in cities like Florence, Bologna, Göttingen, Paris, London, and Edinburgh, to name just a handful of great centers of learning, science may not have come to shape our culture, industry, and greatest ambitions as it has. Science is not infallible, of course, but scientific thinking underlies a great deal of what we do and of how we try to decide what is and isn't so. This makes it worth taking a close look behind the curtain to better see how it does what it does. That includes seeing how our imperfect human brains, those of even the most rigorous thinkers, can fool themselves.

Unfortunately, we must also recognize that some researchers make up data. In the most extreme cases, they report data that were never collected from experiments that were never conducted. They get away with it because fraud is relatively rare among researchers and so peer reviewers are not on their guard. In other cases, an investigator changes a few data points to make the data more closely reflect his or her pet hypotheses. In less extreme cases, the investigator omits certain data points because they don't conform to the hypothesis, or selects only cases that he or she knows will contribute favorably to the hypothesis. A case of fraud occurred in 2015 when Dong-Pyou Han, a former biomedical scientist at Iowa State University in Ames, was found to have fabricated and falsified data about a potential HIV vaccine. In an unusual outcome, he didn't just lose his job at the university but was sentenced to almost five years in prison.

The entire controversy about whether the measles, mumps, and rubella (MMR) vaccine causes autism was propagated by Andrew Wakefield in an article with falsified data that has now been retracted—and yet millions of people continue to believe in the connection. In some cases, a researcher will manipulate the data or delete data according to established principles, but fail to report these moves, which makes interpretation and replication more difficult (and which borders on scientific misconduct).

The search for proof, for certainty, drives science, but it also drives our sense of justice and all our judicial systems. Scientific practice has shown us the right way to proceed with this search.

There are two pervasive myths about how science is done. The first is that science is neat and tidy, that scientists never disagree about anything. The second is that a single experiment tells us all we need to know about a phenomenon, that science moves forward in leaps and bounds after every experiment is published. Real science is replete with controversy, doubts, and debates about what we really know. Real scientific knowledge is gradually established through many replications and converging findings. Scientific knowledge comes from amassing large amounts of data from a large number of experiments, performed by multiple laboratories. Any one experiment is just a brick in a large wall. Only when a critical mass of experiments has been completed are we in a position to regard the entire wall of data and draw any firm conclusions.

The unit of currency is not the single experiment, but the meta-analysis. Before scientists reach a consensus about something, there has usually been a meta-analysis, tying together the different pieces of evidence for or against a hypothesis.

If the idea of a meta-analysis versus a single experiment reminds you of the selective windowing and small sample problems mentioned in Part Two, it should. A single experiment, even with a lot of participants or observations, could still just be an anomaly—that eighty miles per gallon you were lucky to get the one time you tested your car. A dozen experiments, conducted at different times and places, give you a better idea of how robust the phenomenon is. The next time you read that a new face cream will make you look twenty years younger, or about a new herbal remedy for the common cold, among the other questions you should ask is whether a meta-analysis supports the claim or whether it's a single study.

Deduction and Induction

Scientific progress depends on two kinds of reasoning. In deduction, we reason from the general to the specific, and if we follow the rules of logic, we can be certain of our conclusion. In induction, we take a set of observations or facts, and try to come up with a general principle that can account for them. This is reasoning from the specific to the general. The conclusion of inductive reasoning is not certain—it is based on our observations and our understanding of the world, and it involves a leap beyond what the data actually tell us.

Probability, as introduced in Part One, is deductive. We work from general information (such as "this is a fair coin") to a specific prediction (the probability of getting three heads in a row). Statistics is inductive. We work from a particular set of observations (such as flipping three heads in a row) to a general statement (about whether the coin is fair or not). Or as another example, we would use probability (deduction) to indicate the likelihood that a particular headache medicine will help you. If your headache didn't go away, we could use statistics (induction) to estimate the likelihood that your pill came from a bad batch.

Induction and deduction don't just apply to numerical things like probability and statistics. Here is an example of deductive logic in words. If the premise (the first statement) is true, the conclusion must be also:

> Gabriel García Márquez is a human.
> All humans are mortal.
> Therefore (this is the deductive conclusion) Gabriel García
>   Márquez is mortal.

1. Some automobiles are Fords.

2. All Fords are automobiles.

3. The guy who played Han Solo is an automobile!

The type of deductive argument about Márquez is called a syllogism. In syllogisms, it is the *form* of the argument that guarantees that the conclusion follows. You can construct a syllogism with a premise that you know (or think to be) false, but that doesn't invalidate the syllogism—in other words, the logic of the whole thing still holds.

> The moon is made of green cheese.
> Green cheese costs $22.99 per pound.
> Therefore, the moon costs $22.99 per pound.

Now, clearly the moon is *not* made of green cheese, but IF it were, the deduction is logically valid. If it makes you feel better, you can rewrite the syllogism so that this is made explicit:

> IF the moon is made of green cheese
> AND IF green cheese costs $22.99 per pound
> THEN the moon costs $22.99 per pound.

There are several distinct types of deductive arguments, and they're typically taught in philosophy or math classes on formal logic. Another common form involves conditionals. This one is called *modus ponens*. It's easy to remember what it's called with this example (using Poe as in *ponens*):

> If Edgar Allan Poe went to the party, he wore a black cape.
> Edgar Allan Poe went to the party.
> Therefore, he wore a black cape.

Formal logic can take some time to master, because, as with many forms of reasoning, our intuitions fail us. In logic, as in running a race, order matters. Does the following sound like a valid or invalid conclusion?

> If Edgar Allan Poe went to the party, he wore a black cape.
> Edgar Allan Poe wore a black cape.
> Therefore, he went to the party.

While it *might* be true that Poe went to the party, it is not *necessarily* true. He could have worn the cape for another reason (perhaps it was cold, perhaps it was Halloween, perhaps he was acting in a play that required a cape and wanted to get in character). Drawing the conclusion above represents an error of reasoning called the *fallacy of affirming the consequent,* or the *converse error.*

If you have a difficult time remembering what it's called, consider this example:

> If Chuck Taylor is wearing Converse shoes, then his feet are
> covered.
> Chuck Taylor's feet are covered.
> Therefore, he is wearing Converse shoes.

This reasoning obviously doesn't hold, because wearing Converse shoes is not the only way to have your feet covered—you could be wearing any number of different shoe brands, or have garbage bags on your feet, tied around the ankles.

However, you *can* say with certainty that if Chuck Taylor's feet are not covered, he is not wearing Converse shoes. This is called the *contrapositive* of the first statement.

Logical statements don't work like the minus signs in equations—you can't just negate one side and have it automatically negate the other. You have to memorize these rules. It's somewhat easier to do using quasi-mathematical notation. The statements above can be represented this way, where A stands for any premise, such as "If Chuck Taylor is wearing Converse shoes," or "If the moon is made of green cheese" or "If the Mets win the pennant this year." B is the consequence, such as "then Chuck's feet are covered," or "then the moon should appear green in the night sky" or "I will eat my hat."

Using this generalized notation, we say *If A* as a shorthand for "If A is true." We say *B* or *Not B* as a shorthand for "B is true" or "B is not true." So . . .

If A, then B

A

Therefore, B

In logic books, you may see the word *then* replaced with an arrow (→) and you may see the word *not* replaced with this symbol: ~. You may see the word *therefore* replaced with ∴ as in:

If A → B

A

∴ B

Don't let that disturb you. It's just some people trying to be fancy.

Now there are four possibilities for statements like this: A can be true or not true, and B can be true or not true. Each of the possibilities has a special name.

I. Modus ponens. This is also called affirming the antecedent. "Ante" means before, like when you "ante up" in poker, putting money in the pot before any cards are played.

If A → B

A ∴ B

VALID

Example: If that woman is my sister, then she is younger than I am.

That woman is my sister.

Therefore, she is younger than I am.

2. The contrapositive.

If A → B

~ B ∴ ~ A

Example: If that woman is my sister, then she is younger
than I am.

That woman is not younger than I am.

Therefore, she is not my sister.

3. The converse.

If A → B

B ∴ A

This is a *not* a valid deduction.

Example: If that woman is my sister, then she is younger
than I am.

That woman is younger than I am.

Therefore, she is my sister.

This is invalid because there are many women younger than I am
who are not my sister.

4. The inverse.

If A → B

~A ∴ ~B

This is a *not* a valid deduction.

> Example: If that woman is my sister, then she is younger
>   than I am.
> That woman is not my sister.
> Therefore, she is not younger than I am.

This is invalid because many women who are not my sister are still younger than I am.

Inductive reasoning is based on there being evidence that suggests the conclusion is true, but does not guarantee it. Unlike deduction, it leads to uncertain but (if properly done) probable conclusions.

An example of induction is:

> All mammals we have seen so far have kidneys.
> Therefore (this is the inductive step), if we discover a new
>   mammal, it will probably have kidneys.

Science progresses by a combination of deduction and induction. Without induction, we'd have no hypotheses about the world. We use it all the time in daily life.

> Every time I've hired Patrick to do a repair around the
>   house, he's botched the job.
> Therefore, if I hire Patrick to do this next repair, he'll botch
>   this one too.

> Every airline pilot I've met is organized, conscientious, and
> meticulous.
> Lee is an airline pilot. He has these qualities, and he's also
> good at math.
> Therefore, all airline pilots are good at math.

Of course, this second example doesn't necessarily follow. We're
making an inference. With what we know about the world, and the
job requirements for being a pilot—plotting courses, estimating the
influence of wind velocity on arrival time, etc.—this seems reasonable. But consider:

> Every airline pilot I've met is organized, conscientious, and
> meticulous.
> Lee is an airline pilot. He has these qualities, and he also
> likes photography.
> Therefore, all airline pilots like photography.

Here our inference is less certain. Our real-world knowledge
suggests that photography is a personal preference, and it doesn't
necessarily follow that a pilot would enjoy it more or less than a
non-pilot.

The great fictional detective Sherlock Holmes draws conclusions
through clever reasoning, and although he claims to be using
deduction, in fact he's using a different form of reasoning called
abduction. Nearly all of Holmes's conclusions are clever guesses,
based on facts, but not in a way that the conclusion is airtight or
inevitable. In abductive reasoning, we start with a set of observations and then generate a theory that accounts for them. Of the

infinity of different theories that could account for something, we seek the most likely.

For example, Holmes concludes that a supposed suicide was really a murder:

> HOLMES: The wound was on the right side of his head. Van Coon was left-handed. Requires quite a bit of contortion.
>
> DETECTIVE INSPECTOR DIMMOCK: Left-handed?
>
> HOLMES: Oh, I'm amazed you didn't notice. All you have to do is look around this flat. Coffee table on the left-hand side; coffee mug handle pointing to the left. Power sockets: habitually used the ones on the left . . . Pen and paper on the left-hand side of the phone because he picked it up with his right and took down messages with his left . . . There's a knife on the breadboard with butter on the right side of the blade because he used it with his left. It's highly unlikely that a left-handed man would shoot himself in the *right* side of his head. Conclusion: Someone broke in here and murdered him . . .
>
> DIMMOCK: But the gun . . . why—
>
> HOLMES: He was waiting for the killer. He'd been threatened.

Note that Sherlock uses the phrase *highly unlikely*. This signals that he's not using deduction. And it's not induction because he's not going from the specifics to the general—in a way, he's going from one set of specifics (the observations he makes in the victim's flat) to another specific (ruling it murder rather than suicide). Abduction, my dear Watson.

Arguments

When evidence is offered to support a statement, these combined statements take on a special status—what logicians call an argument. Here, the word *argument* doesn't mean a dispute or disagreement with someone; it means a formal logical system of statements. Arguments have two parts: evidence and a conclusion. The evidence can be one or more statements, or premises. (A statement without evidence, or without a conclusion, is not an argument in this sense of the word.)

Arguments set up a system. We often begin with the conclusion—I know this sounds backward, but it's how we typically speak; we state the conclusion and *then* bring out the evidence.

> Conclusion: Jacques cheats at pool.
> Evidence (or premise): When your back was turned, I saw him move the ball before taking a shot.

Deductive reasoning follows the process in the opposite direction.

> Premise: When your back was turned, I saw him move the ball before taking a shot.
> Conclusion: Jacques cheats at pool.

This is closely related to how scientists talk about the results of experiments, which are a kind of argument, again in two parts.

> Hypothesis = H
> Implication = I

H: There are no black swans.

I: If H is true, then neither I nor anyone else will ever see a black swan.

But I is not true. My uncle Ernie saw a black swan, and then took me to see it too.

Therefore, reject H.

A Deductive Argument

The germ theory of disease was discovered through the application of deduction. Ignaz Semmelweis was a Hungarian physician who conducted a set of experiments (twelve years before Pasteur's germ and bacteria research) to determine what was causing high mortality rates at a maternity ward in the Vienna General Hospital. The scientific method was not well established at that point, but his systematic observations and manipulations helped not only to pinpoint the culprit, but also to advance scientific knowledge. His experiments are a model of deductive logic and scientific reasoning.

Built into the scenario was a kind of control condition: The Vienna General had two maternity wards adjacent to each other, the first division (with the high mortality rate) and the second division (with a low mortality rate). No one could figure out why infant and mother death rates were so much higher in one ward than the other.

One explanation offered by a board of inquiry was that the configuration of the first division promoted psychological distress: Whenever a priest was called in to give last rites to a dying woman, he had to pass right by maternity beds in the first division to get to her; this was preceded by a nurse ringing a bell. The combination was believed to terrify the women giving birth and therefore make them more likely victims of this "childbed fever." The priest did not have to pass by birthing mothers in the second division when he delivered last rites because he had direct access to the room where dying women were kept.

Semmelweis proposed a hypothesis and implication that described an experiment:

H: The presence of the ringing bell and the priest increases chances of infection.

I: If the bell and priest are not present, infection is not increased.

Semmelweis persuaded the priest to take an awkward, circuitous route to avoid passing the birthing mothers of the first division, and he persuaded the nurse to stop ringing the bell. The mortality rate did not decrease.

I is not true.
Therefore H is false.

We reject the hypothesis after careful experimentation.

Semmelweis entertained other hypotheses. It wasn't overcrowding, because, in fact, the second division was the more crowded one. It wasn't temperature or humidity, because they were the same in the two divisions. As often happens in scientific discovery, a chance event, purely serendipitous, led to an insight. A good friend of Semmelweis's was accidentally cut by the scalpel of a student who had just finished performing an autopsy. The friend became very sick, and the subsequent autopsy revealed some of the same signs of infection as were found in the women who were dying during child-birth. Semmelweis wondered if there was a connection between the particles or chemicals found in cadavers and the spread of the disease. Another difference between the two divisions that had seemed irrelevant now suddenly seemed relevant: The first division staff were medical students, who were often performing autopsies or cadaver dissections when they were called away to deliver a baby; the second division staff were midwives who had no other duties. It was not common practice for doctors to wash their hands, and so Semmelweis proposed the following:

H: The presence of cadaverous contaminants on the hands of doctors increases chances of infection.

I: If the contaminants are neutralized, infection is not increased.

Of course, an alternative *I* was possible too: If the workers in the two divisions were switched (if midwives delivered in division one and medical students in division two) infection would be decreased. This is a valid implication too, but for two reasons switching the workers was not as good an idea as getting the doctors to wash their hands. First, if the hypothesis was really true, the death rate at the hospital would remain the same—all Semmelweis would have done was to shift the deaths from one division to another. Second, when not delivering babies, the doctors still had to work in their labs in division one, and so there would be an increased delay for both sets of workers to reach mothers in labor, which could contribute to additional deaths. Getting the doctors to wash their hands had the advantage that if it worked, the death rate throughout the hospital would be lowered.

Semmelweis conducted the experiment by asking the doctors to disinfect their hands with a solution containing chlorine. The mortality rate in the first division dropped from 18 percent to under 2 percent.

Logical Fallacies

Illusory Correlation

The brain is a giant pattern detector, and it seeks to extract order and structure from what often appear to be random configurations. We see Orion the Hunter in the night sky not because the stars were organized that way but because our brains can project patterns onto randomness.

When that friend phones you just as you're thinking of them, that kind of coincidence is so surprising that your brain registers it. What it doesn't do such a good job of is registering all the times you *didn't* think of someone and they called you. You can think of this like one of those fourfold tables from Part One. Suppose it's a particularly amazing week filled with coincidences (a black cat crosses your path as you walk by a junkyard full of broken mirrors, make your way up to the thirteenth floor of a building to find the movie *Friday the 13th* playing on a television set there). Let's say you get twenty phone calls that week and two of them were from long-lost friends whom you hadn't thought about for a while, but they called within ten minutes of you thinking of them. That's the top row of your table: twenty calls, two that you summoned using extrasensory signaling, eighteen that you didn't. But wait! We have to fill in the bottom row of the table: How many times were you thinking about people and they *didn't* call, and—here's my favorite—how many times were you *not* thinking about someone and they didn't call?

Was I Thinking About Them Just Before?

		YES	NO	
Someone Phoned	**YES**	2	18	20
	NO	50	930	980
		52	948	1,000

To fill out the rest of the table, let's say there are 52 times in a week that you're thinking about people, and 930 times in a week when you are not thinking about people. (This last one is just a crazy guess, but if we divide up the 168-hour week into ten-minute increments, that's about 980 total thoughts, and we already know that 50 of those were about people who didn't phone you, leaving 930 thoughts about things other than people; this is probably an underestimate, but the point is made with any reasonable number you care to put here—try it yourself.)

The brain really only notices the upper left-hand square and ignores the other three, much to the detriment of logical thinking (and to the encouragement of magical thinking). Now, before you book a trip to Vegas to play the roulette wheel, let's run the numbers. What is the probability that someone will call *given* that you just thought about them? It's only two out of fifty-two, or 4 percent.

That's right, 4 percent of the time when you think of someone they call you. That's not so impressive.

What might account for the 4 percent of the times when this coincidence occurs? A physicist might just invoke the 1,000 events in your fourfold table and note that only two of them (two-tenths of 1 percent) appear to be "weird" and so you should just expect this by chance. A social psychologist might wonder if there was some external event that caused both you and your friend to think of each other, thus prompting the call. You read about the terrorist attacks in Paris on November 13, 2015. Somewhere in the back of your mind, you remember that you and a college friend always talked about going to Paris. She calls you and you're so surprised to hear from her you forget the Paris connection, but she is reacting to the same event, and that's why she picked up the phone.

If this reminds you of the twins-reared-apart story earlier, it should. Illusory correlation is the standard explanation offered by behavioral geneticists for the strange confluence of behaviors, such as both twins scratching their heads with their middle finger, or both wrapping tape around pens and pencils to improve their grip. We are fascinated by the contents of the upper left-hand cell in the fourfold table, fixated on all the things that the twins do in common. We tend to ignore all the things that one twin does and the other doesn't.

Framing of Probabilities

After that phone call from your old college friend, you decide to go to Paris on vacation for a week next summer. While standing in front of the *Mona Lisa*, you hear a familiar voice and look up to see your old college roommate Justin, whom you haven't seen in years.

"I can't believe it!" Justin says. "I know!" you say. "What are the odds that I'd run into you here in Paris, standing right in front of the *Mona Lisa*! They must be millions to one!"

Yes, the odds of running into Justin in front of the *Mona Lisa* are probably millions to one (they'd be difficult to calculate precisely, yet any calculation you do would make clear that this was very unlikely). But this way of framing the probability is fallacious. Let's take a step back. What if you hadn't run into Justin just as you were standing in front of the *Mona Lisa*, but as you were in front of the *Venus de Milo*, in *les toilettes*, or even as you were walking in the entrance? What if you had run into Justin at your hotel, at a café, or the Eiffel Tower? You would have been just as surprised. For that matter, forget about Justin—if you had run into *anyone you knew* during that vacation, *anywhere in Paris,* you'd be just as surprised. And why limit it to your vacation in Paris? It could be on a business trip to Madrid, while changing planes in Cleveland, or at a spa in Tucson. Let's frame the probability this way: Sometime in your adult life, you'll run into someone you know where you wouldn't expect to run into them. Clearly the odds of that happening are quite good. But the brain doesn't automatically think this way—cognitive science has shown us just how necessary it is for us to train ourselves to avoid squishy thinking.

Framing Risk

A related problem in framing probabilities is the failure to frame risks logically. Even counting the airplane fatalities of the 9/11 attacks in the United States, air travel remained (and continues to remain) the safest transportation mode, followed closely by rail transportation.

The chances of dying on a commercial flight or train trip are next to zero. Yet, right after 9/11, many U.S. travelers avoided airplanes and took to the highways instead. Automobile deaths increased dramatically. People followed their emotional intuition rather than a logical response, oblivious to the increased risk. The *rate* of vehicular accidents did not increase beyond baseline, but the sum of people who died in all transportation-related accidents increased as more people chose a less safe mode of travel.

You might pull up a statistic such as this one:

More people died in plane crashes in 2014 than in 1960.

From this, you might conclude that air travel has become much less safe. The statistic is correct, but it's not the statistic that's relevant. If you're trying to figure out how safe air travel is, looking at the total number of deaths doesn't tell you that. You need to look at the death *rate*—the deaths per miles flown, or deaths per flight, or something that equalizes the baseline. There were not nearly as many flights in 1960, but they were more dangerous.

By similar logic, you can say that more people are killed on highways between five and seven p.m. than between two and four a.m., so you should avoid driving between five and seven. But the simple fact is that many times more people are driving between five and seven—you need to look at the *rate* of death (per mile or per trip or per car), not the raw number. If you do, you'll find that driving in the evening is safer (in part because people on the road between two and four a.m. are more likely to be drunk or sleep-deprived).

After the Paris attacks of November 13, 2015, CNN reported that at least one of the attackers had entered the European Union as a refugee, against a backdrop of growing anti-refugee sentiment in Europe. Anti-refugee activists had been calling for stricter border control. This is a social and political issue and it is not my intention to take a stand on it, but the numbers can inform the decision making. Closing the borders completely to migrants and refugees might have thwarted the attacks, which took roughly 130 lives. Denying entry to a million migrants coming from war-torn regions such as Syria and Afghanistan would, with great certainty, have cost thousands of them their lives, far more than the 130 who died in the attacks. There are other risks to both courses of action, and other considerations. But to someone who isn't thinking through the logic of the numbers, a headline like "One of the attackers was a refugee" inflames the emotions around anti-immigrant sentiment, without acknowledging the many lives that immigration policies saved. The lie that terrorists want you to believe is that you are in immediate and great peril.

Misframing is often used by salespeople to persuade you to buy their products. Suppose you get an email from a home-security company with this pitch: "Ninety percent of home robberies are solved with video provided by the homeowner." It sounds so empirical. So scientific.

Start with a plausibility check. Forget about the second part of the sentence, about the video, and just look at the first part: "Ninety percent of home robberies are solved . . ." Does that seem reasonable? Without looking up the actual statistics, just using your real-world knowledge, it seems doubtful that 90 percent of home robberies are solved. This would be a fantastic success rate for any police department. Off to the Internet. An FBI page reports that about 30 percent of robbery cases are "cleared," meaning solved.

So we can reject as highly unlikely the initial statement. It said 90 percent of home robberies are solved with video provided by the homeowner. But that can't be true—it would imply that more than 90 percent of home robberies are solved, because some are certainly solved without home video. What the company more likely means is that 90 percent of solved robberies are from video provided by the homeowner.

Isn't that the same thing?

No, because the sample pool is different. In the first case, we're looking at all home robberies committed. In the second case we're looking only at the ones that were solved, a much smaller number. Here it is visually:

All home robberies in a neighborhood:

Solved home robberies in a neighborhood
(using the 30 percent figure obtained earlier):

So does that mean that if I have a video camera there is a 90 percent chance that the police will be able to solve a burglary at my house?

No!

All you know is that *if* a robbery is solved, there is a 90 percent chance that the police were aided by a home video. If you're thinking that we have enough information to answer the question you're really interested in (what is the chance that the police will be able to solve a burglary at my house if I buy a home-security system versus if I don't), you're wrong—we need to set up a fourfold table like the ones in Part One, but if you start, you'll see that we have information on only one of the rows. We know which percentage of solved crimes had home video. But to fill out the fourfold table, we'd also need to know what proportion of the unsolved crimes had home video (or, alternatively, what proportion of the home videos taken resulted in unsolved crimes).

Remember, P(burglary solved | home video) ≠ P(home video | burglary solved).

The misframing of the data is meant to spark your emotions and cause you to purchase a product that may not have the intended result at all.

Belief Perseverance

An odd feature of human cognition is that once we form a belief or accept a claim, it's very hard for us to let go, even in the face of overwhelming evidence and scientific proof to the contrary. Research reports say we should eat a low-fat, high-carb diet and so we do.

New research undermines the earlier finding—quite convincingly—yet we are reluctant to change our eating habits. Why? Because on acquiring the new information, we tend to build up internal stories to help us assimilate the knowledge. "Eating fats will make me fat," we tell ourselves, "so the low-fat diet makes a lot of sense." We read about a man convicted of a grisly homicide. We see his picture in the newspaper and think that we can make out the beady eyes and unforgiving jaw of a cold-blooded murderer. We convince ourselves that he "looks like a killer." His eyebrows arched, his mouth relaxed, he seems to lack remorse. When he's later acquitted based on exculpatory evidence, we can't shake the feeling that even if he didn't do *this* murder, he must have done another one. Otherwise he wouldn't look so guilty.

In a famous psychology experiment, participants were shown photos of members of the opposite sex while ostensibly connected to physiological monitoring equipment indicating their arousal levels. In fact, they weren't connected to the equipment at all—it was under the experimenter's control. The experimenter gave the participants false feedback to make them believe that they were particularly attracted to a person in one of the photos more than the others. When the experiment was over, the participants were shown that the "reactions" of their own body were in fact premanufactured tape recordings. The kicker is that the experimenter allowed them to choose one photo to take home with them. Logically, they should have chosen the picture that they found most attractive at that moment—the evidence for liking a particular picture had been completely discredited. But the participants tended to choose the picture that was consistent with their initial belief. The experimenters showed that the effect was driven by the sort of self-persuasion described above.

Autism and Vaccines: Four Pitfalls in Reasoning

The story with autism and vaccines involves four different pitfalls in critical thinking: illusory correlation, belief perseverance, persuasion by association, and the logical fallacy we saw earlier, *post hoc, ergo propter hoc* (loosely translated, it means "because this happened after that, that must have caused this").

Between 1990 and 2010, the number of children diagnosed with autism spectrum disorders (ASD) rose sixfold, more than doubling

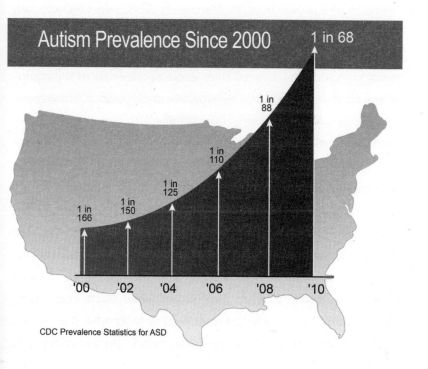

Autism Prevalence Since 2000   1 in 68

1 in 88

1 in 110

1 in 125

1 in 150

1 in 166

'00   '02   '04   '06   '08   '10

CDC Prevalence Statistics for ASD

in the last ten years. The prevalence of autism has increased exponentially from the 1970s to now.

The majority of the rise has been accounted for by three factors: increased awareness of autism (more parents are on the alert and bring their children in for evaluation, professionals are more willing to make the diagnosis); widened definitions that include more cases; and the fact that people are having children later in life (advanced parental age is correlated with the likelihood of having children with autism and many other disorders).

If you allow the Internet to guide your thinking on why autism has increased, you'll be introduced to a world of fiendish culprits: GMOs, refined sugar, childhood vaccines, glyphosates, Wi-Fi, and proximity to freeways. What's a concerned citizen to do? It sure would be nice if an expert would weigh in. Voilà—an MIT scientist comes to the rescue! Dr. Stephanie Seneff made headlines in 2015 when she reported a link between a rise in the use of glyphosate, the active ingredient in the weed killer Roundup, and the rise of autism. That's right, two things rise—like pirates and global warming—so there must be a causal connection, right?

Post hoc, ergo propter hoc, anyone?

Dr. Seneff is a computer scientist with no training in agriculture, genetics, or epidemiology. But she is a *scientist* at the venerable MIT, so many people wrongly assume that her expertise extends beyond her training. She also couches her argument in the language of science, giving it a real pseudoscientific, counterknowledge gloss:

1. Glyphosate interrupts the shikimate pathway in plants.
2. The shikimate pathway allows plants to create amino acids.
3. When the pathway is interrupted, the plants die.

Seneff concedes that human cells don't have a shikimate pathway, but she continues:

4. We have millions of bacteria in our gut ("gut flora").
5. Those bacteria do have a shikimate pathway.
6. When glyphosate enters our system, it disturbs our digestion and our immune function.
7. Glyphosate in humans can also inhibit liver function.

If you're wondering what all this has to do with ASD, you should be. Seneff lays out a case (without citing any evidence) for increased prevalence of digestive problems and immune system dysfunction, but these have nothing to do with ASD.

Others searching for an explanation for the rise in autism rates have pointed to the MMR (measles-mumps-rubella) vaccine, and the antiseptic, antifungal compound thimerosal (thiomersal) it contains. Thimerosal is derived from mercury, and the amount contained in vaccines is typically one-fortieth of what the World Health Organization (WHO) considers the amount tolerable per day. Note that the WHO guidelines are expressed as per-day amounts, and with the vaccine, you're getting it only once.

Although there was no evidence that thimerosal was linked to autism, it was removed from vaccines in 1992 in Denmark and Sweden, and in the United States starting in 1999, as a "precautionary measure." Autism rates have continued to increase apace even with the agent removed. The illusory correlation (as in pirates and global warming) is that the MMR vaccine is typically given between twelve and fifteen months of age, and if a child has autism, the earliest that it is typically diagnosed is between eighteen and twenty-four months of age. Parents tended to focus on the upper left-hand cell of a fourfold table—the number of times a child received a vaccination and was later diagnosed with autism—without considering how many children who were not vaccinated still developed autism, or how many *millions* of children were vaccinated and did not develop autism.

To make matters worse, a now-discredited physician, Andrew Wakefield, published a scientific paper in 1998 claiming a link. The *British Medical Journal* declared his work fraudulent, and six years later, the journal that originally published it, the *Lancet*, retracted it. His medical license was revoked. Wakefield was a surgeon, not an expert in epidemiology, toxicology, genetics, neurology, or any specialization that would have qualified him as an expert on autism.

Post hoc, ergo propter hoc caused people to believe the correlation implied causation. Illusory correlation caused them to focus only on the coincidence of some people developing autism who also had the vaccine. The testimony of a computer scientist and a physician caused people to be persuaded by association. Belief perseverance caused people who initially believed the link to cling to their beliefs even after the evidence had been removed.

Parents continue to blame the vaccine for autism, and many parents stopped vaccinating their children. This led to several outbreaks of measles around the world. All because of a spurious link and the failure of a great many people to distinguish between correlation and causation, and a failure to form beliefs based on what is now overwhelming scientific evidence.

KNOWING WHAT YOU DON'T KNOW

> *. . . as we know, there are known knowns; there are things we know we know. We also know there are known unknowns; that is to say we know there are some things we do not know. But there are also unknown unknowns—the ones we don't know we don't know.*
>
> —U.S. Secretary of Defense Donald Rumsfeld

This is clearly tortured language, and the meaning of the sentence is obscured by that. There's no reason for the repetitive use of the same word, and the secretary might have been clearer if he had said instead, "There are things we know, things we are aware that we do not know, and some things we aren't even aware that we don't know." There's a fourth possibility, of course—things we know that we aren't aware we know. You've probably experienced this—someone asks you a question and you answer it, and then say to yourself, "I'm not even sure how I knew that."

Either way, the fundamental point is sound, you know? What will really hurt you, and cause untold amounts of damage and inconvenience, are the things you think you know but don't (per Mark Twain's/Josh Billings's epigraph at the beginning of this book), and the things that you weren't even aware of that are supremely relevant to the decision you have ahead (the unknown unknowns). Formulating a proper scientific question requires taking an account of what we know and what we don't know. A properly formulated scientific hypothesis is *falsifiable*—there are steps we can take, at least in theory, to test the true state of the world, to determine if our hypothesis is true or not. In practice, this means considering alternative explanations ahead of time, before conducting the experiment, and designing the experiment so that the alternatives are ruled out.

If you're trying out a new medicine on two groups of people, the experimental conditions have to be the same in order to conclude that medicine A is better than medicine B. If all the people in group A get to take their medicine in a windowed room with a nice view, and the people in group B have to take it in a smelly basement lab, you've got a confounding factor that doesn't allow you to conclude the difference (if you find one) was due solely to the medication. The smelly basement problem is a known known. Whether medicine A works better than medicine B is a known unknown (it's why we're conducting the experiment). The unknown unknown here would be some other potentially confounding factor. Maybe people with high blood pressure respond better to medicine A in every case, and people with low blood pressure respond better to medicine B. Maybe family history matters. Maybe the time of day the medication is taken makes a difference. Once you identify a potential confounding factor, it very neatly moves from the category of unknown unknown to known unknown. Then we can modify the experiment, or do additional research that will help us to find out.

The trick to designing good experiments—or evaluating ones that have already been conducted—comes down to being able to generate alternative explanations. Uncovering unknown unknowns might be said to be *the* principal job of scientists. When experiments yield surprising results, we rejoice because this is a chance to learn something we didn't know. The B-movie characterization of the scientist who clings to his pet theory to his last breath doesn't apply to any scientist I know; real scientists know that they only learn when things don't turn out the way they thought they would.

In a nutshell:

1. There are some things we know, such as the distance from the Earth to the sun. You may not be able to generate an answer without looking it up, but you are aware that the answer is known. This is Rummy's *known known*.

2. There are some things that we don't know, such as how neural firing leads to feelings of joy. We're aware that we don't know the answer to this. This is Rummy's *known unknown*.

3. There are some things that we know, but we aren't aware that we know them, or forget that we know them. What is your grandmother's maiden name? Who sat next to you in third grade? If the right retrieval cues help you to recollect something, you find that you knew it, although you didn't realize ahead of time that you did. Although Rumsfeld doesn't mention them, this is an *unknown known*.

4. There are some things that we don't know, and we're not even aware we don't know them. If you've bought a house,

you've probably hired various inspectors to report on the condition of the roof, the foundation, and the existence of termites or other wood-destroying organisms. If you had never heard of radon, and your real estate agent was more interested in closing the deal than protecting your family's health, you wouldn't think to test for it. But many homes do have high levels of radon, a known carcinogen. This would count as an *unknown unknown* (although, having read this paragraph, it is no longer one). Note that whether you're aware or unaware of an unknown depends on your expertise and experience. A pest-control inspector would tell you that he is only reporting on what's visible—it is known to him that there might be hidden damage to your house, in areas he was unable to access. The nature and extent of this damage, if any, is unknown to him, but he's aware that it might be there (a *known unknown*). If you blindly accept his report and assume it is complete, then you're unaware that additional damage could exist (an *unknown unknown*).

We can clarify Secretary Rumsfeld's four possibilities with a four-fold table:

What we know that we know: GOOD—PUT IT IN THE BANK	What we know that we don't know: NOT BAD, WE CAN LEARN IT
What we don't know that we know: A BONUS	What we don't know that we don't know: DANGER—HIDDEN SHOALS

The unknown unknowns are the most dangerous. Some of the biggest human-caused disasters can be traced to these. When bridges collapse, countries lose wars, or home purchasers face foreclosure, it's often because someone didn't allow for the possibility that they don't know everything, and they proceeded along blindly thinking that every contingency had been accounted for. One of the main purposes of training someone for a PhD, a law or medical degree, an MBA, or military leadership is to teach them to identify and think systematically about what they don't know, to turn unknown unknowns into known unknowns.

A final class that Secretary Rumsfeld didn't talk about either are incorrect knowns—things that we think are so, but aren't. Believing false claims falls into this category. One of the biggest causes of bad, even fatal, outcomes is belief in things that are untrue.

BAYESIAN THINKING IN SCIENCE AND IN COURT

Recall from Part One the idea of Bayesian probability, in which you can modify or update your belief about something based on new data as it comes in, or on the prior probability of something being true—the probability you have pneumonia *given* that you show certain symptoms, or the probability that a person will vote for a particular party *given* where they live.

In the Bayesian approach, we assign a subjective probability to the hypothesis (the *prior* probability), and then modify that probability in light of the data collected (the *posterior* probability, because it's the one you arrive at after you've conducted the experiment). If we had reason to believe the hypothesis was true before we tested it, it doesn't take much evidence for us to confirm it. If we had reason to believe the hypothesis unlikely before we tested it, we need more evidence.

Unlikely claims, then, according to a Bayesian perspective, require stronger proof than likely ones. Suppose your friend says she saw something flying right outside the window. You might entertain three hypotheses, *given* your own recent experiences at that window: It is a robin, it is a sparrow, or it is a pig. You can assign probabilities to these three hypotheses. Now your friend shows you a photo of a pig flying outside the window. Your prior belief that pigs fly is so small that the posterior probability is still very small, even with this evidence. You're probably now entertaining new hypotheses that the photo was doctored, or that there was some other kind of trickery involved. If this reminds you of the fourfold tables and the likelihood that someone has breast cancer given a positive test, it should—the fourfold tables are simply a method for performing Bayesian calculations.

Scientists should set a higher threshold for evidence that goes against standard theories or models than for evidence that is consistent with what we know. Following thousands of successful trials for a new retroviral drug in mice and monkeys, when we find that it works in humans we are not surprised—we're willing to accept the evidence following standard conventions for proof. We might be convinced by a single study with only a few hundred participants. But if someone tells us that sitting under a pyramid for three days will cure AIDS, by channeling qi into your chakras, this requires stronger evidence than a single experiment because it is farfetched and nothing like it has ever been demonstrated before. We'd want to see the result replicated many times and under many different conditions, and ultimately, a meta-analysis.

The Bayesian approach isn't the only way that scientists deal with unlikely events. In their search for the Higgs boson, physicists set a threshold (using conventional, not Bayesian, statistical tests) 50,000 times more stringent than usual—not because the Higgs was unlikely (its existence was hypothesized for decades) but because the cost of being wrong is very high (the experiments are very expensive to conduct).

The application of Bayes's rule can perhaps best be illustrated with an example from forensic science. One of the cornerstone principles of forensic science was developed by the French physician and lawyer Edmond Locard: Every contact leaves a trace. Locard stated that either the wrongdoer leaves signs at the scene of the crime or has taken away with her—on her person, body, or clothes—indications of where she has been or what she has done.

Suppose a criminal breaks into the stables to drug a horse the night before a big race. He will leave some traces of his presence at the crime scene—footprints, perhaps skin, hair, clothing fibers, etc. Evidence has been transferred from the criminal to the scene of the crime. And similarly, he will pick up dirt, horsehair, blanket fibers, and such from the stable, and in this way evidence has been transferred from the crime scene to the criminal.

Now suppose someone is arrested the next day. Samples are taken from his clothing, hands, and fingernails, and similarities are found between these samples and other samples taken at the crime scene. The district attorney wants to evaluate the strength of this evidence. The similarities may exist because the suspect is guilty. Or perhaps the suspect is innocent, but was in contact with the guilty party—that contact too would leave a trace. Or perhaps the suspect, quite innocently, was in another barn, interacting innocently with another horse, accounting for the similarities.

Using Bayes's rule allows us to combine objective probabilities, such as the probability of the suspect's DNA matching the DNA found at the crime scene, with personal, subjective views, such as the credibility of a witness, or the honesty and track record of the CSI officer who had custody of the DNA sample. Is the suspect someone who has done this before, or someone who knows nothing about horse racing, has no connection to anyone involved in the race, and has a very good alibi? These factors help us to determine a prior, subjective probability that the suspect is guilty.

If we take literally the assumption in the American legal system that one is innocent until proven guilty, then the prior probability of a suspect being guilty is zero, and any evidence, no matter how damning, won't yield a posterior probability above zero, because you'll always be multiplying by zero. A more reasonable way to establish the prior probability of a suspect's innocence is to consider anyone in the population equally likely. Thus, if the suspect was apprehended in a city of 100,000 people, and investigators have reason to believe that the perpetrator was a resident of the city, the prior odds of the suspect being guilty are 1 in 100,000. Of course, evidence can narrow the population—we may know, for example, that there were no signs of forced entry, and so the suspect had to be one of fifty people who had access to the facility.

Our prior hypothesis (*a priori* in Latin) is that the suspect is guilty with a probability of .02 (one of fifty people who had access). Now let's suppose the perpetrator and the horse got in a scuffle, and human blood was found at the scene. Our forensics team tells us that the probability that the suspect's blood matches the blood found at the scene is .85. We construct a fourfold table as before. We fill in the bottom row under the table first: The suspect has a one in fifty chance to be guilty (the *Guilty: Yes* column), and a forty-nine in fifty chance to be innocent. The lab told us that there's a .85 probability of a blood match, so we enter that in the upper left: the probability that the suspect is guilty *and* the blood matches. That means the lower left cell has to be .15 (the probabilities have to add up to one). The .85 blood match means something else: that there's a .15 chance the blood was left by someone else, not our suspect, which would absolve him and render him not guilty. There's a .15 chance that one of the people in the right-hand column will match, so we multiply 49 × .15 to get 7.35 in the upper right cell. We subtract that from the forty-nine in order to find the value for the bottom right cell.

		Suspect Guilty		
		YES	NO	
Blood Match	YES	0.85	7.35	8.2
	NO	0.15	41.65	41.8
		1	49	50

Now we can calculate the information we want the judge and jury to evaluate.

$$P(\text{Guilty} \mid \text{Match}) = .85/8.2 = .10$$
$$P(\text{Innocent} \mid \text{Match}) = 7.35/8.2 = .90$$

Given the evidence, it is about nine times more likely that our suspect is innocent than guilty. We started out with him having a .02 chance of being guilty, so the new information has increased his guilt by a factor of five, but it is still more likely that he is innocent.

Suppose, however, some new evidence comes in—horsehair found on the suspect's coat—and the probability that the horsehair belongs to the drugged horse is .95 (only five chances in one hundred that the hair belongs to a different horse). We can chain our Bayesian probabilities together now, filling out a new table. In the bottom margin, we enter the values we just calculated, .10 and .90. (Statisticians sometimes say that yesterday's posteriors are today's priors.) If you'd rather think of these numbers as "one chance in ten" and "nine chances in ten," go ahead and enter them as whole numbers.

Suspect Guilty

		YES	NO	
Blood Match	**YES**	0.95	0.45	1.4
	NO	0.05	8.55	8.6
		1	9	10

We know from our forensics team that the probability of a match for the hair sample is .95. Multiplying that by one, we get the entry for the upper left, and subtracting that from one we get the entry for the lower left. If there is a .95 chance that the sample matches the victimized horse, that implies that there is a .05 chance that the sample matches a different animal (which would absolve the suspect) so the upper right-hand cell is the product of .05 and the marginal total of 9 = .45. Now when we perform our calculations, we see that

P(Guilty | Evidence) = .68   P(Evidence | Guilty) = .95
P(Innocent | Evidence) = .32   P(Evidence | Innocent) = .05

The new evidence shows us that it is about twice as likely that the suspect is guilty as that he is innocent, given the evidence. Many attorneys and judges do not know how to organize the evidence like this, but you can see how helpful it is. The problem of mistakenly thinking that P(Guilty | Evidence) = P(Evidence | Guilt) is so widespread it has been dubbed *the prosecutor's fallacy*.

If you prefer, the application of Bayes's rule can be done mathematically, rather than using the fourfold table, and this is shown in the appendix.

Four Case Studies

Science doesn't present us with certainty, only probabilities. We don't know for 100 percent sure that the sun will come up tomorrow, or that the magnet we pick up will attract steel, or that nothing travels faster than the speed of light. We think these things very likely, but science yields only the best Bayesian conclusions we can have, given what we know so far.

Bayesian reasoning asks us to consider probabilities in light of what we know about the state of the world. Crucial to this is engaging in the kind of critical thinking described in this field guide. Critical thinking is something that can be taught, and practiced, and honed as a skill. Rigorous study of particular cases is a standard approach because it allows us to practice what we've learned in new contexts—what learning theorists call *far transfer*. Far transfer is the most effective way we know to make knowledge stick.

There is an infinite variety of ways that faulty reasoning and misinformation can sneak up on us. Our brains weren't built to excel at this. It's always been a part of science to take a step back and engage in careful, systematic reasoning. Case studies are presented as stories, based on true incidents or composites of true incidents, and of course, we are a story-loving species. We remember the stories and the interesting way they loop back to the fundamental concepts. Think of the following as problem sets we can all explore together.

Shadow the Wonder Dog Has Cancer (or Does He?)

We got our dog Shadow, a Pomeranian-Sheltie mix, from a rescue shelter when he was two years old. He got his name, we learned, because he would follow us from room to room around the house during the day, never far away. As often happens with pets, our rhythms synchronized—we would fall asleep and wake up at the same time, get hungry around the same time, feel like getting exercise at the same time. He traveled with us often on business trips to other cities, becoming acclimated to planes, trains, and automobiles.

When Shadow was thirteen, he began having trouble urinating, and one morning we found blood in his urine. Our vet conducted an ultrasound examination and found a growth on his bladder. The only way to tell whether it was cancerous was to perform two surgical procedures that the oncologist was urging: a cystoscopy, which would run a miniature camera through his urethra into the bladder, and a biopsy to sample the mass and study it under the microscope. The general practitioner cautioned against this because of the risks of general anesthesia in a dog Shadow's age. If it did turn out to be cancerous, the oncologist would want to perform surgery and start chemotherapy. Without any further tests, the doctors were still pretty certain that this was bladder cancer, known as transitional cell carcinoma (TCC). On average, dogs live only six months following this diagnosis.

As my wife and I looked into Shadow's eyes, we felt utterly helpless. We didn't know if he was in pain, and if so, how much more he was facing, either from the treatment or from the disease. His care was entirely in our hands. This made the decision particularly emotional, but that didn't mean we threw rationality out the window. You can think critically even when the decision is emotional. Even when it's your dog.

This is a typical medical scenario for people or pets: two doctors, two different opinions, many questions. What are the risks of surgery? What are the risks of the biopsy? How long is Shadow likely to live if we give him the operation and how long is he likely to live if we don't?

In a biopsy, a small needle is used to collect a sample of tissue that is then sent to a pathologist, who reports on the likelihood that it is cancerous or not. (Pathology, like most science we've seen, does not deal in certainties, just likelihoods and the probability that the sample contains cancer, which is then applied to the probability that the unsampled parts of the organ might also contain cancer; if you're looking for certainty, pathology is not the place to look.) Patients and pet owners almost never ask about the risk of biopsy. For humans, these statistics are well known, but they are less well tracked in veterinary medicine. Our vet estimated that there was a 5 percent chance of life-threatening infection, and a 10 percent chance that some cancerous material (if indeed the mass was cancerous) would be "shed" into the abdomen on the needle's way out, seeding further cancer growth. An additional risk was that biopsies leave behind scar tissue that makes it more difficult to operate later if that's what you decide to do. The anesthesia needed for the procedure could kill Shadow. In short, the diagnostic procedure could make him worse.

Our vet presented us with six options:

1. Biopsy through the abdominal wall in the hope of obtaining a more definitive diagnosis.

2. Diagnostic catheterization (using a catheter to traumatize a portion of the mass, allowing cells to exfoliate and then be examined).

3. Biopsy using the same cystoscopic camera they wanted to use anyway to better image the mass (through the urethra).

4. Major surgery right now to view the mass directly, and remove it if possible. The problem with this is that most bladder cancers return within twelve months because the surgeons are unable to remove every cancerous cell, and the ones left behind typically keep on growing at a rapid rate.

5. Do nothing.

6. Put Shadow to sleep right now, in recognition of the fact that it most probably is bladder cancer, and he doesn't have long to live anyway.

We asked about what the treatment options were if it was found to be cancer, and what they might be if it was not cancer. Too often, patients focus on the immediate, upcoming procedure without regard for what the next steps might be.

If the mass was cancerous, the big worry was that the tumor could grow and eventually block one of the tubes that brings urine into the bladder from the kidneys, or that allows urine to leave the bladder and end up on a lawn or fire hydrant of choice. If that blockage occurs, Shadow could experience great pain and die within a day. Along the way to this, there could be temporary blockages as a result of swelling. Because of the position of the bladder within the body, and the angle of ultrasound, it was difficult to tell how close the mass was to these tubes (the ureter and urethra).

So what about the six options presented above—how to decide which (if any) to choose? We ruled out two of them: putting Shadow to sleep and doing nothing. Recall that the oncologist was pushing for surgery because that is their gold standard, their protocol for such cases. We asked for some statistics and she said she'd have to do some research and get back to us. Later, she said that there was a 20 percent chance that the surgery would end badly, killing Shadow right away. So we ruled out the major surgery because we weren't even sure yet if the mass was cancerous.

We asked for life-expectancy statistics on the various remaining scenarios. Unfortunately, most such statistics are not kept by the veterinary community, and in any case, those that are kept skew toward short life expectancy because many pet owners choose euthanasia. That is, many owners opt to put their pets down before the disease progresses because of concerns about either the animal's quality of life or the owners' quality of life: Dogs with TCC often experience incontinence (we had already noticed that Shadow was leaving us little surprises around the house). We didn't have a definitive diagnosis yet, but based on the sparse statistics that existed, it looked as though Shadow would live three months *with or without treatment.* Three months if we do nothing, three months if we give him chemo, three months if we give him surgery. How could that be? Ten years ago, we found out, vets would recommend euthanasia on first diagnosis of TCC. And at the first sign of chronic incontinence, owners would put their dogs down. So owners were typically ending their dogs' lives before the cancer did, and this made the statistics unreliable.

We did some research on our own, using "transitional cell

carcinoma" and "dog *or* canine" as the search terms. We found out that there was a 30 percent chance Shadow could improve simply by taking a nonsteroidal anti-inflammatory called Piroxicam. Piroxicam has its own side effects, including stomach upset, vomiting, loss of appetite, and kidney and liver trouble. We asked the vet about it and she agreed that it made sense to start him on Piroxicam no matter what else we were doing.

From the Purdue University website—Purdue has one of the leading veterinary medical centers—we were able to obtain the following survival statistics:

1. Median survival with major surgery = 109 days
2. Median survival with chemotherapy = 130 days
3. Median survival with Piroxicam = 195 days

The range of survival times in all of these studies, however, varied tremendously from dog to dog. Some dogs died after only a few days, while others lived more than two years.

We decided that the most rational choice was to start Shadow on the Piroxicam because its side effects were relatively minor, compared to the others, and to get the cystoscopy in order to give the doctor a better look at the mass and the associated biopsy to give us more to go on. Shadow would have to be lightly anesthetized, but it was only for a short time and the doctors were confident that he would emerge fine.

Two weeks later, the cystoscopy showed that the mass was in fact very close to the ureters and the urethral openings—so close, in fact, that surgery wouldn't help if the mass was cancerous because too much of the tumor would be left behind. The pathologist wasn't able to tell if the tissue was cancerous or not because the procedure ended up not getting a large enough sample. So after all that, we still didn't have a diagnosis. Yet the statistics above suggested that if Shadow was among the 30 percent of dogs for whom Piroxicam worked, that would yield the best life expectancy. We wouldn't have to subject him to the discomforts of surgery or chemo, and we could just enjoy our time together at home.

There are many instances, with both pets and humans, that a treatment doesn't statistically improve your life expectancy. Taking a statin if you are not in a high-risk group or surgically removing the prostate for cancer if you do not have fast-moving prostate cancer are both treatments with negligible impact on life expectancy. It sounds counterintuitive, but it's true: Not all treatments actually help. It's clear that Shadow would be better off without the surgery (so that we could avoid the 20 percent chance it would kill him) and the chemo wouldn't buy him any time, statistically.

Shadow responded to the Piroxicam very well and within three days he was back to himself—energetic, in a good mood, happy. By one week he had no more difficulty urinating. We saw occasional minor amounts of blood in the urine, but we were told this was normal after biopsy. Then, 161 days after the initial suspicion (which was never confirmed) of TCC, his kidneys started to fail. We checked him into a specialty oncology clinic. The doctors weren't sure whether the organ failure might be related to TCC, or why it was occurring now. They prescribed medications to address common kidney conditions and ran dozens of tests without getting any closer to understanding what was happening. Shadow grew increasingly uncomfortable and stopped eating. We put him on an IV drip painkiller and two days later, when we took him off for just a few minutes to see how he was doing, he was clearly in pain. We talked to his current and former doctors, carefully describing the situation, its progression, and his condition. All agreed it was time to let him go. We had Shadow's company—and he had ours—for a month longer than the average chemotherapy patient, and during that month he was able to avoid hospitals, catheters, IV lines, and scalpels.

We went to the oncology hospital—the staff knew us well because we had been visiting Shadow there every day in between his tests and treatments—and arranged for him to be put to sleep. He was in pain, and we felt that we had perhaps waited one or two days too long. It was awful to see that large personality suddenly drift away and disappear. We found comfort knowing that we had considered every stage of his care and that he had as good a life as we were able to give him for as long as possible. Perhaps the most difficult emotion that people experience after a disease ends a life is regret over the choices made. We were able to say good-bye to Shadow with no regrets over our decisions. We let our critical thinking, our use of Bayesian reasoning, guide us.

Were Neil Armstrong and Buzz Aldrin Thespians?

Moon-landing deniers point to a number of inconsistencies and unanswered questions. "There should have been more than a two-second delay in communications between the Earth and the moon, because of its distance." "The quality of the photographs is implausibly high." "There are no stars in the sky in any of the photos." "How could the photos of the American flag show ripples in it, as though waving in the air, if the moon has no atmosphere?" The capper is a report by an aerospace worker, Bill Kaysing, who wrote that the probability of a successful landing on the moon was .0017 percent (note the precision of this estimate!). Many more such claims exist. Part of what keeps counterknowledge going is the sheer number of unanswered questions that keep popping up, like a game of Whac-A-Mole. If you want to convince people of something that's not true, it's apparently very effective to simply snow them with one question after another, and hope that they will be sufficiently impressed—and overwhelmed—that they won't bother to look for explanations. But even 1,000 unanswered questions don't necessarily mean that something didn't happen, as any investigator knows. The websites dedicated to the moon landing denial don't cite the evidence for it, nor do they publish rebuttals to their claims.

In the case of the moon landing, each of these (and the other claims) is easily refuted. There *was* a two-second delay in Earth-moon communications that can be easily heard on the original tapes, but some documentary films and news reports edited out the delay in the interest of presenting a more compelling broadcast. The quality of the photographs is high because the astronauts used a high-resolution Hasselblad camera with 70mm high-resolution film. There are no stars in the lunar sky because most of the images we saw were taken during lunar daytime (otherwise, we wouldn't have been able to see the astronauts). The flag doesn't show ripples: Aware that there was no atmosphere, NASA prepared the flag with a t-bar to support its top edge and the "ripples" are simply folds in the fabric. With no wind to blow the flag, its creases stay in place. This claim is based on still photos in which there appears to be a rippling effect, but moving film images show that the flag is not blowing, it's static.

But what about the report of an aerospace worker that a moon landing was highly improbable? First, the "aerospace worker" was not trained in engineering or science; he was a writer with a BA in English who happened to work for Rocketdyne. The source of his estimate appears to be from a Rocketdyne report from the 1950s, back when space technology was still in its infancy. Although there are still unanswered questions (e.g., why are some of the original telemetry recordings missing?), the weight of evidence overwhelmingly points to the moon landing being real. It's not certainty, it's just very, very likely. If you're going to use spuriously obtained probability estimates to claim that past events didn't happen, you'd have to similarly conclude that human beings don't really exist: It's been claimed that the chances of life forming on Earth is many billions to one. Like many examples of counterknowledge, this uses the language of science—in this case probability—in a way that utterly debases that fine language.

Statistics Onstage (and in a Box)

David Blaine is a celebrity magician and illusionist. He also claims to have completed great feats of physical endurance (at least one was recognized by the *Guinness Book of World Records*). The question for a critical thinker is: Did he actually demonstrate physical endurance or was he using a clever illusion? Certainly, as a skilled magician, it would be easy for him to fake the endurance work.

In a TED talk with more than 10 million views, he claims to have held his breath for seventeen minutes underwater, and tells us how he trained himself to do it. Other claims are that he froze himself in a block of ice for a week, fasted in a glass box for forty-four days, and was buried alive in a coffin for a week. Are these claims true? Are they even plausible? Are there alternative explanations?

In his videos, Blaine has a down-to-earth manner; he doesn't speak quickly, he doesn't seem slick. He's believable because his speech sounds so awkward that it's difficult to imagine he's calculated just what to say and how to say it. But bear in mind: Professional magicians typically calculate and plan everything they say. Every single move, every apparently spontaneous scratch of the head, is typically rehearsed over and over again. The illusion they're trying to create—the feat of magic—works because the magician is expert at misdirecting your attention and subverting your assumptions about what's spontaneous and what's not.

So how do we apply critical thinking to his endurance performances?

If you're thinking about hierarchies of source quality, you'll focus on the fact that he has a TED talk and TED talks are fact-checked and very tightly curated. Or are they? Well, actually, there are more than 5,000 TED-branded events, but only two are vetted—TED and TED-Global. Blaine's video comes from a talk he delivered at TEDMED, one of the more than 4,998 conferences that are run by enthusiasts and volunteers and are not vetted by the TED organization. This doesn't mean it's not true, just that we can't rely on the reputation and authority of TED to establish its truth. Recall TMZ and the reporting of Michael Jackson's death—they're going to be right some of the time, and maybe even a lot of the time, but you can't know for sure.

Before looking at the underwater breath holding, let's look more closely at a couple of Blaine's other claims. For starters, Fox television reported his ice-block demonstration to be a hoax. A trap-door beneath the chamber he was in led to a warm and comfortable room, Fox reported, while a body double took his place in the ice block. How did he get away with this trick? A lot of what magicians practice over and over again is getting the audience to accept things that are a bit out of the ordinary. There are some telltale clues that all was not as it seems. First, why is he wearing a mask? (You might assume that it's because it's part of the show, or because it makes him look fierce. The real reason might be because it makes it easier to fool you with a body double.) Why do they need to spray sheets of water over the ice at periodic intervals? (Blaine says it's to prevent the ice from melting; maybe it's so that he can change places with the body double during the brief moment you can't see through the ice.) What about the physiological monitoring equipment on his body, reporting his heart rate and body temperature—surely that's real, isn't it? (Who says that the equipment is actually hooked up to him? Perhaps it wasn't and was instead being fed by a computer.)

If Blaine was lying about the ice block—claiming it was a feat of endurance when in reality it was just conjuring, a magic trick—why not lie about other feats of endurance too? As a performer with a large audience, he would want to ensure that his demonstrations work every time. Using illusions and tricks may be more reliable, and safer, than trying to push endurance limits. But even if it did involve a trick, perhaps it's too harsh to call it a lie—it's all part of the show, isn't it? No one really believes that magicians are calling upon unseen forces; we know that they rehearse like the dickens and use misdirection. Who cares? Well, most reputable magicians, when asked, will come clean and admit that what they are doing are rehearsed illusions, not demonstrations of the black arts. Glenn Falkenstein, for example, performed a mind-reading act that was among the most impressive ever seen. But at the end of each show, he was quick to point out that there was no actual mind-reading involved. Why? Out of a sense of ethics. The world is full of people who believe things that aren't true, and believe many things that are ridiculous, he said. Millions of people who have a poor understanding of cause and effect waste their money and energy on psychics, astrologers, gambling, and "alternative" therapies with no proven efficacy. Being forthright about how this sort of entertainment is accomplished is important, he said, so that people are not led to believe things that aren't so.

In another demonstration, Blaine claims to have stuck a needle clear through his hand. Was this an illusion or did he really do it? In videos, it certainly looks real, but of course, that's what magic is all about. (Search YouTube and you'll find videos showing how it can be done with specialized apparatus.) What about the forty-four-day fast in a glass box? There was even a peer-reviewed paper in the *New England Journal of Medicine* about that, and in terms of information sources, that's about as good as it gets. Upon closer examination, however, the physicians who authored that paper only examined Blaine after the fast, not before or during, and so they can't provide independent verification that he actually fasted. Was this question ever raised during peer review? The current editor of the journal searched his office archives but the records had been destroyed, since the article was published a decade before my inquiry. The lead author on the article told me in an email that based on the hormones she measured after the event, he was indeed fasting, but it's possible as well that he was sneaking in some food; she couldn't comment on that. She did point me to an article by a colleague of hers in another peer-reviewed journal, in which a physician *did* monitor Blaine throughout the fast (the article didn't show up in my PubMed or Google Scholar searches because David Blaine was not mentioned in the article by name). Relevant is the following passage from the article, which appeared in the journal *Nutrition*:

Immediately before the start of the fast, DB appeared to have a muscular build that was consistent with the body mass index, body composition figures, and upper arm muscle circumference, which are reported below. On the evening of Saturday, September 6, 2003, DB entered a transparent Perspex box, measuring 2.1 x 2.1 x 0.9 m, which was suspended in air for the next 44 d, close to Tower Bridge, London. Continuous detailed video monitoring was available to one of the investigators (ME, office and at home), who was able to assess the clinical state and physical activity of DB. DB, who was 30 y old, had consumed before the event, a diet that was estimated, but not verified, to have increased his weight by as much as 6–7 kg. He also took some multivitamin tablets for a few days before the event, which he stopped on entry into the box. He felt weaker and more lethargic as the event progressed. From about 2 wk onward he experienced some dizziness and faintness on standing up quickly, and on some occasions, temporary visual problems, as if "blacking out." He also developed transient sharp shooting pains in his limbs and trunk, abdominal discomfort, nausea, and some irregular heart beats. . . . A small amount of bleeding from his nose occurred on the fifth day after entry to the box and this recurred later. There were no other obvious signs or symptoms of a bleeding tendency. There were also no signs of edema before or at the end of the fast. In addition, there were no clinical signs of thiamine deficiency. DB, who was initially a muscular looking man, was visibly thinner on exit from the box. His blood pressure taken almost immediately before the event began was 140/90 mmHg while lying and 130/80 mmHg while standing, and at the end it was 109/74 mmHg while lying (pulse 89 beats/min) and 109/65 mmHg while standing (pulse 119 beats/min).

From this report, it does sound as though he really did fast. A skeptic might discount his reports of pain and nausea as showmanship, but it is difficult to fake irregular heartbeat and weight loss.

But it's the breath holding, televised on Oprah Winfrey's show, that is the focus of Blaine's TEDMED talk. In it, Blaine uses a lot of scientific and medical terminology to prop up the narrative that this was a medically based endurance demonstration, not a mere trick. Blaine describes the research he did:

"I met with a top neurosurgeon and I asked him how long . . . anything over six minutes you have a serious risk of hypoxic brain damage . . . perflubron." Blaine mentions liquid breathing; a hypoxic tent to build up red blood cell count; pure O_2. That got him to fifteen minutes. He goes on to elaborate a training regimen in which he gradually built up to seventeen minutes. He throws out terms like "blood shunting" and "ischemia." Did Blaine actually do what he said he did? Was the medical jargon he threw out for real, or just pseudoscientific babble he invoked to overwhelm, to make us *think* he knew what he was talking about?

As always, we start with a plausibility check. If you've ever tried holding your breath, you probably held out for half a minute—maybe even an entire minute. A bit of research reveals that professional pearl divers routinely hold their breaths for seven minutes. The world record for breath holding *before* Blaine's was just under seventeen minutes. As you continue to read up on the topic, you'll discover that there are two kinds of breath-holding competitions: plain old, regular old breath holding, like you and your older brother did in the community pool when you were kids, and *aided* breath holding, in which competitors are allowed to do things like inhale 100 percent pure oxygen for half an hour prior to competing. This is sounding more plausible, but how far can you get with aided breath holding—can it actually bridge the gap between a few minutes and seventeen minutes? At this point, you might try to learn what the experts have to say—pulmonologists (who would know something about lung capacity and the breathing reflex) and neurologists (who would know how long the brain can last without an influx of oxygen). The two pulmonologists I checked with described a training regimen much like the one Blaine describes in his video; both felt that with these "tricks" or special measures, seventeen minutes of breath holding would be possible. Indeed, Blaine's record was broken in 2012 by Stig Severinsen, who held his breath for twenty minutes and ten seconds (after inhaling pure oxygen, of course), and who then broke his own record a month later, achieving twenty-two minutes. David Eidelman, MD, a pulmonary specialist and dean of the McGill Medical School said: "I agree that it does sound hard to believe. . . . However, by inhaling oxygen first, fasting, and using yoga-type techniques to lower metabolic rate while holding one's breath underwater, it seems that this is possible. So, while I retain some skepticism, I do not think I can prove it is impossible."

Charles Fuller, MD, a pulmonary specialist at UC Davis, adds, "There is sufficient evidence to indicate that Blaine is being truthful, as this event is physiologically feasible. Given the caveat that Blaine is a magician and there could have been other contributing factors to his successful seventeen-minute breath hold, there is also ample physiological evidence that this feat could have been accomplished. There is a subset of people in the breath-hold world who vie for a record officially known as 'pre-oxygenated static apnea.' In this event, breath holding is sponsored by Guinness World Records, as sports divers consider this cheating. Breath-hold duration is measured after hyper-ventilating (blowing of carbon dioxide) for thirty minutes while breathing 100 percent pure oxygen. Further, the event is typically held in a warm pool (which reduces metabolic oxygen demand), with the head held just below the surface, which induces the human dive reflex (further depressing metabolic oxygen demand). In other words, all 'tricks' which extend the human capacity for conscious breath hold. Most importantly, prior to Blaine, the record was just under his sev-enteen minutes [by an athlete who was *not* a magician], and there are additional individuals who have now been recorded for longer breath holds in excess of twenty minutes. Thus, ample evidence that this feat could have been accomplished as claimed."

So far, Blaine's story seems plausible, and his talk hits all the right notes. But what about brain damage? Blaine himself men-tioned this as a problem. You've no doubt heard that if the brain loses oxygen for even three minutes, irreparable damage and brain death can occur. If you're not breathing for seventeen minutes, how do you prevent brain death? A good question for a neurologist.

Scott Grafton, MD: "Oxygen doesn't stay in the blood all by itself. Think oil and water. It will quickly diffuse out of the liquid blood—it needs to bind to something. Blood carries red cells. Each red cell is loaded with hemoglobin (Hgb) molecules. These hemoglobin mol-ecules can potentially bind up to four oxygen molecules. Each time a red cell passes through the lungs, the number of Hgb molecules with oxygen bound to them increases. The stronger the concentration of oxygen in the air, the more Hgb molecules will bind to it. So load 'em up! Breathe 100 percent oxygen for thirty minutes so that total oxygen binding is as close to 100 percent saturation as you can get.

"Each time the red cell passes through the brain, oxygen will have a probability of unbinding from the molecule, diffusing across cell membranes to enter the brain tissue, where it binds to other molecules that use it in oxidative metabolism. The probability of a given oxygen molecule unbinding from hemoglobin and diffusing is a function of the relative difference of oxygen concentration on either side of the membranes."

In other words, the more oxygen the brain needs, the more likely it will be to pull oxygen out of hemoglobin. By breathing pure oxygen for thirty minutes, the competitive breath holder will maximize the amount of oxygen in the brain *and* in the blood. Then, once the breath-holding event starts, oxygen levels in the brain will decrease as they normally do over time, and the competitive breath holder will very efficiently pull out whatever oxygen happens to be left in hemoglobin to oxygenate the brain.

Grafton continues, "Not all the hemoglobin molecules are loaded up with oxygen on each pass through the lungs, and not all of them unload on each pass through the organs. It takes quite a few passes to unload all of them. When we say that brain death happens quickly due to a lack of oxygen, it is usually in the context of a lack of circulation (heart attack) when the heart is no longer delivering blood to the brain. Stop the pump, and no red cells are available to offer up oxygen and brain tissue dies fast. In a person submerged, there is a race between brain injury and pump failure.

"One key trick: The muscles need to be at rest. Muscles are loaded with myoglobin, which holds on to oxygen four times more strongly than red-cell hemoglobin does. If you're using your muscles, this will accelerate the loss of oxygen overall. Keep muscle demand low." This is the *static* in the static apnea that Dr. Fuller mentioned.

So from a medical standpoint, David Blaine's claim appears plausible. That might be the end of the story, except for this. An article in the *Dallas Observer* claims the breath holding was a trick, and that Blaine—a master illusionist—used a well-hidden breathing tube. There's nothing about this in other mainstream media, which doesn't mean the *Observer* is wrong, of course, but why is this paper the only one reporting it? Perhaps a magician who performs a trick but claims it wasn't one is not big news.

Reporter John Tierney traveled to Grand Cayman Island to write about Blaine's preparation for the breath holding for an article in the *New York Times,* and then wrote about the *Oprah* appearance in his blog a week later. Tierney makes a lot out of Blaine's heart rate, as reported on a monitor next to his tank on the Winfrey show, but as with the ice-block demonstration, there's no evidence this monitor was actually connected to Blaine, and it might really have been more for showmanship—to make the audience think that the conditions were really rough (a standard practice for magicians). Neither Tierney nor a physician involved in the training mention how closely they were monitoring Blaine during practice trials in the Caymans—it's possible that they took him at his word that he didn't have any apparatus. Perhaps the real motive of this training was that Blaine figured if he could fool *them,* he could fool a television audience. Tierney writes, "I was there at the pool along with some free-divers who are experts at static apnea (holding your breath while remaining immobile). Dr. Ralph Potkin, a pulmonologist who studies breath holding and is the team physician for the United States free-diving team, attached electrodes to Blaine's body during the session and measured his heart, blood, and breathing as Blaine kept his head submerged in the water for sixteen minutes.

"I've always been skeptical of cons—I did a long piece on James Randi a while back, and was with him in Detroit when he was exposing an evangelist named Peter Popoff—but I saw no reason to doubt Blaine's feat. His breath hold in front of me was done in clear water in the shallow end of the very ordinary swimming pool at our hotel, with experts in breath holding a few feet away watching him all the time. His nose and mouth were clearly below the water—but just a couple of inches, so they were visible at all times. You tell me how he snuck a breathing tube in there so that no one noticed it or any bubbles. Magicians fool people by distracting them with motions and patter, but the whole point of static apnea is to remain absolutely motionless in order to conserve oxygen, which is what David did. (It's remarkable what a difference that makes—the trainers who were working with David did a short session with me and my photographer. We pre-breathed air instead of oxygen, but we were amazed at how long we went—I got up to 3 minutes and 41 seconds, and the photographer even longer.)"

So now the *Dallas Observer* says it was faked, and a *New York Times* reporter seems to believe it wasn't. What do professional magicians think? I spoke to four. One said, "It's got to be a trick. A lot of his demonstrations are known, at least within the magic community, to use camera trickery and [a] very involved setup. It would be very easy for him to have a breathing tube that allows him to take oxygen in, and to exhale carbon dioxide, without making bubbles in the water. And if he practices, he wouldn't need to be doing it all that often—he could actually hold his breath for a minute or two at a time in between tube breaths. And there could be other camera trickery—he might not actually be in the water! Projection or green screen could make it appear that he was."

The second magician, who had worked with Blaine a decade earlier, added, "His hero is Houdini, who became famous for doing stunts. Houdini made a reputation in part by doing things that people did in the 1920s—flagpole sitting and so on. Some do require endurance and some are faked slightly; some aren't as hard as you'd think, but most people never try them. I don't see why Blaine would fake the block-of-ice trick—that one is simple because of the igloo effect—it's not actually that cold in there. It *looks* impressive. If he was in a freezer that would be different.

"But seventeen-minute breath holding? If he can super-oxygenate his blood, that can help. I know that he does train and does some things that are remarkable. But I'm sure the breath-holding trick is partly enabled. He does hold his breath, I think, but not 100 percent of the time. It's quite easy to fake. He probably has [a] breathing tube and other apparatus.

"Note that a lot of his magic is on TV and there are edits at key points. We assume it's real information and we're seeing everything because that's how our brains construct reality. But as a magician, I see the edits and wonder what was happening during the missing footage."

A third magician added, "Why would you go to all the trouble to train if, as an illusionist, you can do it with equipment? Using equipment creates a more reliable, replicable performance that you can do again and again. Then, you just need to act as though you're in pain, dizzy, disoriented, and as though you've pushed your body beyond all reasonable limits. As an entertainer, you wouldn't want to leave anything to chance—there's too much at stake."

The fact that no one reports seeing David Blaine use a breathing tube does not constitute evidence that he didn't, because it is precisely the *job* of illusionists to play tricks with what you see and what you think you see. And the illusion is even more powerful when it happens to you. I've had the magician Glenn Falkenstein read off the serial numbers of dollar bills in my wallet from across the room while he was blindfolded. I've had the magician Tom Nixon place the seven of diamonds in my hand, yet a few minutes later it had become a completely different card without my being aware of him touching me or the card. I know he's switching the card at some point, but even after having the trick done to me five times, and watching it performed on other people many more times, I still don't know when the switch occurs. That is part of the magician's genius, and part of the entertainment. I don't think for a moment that Falkenstein or Nixon possess occult powers. I know it's entertainment, and they sell it as that.

The fourth magician I asked about it was James Randi, the professional skeptic I (and John Tierney) mentioned earlier, who replicates alleged psychic phenomena through his deft use of illusions and magic tricks. Here's what he wrote via email:

> I recall that when David Blaine first showed up on television performing his stunts, I voluntarily contacted him with a friendly warning that he was—in my opinion as a conjuror—taking chances of personal physical damage. We exchanged friendly correspondence on this matter, until I was abruptly informed that his newly engaged management agency had changed his email address and that he'd been instructed not to correspond further with me. I of course accepted this decision, while hoping that Mr. Blaine would heed my well-intended suggestions.
>
> I have not been in touch with David Blaine since that time. I was alarmed to see the unwise statements he made on the TED appearance, and I have respected the—to me—rather unwise slant that his agency has chosen to give to his claims, but I have respected his privacy.
>
> He let his agency even terminate his connection with me, perhaps because I might have tried to keep him honest. Can't have too much of that quality, of course.

The weight of our fact-checking suggests that the seventeen-minute breath hold is very plausible. That doesn't guarantee that

Blaine didn't use a breathing tube. Whether you believe Blaine pulled off the stunt legitimately is up to you—each of us has to make our own decision. As with any magician, we can't be sure what's true and what's not—and that is the world of ambiguity that magicians spend their professional lives trying to create. In critical thinking, one looks for the most parsimonious account, but in some cases, as here, it is difficult or impossible to choose between the possible explanations or to figure which is more parsimonious. Does it even matter? Well, yes. As Falkenstein said, people who have a poor understanding of cause and effect, or an insufficient understanding of chance and randomness, are easily duped by claims such as these, leading them to too readily accept others. Not to mention the many amateurs who may try to replicate these spectacles, despite the ubiquitous warning of "do not try this at home." The uneducated are easy targets. The difference between doing this by training and doing it by illusion is the difference between being duped and not being duped.

Statistics in the Universe

When you hear names like hydrogen, oxygen, boron, tin, and gold, what do you think of? These are chemical elements of the periodic table, usually taught in middle or high school. They were called elements by scientists because they were believed to be fundamental, indivisible units of matter (from the Latin *elementum,* matter in its most basic form). The Russian scientist Dmitri Mendeleev noticed a pattern in the properties of elements and organized them into a table that made these properties easier to visualize. In the process, he was able to see gaps in the table for elements that had not yet been discovered. Eventually all of the elements between 1 and 118 have either been discovered in nature or synthesized in the laboratory, supporting the theory underlying the table's arrangement.

Later, scientists discovered that the chemical elements were not actually indivisible; they were made of something that the scientists called atoms, from the Greek word *atomos*, for "indivisible." But they were wrong about the indivisibility of those, too—atoms were later discovered to be made up of subatomic particles: protons, neutrons, and electrons. These were also initially thought to be indivisible, but then—you guessed it—that was found to be incorrect. The so-called Standard Model of Particle Physics was formulated in the 1950s and '60s, and theorized that electrons are indivisible, but protons and neutrons are composed of smaller subatomic particles. With the discovery of quarks in the 1970s, this model was confirmed. To further complicate terminology, protons and electrons are a type of *fermion*, and neutrons are a type of *boson* (photons are also a type of boson). The different categories are necessary because the two different types of particles are governed by different laws. Fermions and bosons have been given the name *elementary particle* because it is believed that they are truly indivisible (but time will tell).

According to the Standard Model, there are seventeen different types of elementary particles—twelve kinds of fermions and five kinds of bosons. The Higgs boson, which received a great deal of press in 2012 and 2013, has been the last remaining piece of the Standard Model to be proven—the other sixteen have already been discovered. If it exists, the Higgs would help to explain how matter obtains mass, and fill in a key hole in the theory used to explain the nature of the universe, a hole in the theory that has existed for more than fifty years.

How do we know if we've found it? When particles collide at great . . . Oh, forget it. I'll let a physicist explain it. Here's Professor Harrison Prosper, describing this plot and the little "blip" next to the arrow corresponding to 125 gigaelectronvolts (GeV) on the horizontal axis:

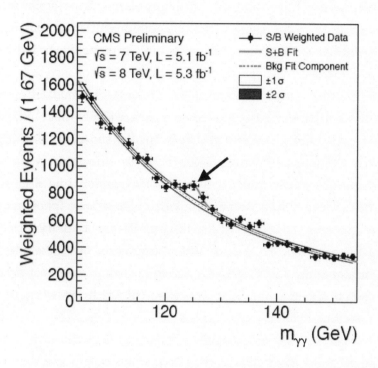

The graph shows "a spectrum arising from proton-proton collisions that resulted in the creation of a pair of photons (gammas in high energy argot)," Prosper says. "The Standard Model predicts that the Higgs boson should decay (that is, break up) into a pair of photons. (The Higgs is predicted to decay in other ways too, such as a pair of Z bosons.) The bump in the plot at around 125 GeV is evidence for the existence of some particle of a definite mass that

247

decays into a pair of photons. That something, as far as we've been able to ascertain, is likely to be the Higgs boson."

Not all physicists agree that the experiments are conclusive. Louis Lyons explains, "The Higgs . . . can decay to different sets of particles, and these rates are defined by the S.M. [Standard Model]. We measure these ratios, but with large uncertainties with the present data. They are consistent with the S.M. predictions, but it could be much more convincing with more data. Hence the caution about saying we have discovered the Higgs of the S. M."

In other words, the experiments are so costly and difficult to conduct, that physicists want to avoid a false alarm—they've been wrong before. Although CERN officials announced in 2012 that they had found it, many physicists feel the sample size was too small. There is so much at stake that the physicists have set for themselves a standard of proof, a statistical threshold, that is much stricter than the 1 in 20 used in other fields—1 in 3.5 million. Why such an extreme evidence requirement? Prosper says, "Given that the search for the Higgs took some forty-five years, tens of thousands of scientists and engineers, billions of dollars, not to mention numerous divorces, huge amounts of sleep deprivation, tens of thousands of bad airline meals, etc., etc., we want to be sure as is humanly possible that this is real."

Physicist Mads Toudal Frandsen adds, "The CERN data is generally taken as evidence that the particle is the Higgs particle. It is true that the Higgs particle can explain the data but there can be other explanations; we would also get this data from other particles. The current data is not precise enough to determine exactly what the particle is. It could be a number of other known particles." Recall the discussion earlier in the *Field Guide* about alternative explanations. Physicists are on the alert for this.

If the plot is showing evidence of a different kind of particle, something that is *not* the Higgs, this could substantially change our view of how the universe was created. And if it does exist, some physicists, such as Stephen Hawking, fear that this could spell the end of the universe as we know it. The fear is that a quantum fluctuation could create a vacuum bubble that rapidly and continually expands until it wipes out the universe. And if you think physicists don't have a sense of humor, Joseph Lykken, a physicist and director of the Fermi National Accelerator Laboratory in Illinois, noted that it won't happen for a long, long time—10^{100} years from now—"so probably you shouldn't sell your house and you should continue to pay your taxes."

Not everyone is happy with the discovery, and not because it may signal the end of the world—it's because finding something in science that the standard theories predict doesn't open the door for new inquiry. An anomalous, unexplained result is most interesting to scientists because it means their model and understanding was at best incomplete, and at worst, completely wrong—presenting a great opportunity for new learning. In one of the many intersections between art and science, the conductor Benjamin Zander says that when a musician makes a mistake, rather than swearing or saying "oops" or "I'm sorry," she should say, "Now *that's* interesting!" Interesting because it represents an opportunity for learning. It could be that the discovery of the Higgs boson answers all the questions we had. Or, as *Wired* writer Signe Brewster says, "It could lead to an underlying principle that physicists have missed until now. The end goal, as always, is to find a string that, when tugged, rings a clarion bell that draws physicists toward something new." As Einstein reportedly said, if you know how it's going to turn out, it's not science, it's engineering.

Scientists are curious, lifelong learners, eager to find the next challenge. There are some who fear that the discovery of the Higgs may explain so much that it ends the ride. Others are so filled with wonder and the complexities of life and the universe that they are confident we'll never figure it all out. I am among the latter.

As of this writing, tantalizing evidence has emerged from CERN of a new particle that might be a graviton, or a heavier version of the Higgs boson. But the most probable explanation for these surprising new bumps in the data flow is that it is a coincidence—the findings have a 1 in 93 chance of being a fluke, far more likely than the 1 in 3.5 million probability used for the Higgs. But there are qualitative considerations. "What is nice is that it is not a particularly crazy signal, in a quite clean channel," physicist Nima Arkani-Hamed told the *New York Times*. "So, while we are nowhere near moving champagne even vaguely close to the fridge, it is intriguing." Nobody knows yet what it is, and that's just fine with Lykken and many others who love the thrill of the chase.

Science, history, and the news are full of things that we knew, or thought we did, until we discovered we were wrong. An essential component of critical thinking is knowing what we don't know. A guiding principle, then, is simply that we know what we know until we don't any longer. The purpose of the *Field Guide* was to help you to think things through, and to give you greater confidence both in what you think you know, and what you think you don't, and—hopefully—to be able to tell the difference between them.

DISCOVERING YOUR OWN

In George Orwell's *1984*, the Ministry of Truth was the country's official propaganda agency, charged with falsifying historical records and other documents to reflect the administration's agenda. The Ministry also advanced counterknowledge when it served their purposes, such as $2 + 2 = 5$.

Nineteen Eighty-four was published in 1949, half a century before the Internet became our de facto information source. Today, like in *1984,* websites can be altered so that the average person doesn't know that they have been; every trace of an old piece of information can be rewritten, or (in the case of Paul McCartney and Dick Clark) kept out of reach. Today, it can be very difficult for the average Web surfer to know if a site is reporting genuine knowledge or counterknowledge. Unfortunately, sites that advertise that they are telling the truth are often the ones that aren't. In many cases, the word "truth" has been co-opted by people who are propagating counterknowledge or fringe viewpoints that go against what is conventionally accepted as truth. Even site names can be deceptive.

Can we trust experts? It depends. Expertise tends to be narrow. An economist in the highest echelons of the government may not have any special insight into what social programs will be effective for curbing crime. And experts sometimes become co-opted by special interests, and, of course, they make mistakes.

An anti-science bias has entered public discourse and the Web. A lot of things that should be scientific or technical problems—like where to put a power plant and how much it should cost—are political. When that happens, the decision-making process is subverted, and the facts that matter are often not the ones that are under consideration. Or we say that we want to cure an intractable human disease, but mock the first step when tens of millions of dollars are spent studying aphids. The reality is that science progresses by gaining an understanding of basic cellular physiology. With the wrong frame the research looks trivial; with the right frame it can be seen for the potential it truly has to be transformative. Money put into human clinical trials might end up being able to treat the symptoms of a few hundred thousand people. That same money put into basic-level scientific research has the potential to find the cure for *dozens* of diseases and *millions* of people because it is dealing with mechanisms common to many different types of bacteria and viruses. The scientific method is the ground from which all the best critical thinking rises.

In addition to an anti-science bias, there is an anti-skepticism bias when it comes to the Internet. Many people think, "If I found it online it must be true." With no central authority charged with the responsibility of monitoring and regulating websites and other material found online, the responsibility for verifying claims falls on each of us. Fortunately, some websites have cropped up that help. Snopes.com and similar sites are dedicated to exposing urban legends and false claims. Companies such as Consumer Reports run independent laboratories to provide an unbiased assessment of different products, regardless of what their manufacturers claim. Consumer Reports has been around for decades, but it is no great leap to expect that other critical-thinking enterprises will flourish in the twenty-first century. Let's hope so. But whatever helpful media is out there, each of us will still have to apply our judgment.

The promise of the Internet is that it is a great democratizing force, allowing everyone to express their opinions, and everyone to have immediate access to all the world's information. Combine these two, as the Internet and social media do, and you have a virtual world of information and misinformation cohabiting side by side, staring back at you like identical twins, one who will help you and the other who will hurt you. Figuring out which one to choose falls upon all of us, and it requires careful thinking and one thing that most of us feel is in short supply: time. Critical thinking is not something you do once with an issue and then drop it. It's an active and ongoing process. It requires that we all think like Bayesians, updating our knowledge as new information comes in.

Time spent evaluating claims is not just time well spent, it should be considered part of an implicit bargain we've all made. Information gathering and research that used to take anywhere from hours to weeks now takes just seconds. We've saved incalculable numbers of hours of trips to libraries and far-flung archives, of hunting through thick books for the one passage that will answer our questions. The implicit bargain that we all need to make explicit is that we will use just *some* of that time we saved in information acquisition to perform proper information verification. Just as it's difficult to trust someone who has lied to you, it's difficult to trust your own knowledge if half of it turns out to be counterknowledge. The fact is that right now counterknowledge flourishes on Facebook and on Twitter and on blogs . . . on all the semi-organized platforms.

We're far better off knowing a moderate number of things with certainty than a large number of things that might not be so. Counterknowledge and misinformation can be costly, in terms of lives and happiness, and in terms of the time spent trying to undo things that didn't go the way we thought they would. True knowledge simplifies our lives, helping us to make choices that increase our happiness and save time. Following the steps in this *Field Guide* to evaluate the myriad claims we encounter is how we can stay two steps ahead of the millions of lies that are out there on the Web, and ahead of the liars and just plain incompetents who perpetrate them.

APPENDIX
APPLICATION OF BAYES'S RULE

Bayes's rule can be expressed as follows; $P(A \mid B) = \dfrac{(P(B \mid A) \times P(A))}{(P(B))}$

For the current problem, let's use the notation that G refers to the prior probability that the suspect is guilty (before we know anything about the lab report) and E refers to the evidence of a blood match. We want to know P(G|E). Substituting in the above, we put in G for A and E for B to obtain:

$$P(G \mid E) = \frac{(P(E \mid G) \times P(G))}{(P(E))}$$

To compute Bayes's rule and solve for P(G | E), it may be helpful to use a table. The values here are the same as those used in the fourfold table on page 220.

COMPUTATION OF BAYES'S RULE

Hypothesis (H) (1)	Prior Probability P(G) (2)	Evidence Probability P(E | G) (3)	Product (4) = (2)(3)	Posterior Probabilities P(G | E) (6) = (4)/Sum
Guilty	.02	.85	.017	.104
Innocent	.98	.15	.147	.896

Sum = .164
= P(D)

Then, rounding, P(Guilty | Evidence) = .10
P(Innocent | Evidence) = .90

GLOSSARY

This list of definitions is not exhaustive but rather a personal selection driven by my experience in writing this book. Of course, you may wish to apply your own independent thinking here and find some of the definitions deserve to be challenged.

Abduction. A form of reasoning, made popular by Sherlock Holmes, in which clever guesses are used to generate a theory to account for the facts observed.

Accuracy. How close a number is to the true quantity being measured. Not to be confused with precision.

Affirming the antecedent. Same as *modus ponens* (see entry below).

Amalgamating. Combining observations or scores from two or more groups into a single group. If the groups are similar along an important dimension—homogeneous—this is usually the right thing to do. If they are not, it can lead to distortions of the data.

Average. This is a summary statistic, meant to characterize a set of observations. "Average" is a nontechnical term, and usually refers to the *mean* but could also refer to the *median* or *mode*.

Bimodal distribution. A set of observations in which two values occur more often than the others. A graph of their frequency versus their values shows two peaks, or humps, in the distribution.

Conditional probability. The probability of an event occurring *given* that another event occurs or has occurred. For example, the probability that it will rain today *given* that it rained yesterday. The word "given" is represented by a vertical line like this: | .

Contrapositive. A valid type of deduction of the form:

If A, then B
Not B
Therefore, not A

Converse error. An invalid form of deductive reasoning of the form:

If A, then B
B
Therefore, A

Correlation. A statistical measure of the degree to which two variables are related to each other, it can take any value from −1 to 1. A perfect correlation exists (correlation = 1) when one variable changes perfectly with another. A perfect negative correlation exists when one variable changes perfectly opposite the other (correlation = −1). A correlation of 0 exists when two variables are completely unrelated.

A correlation shows only that two (or more) variables are linked, not that one causes the other. Correlation does not imply causation.

A correlation is also useful in that it provides an estimate for how much of the variability in the observations is caused by the two variables being tracked. For example, a correlation of .78 between height and weight indicates that 78 percent of the differences in weight across individuals are linked to differences in height. The statistic doesn't tell us what the remaining 22 percent of the variability is attributed to—additional experimentation would need to be conducted, but one could imagine other factors such as diet, genetics, exercise, and so on are part of that 22 percent.

***Cum hoc, ergo propter hoc* (with this, therefore because of this).** A logical fallacy that arises from thinking that just because two things co-occur, one must have caused the other. Correlation does not imply causation.

Cumulative graph. A graph in which the quantity being measured, say sales or membership in a political party, is represented by the total to date rather than the number of new observations in a time period. This was illustrated using the cumulative sales for the iPhone [page 48].

Deduction. A form of reasoning in which one works from general information to a specific prediction.

Double y-axis. A graphing technique for plotting two sets of observations on the same graph, in which the values for each set are represented on two different axes (typically with different scales). This is only appropriate when the two sets of observations are measuring unlike quantities, as in the graph on page 40. Double y-axis graphs can be misleading because the graph maker can adjust the scaling of the axes in order to make a particular point. The example used in the text was a deceptive graph made depicting practices at Planned Parenthood.

Ecological fallacy. An error in reasoning that occurs when one makes inferences about an individual based on aggregate data (such as a group mean).

Exception fallacy. An error in reasoning that occurs when one makes inferences about a group based on knowledge of a few exceptional individuals.

Extrapolation. The process of making a guess or inference about what value(s) might lie beyond a set of observed values.

Fallacy of affirming the consequent. See *Converse error.*

Framing. The way in which a statistic is reported—for example, the context provided or the comparison group or amalgamating used—can influence one's interpretation of a statistic. Looking at the total number of airline accidents in 2016 versus 1936 may be misleading because there were so many more flights in 2016 versus 1936—various adjusted measures, such as accidents per 100,000 flights or accidents per 100,000 miles flown, provide a more accurate summary. One works to find the true frame for a statistic, that is the appropriate and most informative one. Calculating proportions rather than actual numbers often helps to provide the true frame.

GIGO. Garbage in, garbage out.

Incidence. The number of new cases (e.g., of a disease) reported in a specified period of time.

Inductive. A form of inferential reasoning in which a set of particular observations leads to a general statement.

Interpolation. The process of estimating what intermediate value lies between two observed values.

Inverse error. An invalid type of deductive reasoning of the form:

If A, then B
Not A
Therefore, not B

Mean. One of three measures of the average (the central tendency of a set of observations). It is calculated by taking the sum of all observations divided by the number of observations. It's what people usually are intending when they simply say "average." The other two kinds of averages are the median and the mode. For example, for $\{1, 1, 2, 4, 5, 5\}$ the mean is $(1 + 1 + 2 + 4 + 5 + 5) \div 6 = 3$. Note that, unlike the mode, the mean isn't necessarily a value in the original distribution.

Median. One of three measures of the average (the central tendency of a set of observations). It is the value for which half the observations are larger and half are smaller. When there is an even number of observations, statisticians may take the mean of the two middle observations. For example, for

{10, 12, 16, 17, 20, 28, 32}, the median is 17. For {10, 12, 16, 20, 28, 32}, the median would be 18 (the *mean* of the two middle values, 16 and 20).

Mode. One of three measures of the average (the central tendency of a set of observations). It is the value that occurs most often in a distribution. For example, for {100, 112, 112, 112, 119, 131, 142, 156, 199} the mode is 112. Some distributions are bimodal or multimodal, meaning that two or more values occur an equal number of times.

Modus ponens. A valid type of deductive argument of the form:

    If A, then B
    A
    Therefore, B

***Post hoc, ergo propter hoc* (after this, therefore because of this).** A logical fallacy that arises from thinking that just because one thing (Y) occurs after another (X), that X *caused* Y. X and Y might be correlated, but that does not mean a causative relation exists.

Precision. A measure of the level of resolution of a number. The number 909 is precise to zero decimal points and has only a resolution of the nearest whole number. The number 909.35 is precise to two decimal places and has a resolution of 1/100th of a unit. Precision is not the same as accuracy—the second number is more precise, but if the true value is 909.00, the first number is more accurate.

Prevalence. The number of existing cases (e.g., of a disease).

Scatter plot. A type of graph that represents all points individually. For example, opposite is a scatter plot of the data presented on page 39.

School Funding v. SAT Scores

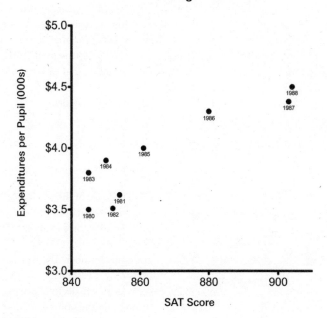

Subdividing. Breaking up a set of observations into smaller groups. This is acceptable when there is heterogeneity within the data and the larger group is composed of entities that vary along an important dimension. But subdividing can be used deceptively to create a large number of small groups that do not differ appreciably along a variable of interest.

Syllogism. A type of logical statement in which the conclusion must necessarily follow from the premises.

Truncated axis. Starting an x- or y-axis on a value other than the lowest one possible. This can sometimes be helpful in allowing viewers to see more clearly the region of the graph in which the observations occur. But used manipulatively, it can distort the reality. The graph shown in this glossary under the entry for "scatter plot" uses two truncated axes effectively and does not give a false impression of the data. The graph shown on page 29 by Fox News does give a false impression of the data, as shown in the redrawn version on page 30.

NOTES

INTRODUCTION: THINKING, CRITICALLY

x **avoid learning a whole lot of things that aren't so** After Huff, D. (1954/1993). *How to Lie with Statistics*. New York: W.W. Norton, p. 19. And, as you'll read later, he probably was echoing Mark Twain, or Josh Billings, or Will Rogers, or who knows who.

xi **Misinformation has been a fixture of human life** Abraham provides misinformation about the identity of his wife, Sarah, to King Abimelech to protect himself. The Trojan horse was a kind of misinformation, appearing as a gift but containing soldiers.

PART ONE: EVALUATING NUMBERS

3 **People choose what to count** This sentence is nearly a direct quote from Best, J. (2005). Lies, calculations and constructions: beyond *How to Lie with Statistics*. *Statistical Science, 20*(3), 210–214.

5 **More people have cell phones than toilets** Wang, Y. (2013, March 25). More people have cell phones than toilets, U.N. study shows. http://news feed.time.com/2013/03/25/more-people-have-cell-phones-than-toilets-u-n -study-shows/.

6 **150,000 girls and young women die of anorexia each year** Steinem, G. (1992). *Revolution from Within*. New York: Little, Brown. Wolf, N. (1991). *The Beauty Myth*. New York: William Morrow.

6 **Add in women from twenty-five to forty-four and you still only get 55,000** This example came to my attention from Best, J. (2005). Lies, calculations and constructions: beyond *How to Lie with Statistics*. *Statistical Science, 20*(3), 210–214. The statistics are available at www.cdc.gov.

6 **anorexia deaths in one year cannot be three times the number of *all* deaths** Maybe you're in the accounts payable department of a big

corporation. An employee put in for reimbursement of gasoline for the business use of his car, $5,000 for the month of April. Start with a little world knowledge: Most cars get better than twenty miles per gallon these days (some get several times that). You also know that the fastest you can reasonably drive is seventy miles per hour, and that if you were to drive ten hours a day, all on the freeway, that would mean 700 miles a day. Keep that up for a standard 21.5-day work month and you've got 15,050 miles. In these kinds of rough estimates, it's standard to use round numbers to make things easier, so let's call that 15,000. Divide that by the fuel economy of 20 mpg and, by a rough estimate, your employee needed 750 gallons of gas. You look up the average national gas price for April and find that it's $2.89. Let's just call that $3.00 (again, rounding, and giving your employee the benefit of the doubt—he may not have managed to get the very best price every time he filled up). $3/gallon times 750 gallons = $2,250. The $5,000 on the expense report doesn't look even remotely plausible now. Even if your employee drove twenty hours a day, the cost wouldn't be that high. https://www.fueleconomy.gov/feg/best/bestworstNF.shtml, retrieved August 1, 2015. http://www.fuelgaugereport.com/.

6 **a telephone call has decreased by 12,000 percent** Pollack, L., & Weiss, H. (1984). Communication satellites: countdown for Intelsat VI. *Science, 223*(4636), 553.

6 **one of 12,000 percent seems wildly unlikely** I suppose you could spin a story that makes this true. Maybe a widget used to cost $1, and now, as part of a big promotion, a company is not just willing to give it to you for free, but to *pay* you $11,999 to take it (that's a 12,000 percent reduction). This happens in real estate and big business. Maybe an old run-down house needs to be razed before a new one can be built; the owner may be paying huge property taxes, the cost of tearing down the house is high, and so the owner is willing to pay someone to take it off of his or her hands. At one point in the late 1990s, several large, debt-ridden record companies were "selling" for $0, provided the new owner would assume their debt.

6 **200 percent reduction in customer complaints** Bailey, C., & Clarke, M. (2008). Aligning business leadership development with business needs: the value of discrimination. *Journal of Management Development, 27*(9), 912–934.

    Other examples of a 200 percent reduction: Rajashekar, B. S., & Kalappa, V. P. (2006). Effects of planting seasons on seed yield & quality of tomato varieties resistant to leaf curl virus. *Seed Research, 34*(2), 223–225. http://www.bostoncio.com/AboutRichardCohen.asp.

7 **50 percent reduction in salary** Illustration © 2016 by Dan Piraro based on an example from Huff, ibid.

7 **making this distinction between percentage point and percentages clear** I'm grateful to James P. Scanlan, attorney-at-law, Washington, D.C., who answered my query to the membership of the American Statistical Association, and provided me with this misuse.

8 **closing of a Connecticut textile mill and its move to Virginia** This example comes from Spirer, L., Spirer, H. F., & Jaffe, A. J. (1987). *Misused Statistics*, New York: Marcel Dekker, p. 194.

Miller, J. (1996, Dec. 29). High costs are blamed for the loss of a mill. *New York Times*, Connecticut Section.

And n. a. (1997, Jan. 12). Correction, *New York Times*, Connecticut Section.

8 **legislation that denied additional benefits** McLarin, K. J. (1993, Dec. 5). New Jersey welfare's give and take; mothers get college aid, but no extra cash for newborns. *New York Times*.

See also: Henneberger, M. (1995, April 11). Rethinking welfare: deterring new births—a special report; state aid is capped, but to what effect? *New York Times*.

8–9 **births to welfare mothers had already fallen by 16 percent** Ibid.

9 **no reason to report the new births** Ibid.

16 **Although they are mathematically equivalent** Koehler, J. J. (2001). The psychology of numbers in the courtroom: how to make DNA-match statistics seem impressive or insufficient. *Southern California Law Review, 74,* 1275–1305.

And Koehler, J. J. (2001). When are people persuaded by DNA match statistics? *Law and Human Behavior, 25*(5), 493–513.

17 **On average, humans have one testicle** Attributed to mathematics professor Desmond MacHale of University College, Cork, Ireland.

17 **temperatures ranging from 15 degrees to 134 degrees** http://en.wikipedia .org/wiki/Death_Valley.

18 **the amount of money spent on lunches in a week** As an example, suppose six adults spend the following amounts on lunch {$12, $10, $10, $12, $11, $11} and six children spend the following {$4, $3.85, $4.15, $3.50, $4.50, $4}. The median (for an even number of observations, the median is sometimes taken as the mean between the two middle numbers, or in this case, the mean of 4.5 and 10) is $7.25. The mean and median are amounts that no one actually spends.

19 **During the 2004 U.S. presidential election** See Gelman, A. (2008). *Red State, Blue State, Rich State, Poor State.* Princeton, NJ: Princeton University Press.

20 **the average life expectancy for males and females** These numbers are for white males and females. Non-white figures for 1850 are not as readily

available. http://www.infoplease.com/ipa/A0005140.html. An additional source of concern is that the U.S. numbers for 1850 are for the state of Massachusetts only, according to the Bureau of the Census.

21 **the average family** The title of this section, and the discussion, follows the work of Jenkins and Tuten very closely:

Jenkins, J., & Tuten, J. (1992). Why isn't the average child from the average family? And similar puzzles. *American Journal of Psychology, 105*(4), 517–526.

22 **the average number of siblings** Stick-figure children from Etsy, https://www.etsy.com/listing/221530596/stick-figure-family-car-van-bike-funny; small and large house drawn by the author; medium house from http://www.clipartbest.com/clipart-9TRgq8pac.

24 **average investor does not earn the average return** A simulation, see Tabarrok, A. (2014, July 11). Average stock market returns aren't average. http://marginalrevolution.com/marginalrevolution/2014/07/average-stock-market-returns-arent-average.html. Accessed October 14, 2014.

26 **poster presented at a conference by a student researcher** Tully, L. M., Lincoln, S. H., Wright, T., & Hooker, C. I. (2013). Neural mechanisms supporting the cognitive control of emotional information in schizophrenia. Poster presented at the 25th Annual Meeting of the Society for Research in Psychopathology. https://www.researchgate.net/publication/266159520_Neural_mechanisms_supporting_the_cognitive_control_of_emotional_information_in_schizophrenia.

I first found this example at www.betterposters.blogspot.com.

27 **gross sales of a publishing company** http://pelgranepress.com/index.php/tag/biz/.

29 **Fox News broadcast the following graph** I've redrawn this for the sake of clarity. For the original, see http://cloudfront.mediamatters.org/static/images/item/fbn-cavuto-20120731-bushexpire.jpg.

30 **Discontinuity in vertical or horizontal axis** Spirer, Spirer, & Jaffe, op. cit., pp. 82–84.

33–34 **Choosing the proper scale and axis** Example from Spirer, Spirer, & Jaffe, op. cit., p. 78.

35 **Many things change at a constant rate** Spirer, Spirer, & Jaffe, op. cit., p. 78.

36 **life expectancy of smokers versus nonsmokers at age twenty-five** These data taken from Jha, P., et al. (2013). 21st-century hazards of smoking and benefits of cessation in the United States. *New England Journal of Medicine, 368*(4), 341–350, Figure 2A for women. Survival probabilities were scaled from the National Health Interview Survey to the U.S. rates of death from all causes at these ages for 2004 with adjustment for differences in age, educational level, alcohol consumption, and adiposity (body-mass

index). I'm grateful to Prabhat Jha for her correspondence about interpreting this.

This form of presentation is based on that of Wainer, H. (1997). *Visual Revelations: Graphical Tales of Fate and Deception from Napoleon Bonaparte to Ross Perot.* New York: Copernicus/Springer-Verlag.

38 **expenditures per public school student and those students' scores on the SAT** This example from Wainer, H. (1997). *Visual Revelations: Graphical Tales of Fate and Deception from Napoleon Bonaparte to Ross Perot.* New York: Copernicus/Springer-Verlag, p. 93. The original appeared in *Forbes* (May 14, 1990).

Of course, there are other variables. Are the spending increases reported in actual or inflation-adjusted dollars? Was the time frame 1980–88 chosen to make that point, and would a different time frame make a different point?

39 **The correlation also provides a good estimate** There is some controversy about whether to use *r* or *r-squared*. For the defense of *r*, see: D'Andrade, R., & Dart, J. (1990). The interpretation of r versus r² or why percent of variance accounted for is a poor measure of size of effect. *Journal of Quantitative Anthropology, 2,* 47–59.

Ozer, D. J. (1985). Correlation and the coefficient of determination. *Psychological Bulletin, 97*(2), 307–315.

40 **services provided by the organization Planned Parenthood** Roth, Z. (2015, Sept. 29). Congressman uses misleading graph to smear Planned Parenthood. msnbc.com.

Politifact explored this issue further, examining the data between the endpoints and furnishing additional contextual information to go along with the usual graph-centered criticism. See https://perma.cc/P8NY-YP49.

47 **presentation on iPhone sales** http://qz.com/122921/the-chart-tim-cook-doesnt-want-you-to-see/; http://www.tekrevue.com/tim-cook-trying-prove-meaningless-chart/.

49 **feature spurious co-occurrences** http://www.tylervigen.com/spurious-correlations.

51 **Randall Munroe in his Internet cartoon *xkcd*** https://xkcd.com/552/.

52 **visual system is pitted against your logical system** This example is based on one in Huff, ibid.

53 **Any model of consumer behavior on a website** This is nearly a direct quote from De Veaux, R. D., & Hand, D. J. (2005). How to lie with bad data. *Statistical Science, 20*(3), 231–238, p. 232.

54 **Colgate's biggest competitor was named nearly as often** I thank my student Vivian Gu for this example.

Derbyshire, D. (2007, Jan. 17). Colgate gets the brush off for "misleading" ads. *The Telegraph*. Retrieved from http://www.telegraph.co.uk/news/uknews/1539715/Colgate-gets-the-brush-off-for-misleading-ads.html.

54 **C-SPAN advertises that they are "available"** http://www.c-span.org/about/history/.

54 **doesn't mean that even one person is watching** Nielsen reports that Americans, on average, receive 189 channels but watch only 17 of them. http://www.nielsen.com/us/en/insights/news/2014/changing-channels-americans-view-just-17-channels-despite-record-number-to-choose-from.html.

55 **water use in the city of Rancho Santa Fe** Boxall, B. (2014, Dec. 2). Rancho Santa Fe ranked as state's largest residential water hog. *Los Angeles Times*. http://www.latimes.com/local/california/la-me-water-rancho-20141202-story.html.

Lovett, I. (2014, Nov. 29). "Where grass is greener, a push to share drought's burden." *New York Times*. http://www.nytimes.com/2014/11/30/us/where-grass-is-greener-a-push-to-share-droughts-burden.html.

56 **flying is actually safer now** http://www.flightsafety.org; Grant, K. B. (2014, Dec. 30). Deadly year for flying—but safer than ever. http://www.cnbc.com/id/102301598.

58 **Newton's law of cooling** For an initial temperature of 155 degrees Fahrenheit, the formula is

$$f(t) = 80e^{-0.08t} + 75.$$

61 **C-SPAN is available in 100 million homes** Bedard, P. (2010, June 22). "Brian Lamb: C-SPAN now reaches 100 million homes." *U.S. News & World Report*. www.usnews.com/news/blogs/washington-whispers/2010/06/22/brian-lamb-c-span-now-reaches-100-million-homes. Retrieved November 22, 2010.

61 **90 percent of the population is within twenty-five miles** Based on Huff, op. cit., p. 48.

62 **3,482 active-duty U.S. military personnel who died in 2010** https://www.cbo.gov/sites/default/files/113th-congress-2013-2014/workingpaper/49837-Casualties_WorkingPaper-2014-08_1.pdf.

62 **total of 1,431,000 people in the military** https://www.census.gov/prod/2011pubs/12statab/defense.pdf.

62 **death rate in 2010** http://www.cdc.gov/nchs/fastats/deaths.htm.

62 **general population of the United States includes** Based on an example from Huff, op. cit., p. 83.

63 **increase in the number of doctors** I thank my student Alexandra Ghelerter for this example. Barnett, A. (1994). How numbers are tricking you. Retrieved from http://www.sandiego.edu/statpage/barnett.htm.

65   **nuances often tell a story** This is Best's term.

66   **there are *six* different indexes** Davidson, A. (2015, July 1). The economy's missing metrics. *New York Times Magazine.*

66   **July 2015 that the unemployment rate dropped** Shell, A. (2015, July 2). Wall Street weighs Fed's next move after jobs data. *USA Today Money.* http://americasmarkets.usatoday.com/2015/07/02/wall-street-gets-what-it -wants-in-june-jobs-count/.

66   **reported the reason for the apparent drop** Schwartz, N. D. (2015, July 3). Jobless rate fell in June, with wages staying flat. *New York Times,* B1.

71   **batting averages for the 2015 season** Stats from http://mlb.mlb.com/stats /sortable.jsp#elem=[object+Object]&tab_level=child&click_text=Sortable +Player+hitting&game_type=%27R%27&season=2015&season_type=ANY &league_code=%27MLB%27&sectionType=sp&statType=hitting&page= 1&ts=1457286793822&playerType=QUALIFIER&timeframe=.

73   **top three causes of death in 2013** http://www.cdc.gov/nchs/fastats/leading -causes-of-death.htm.

79   **attitudes do not seem to fall upon racial lines** This is entirely hypothetical.

79   **Another hurdle: You want age variability** This is from Huff, op. cit., p. 22.

81   **71 percent of *which* British?** Ibid.

81   **answer falsely just to shock the pollster** Many years ago, Chicago columnist Mike Royko encouraged readers to lie to exit pollers on Election Day in the hope that inaccurate data and being made to look foolish would end the practice of TV commentators calling the result of an election before all the votes were counted. I have no data on how many people lied to the exit pollers because of Royko's column, but the fact that exit polls are still a thing suggests it wasn't enough.

82   **the price you pay for not hearing from everyone** Taken from http://www .aapor.org/Education-Resources/Election-Polling-Resources/Margin-of-Sampling-Error-Credibility-Interval.aspx.

82   **Note that these ranges overlap** This is a good rule of thumb, but in some cases this quick method will be inaccurate. See Schenker, N., & Gentleman, J. F. (2001). On judging the significance of differences by examining the overlap between confidence intervals. *American Statistician, 55*(3), 182–186.

83   **Five times out of a hundred** I'm intentionally not making a distinction here between frequentist and Bayesian probability estimates, a distinction that comes up in Part Two.

84   **Margin of error** (image) From Wikipedia.

84   **formula for calculating the margin of error** For large populations, the 95 percent confidence interval can be estimated as $\pm 1.96 \times \text{sqrt}\ [p(1-p)/n]$. To

obtain a 99 percent confidence interval, multiply by 2.58 instead of 1.96. Yes, the interval is *larger* when you're more confident (which should make sense; if you want to be more sure that the range you quote includes the true value, you need a larger range). For smaller populations, the formula is to first compute the standard error:

sqrt [{(Observed proportion) × [l – (Observed proportion)}/sample size]

The width of the 95 percent confidence interval then is ±2 × standard error.

For example, if you sampled fifty overpasses in a large city, you might have found that 20 percent of them needed repair. You calculate the standard error as:

sqrt [(.2 × .8)/50] = sqrt (.l6/50) = .057.

So the width of your 95 percent confidence interval is ±2 × .057 = ±.11 or ±11%. Thus the 95 percent confidence interval is that 20 percent of the overpasses in this town need repair, plus or minus 11 percent. In a news report, the reporter might say that the survey showed 20 percent of overpasses need repair, with a margin of error of 11 percent. To increase the precision of your estimate, you need to sample more. If you go to 200 overpasses (assuming you obtain the same 20 percent figure), your margin of error reduces to about six percent.

85 **this conventional explanation is wrong** Lusinchi, D. (2012). "President" Landon and the 1936 *Literary Digest* poll: were automobile and telephone owners to blame? *Social Science History, 36*(1), 23–54.

85 **An investigation uncovered serious flaws** Clement, S. (2013, June 4). Gallup explains what went wrong in 2012. *Washington Post.* https://www .washingtonpost.com/news/the-fix/wp/2013/06/04/gallup-explains-what -went-wrong-in-2012/.

   http://www.gallup.com/poll/162887/gallup-2012-presidential-election -polling-review.aspx.

88 **trying to figure out what proportion of jelly beans** Taken from http:// www.ropercenter.uconn.edu/support/polling-fundamentals-total-survey -error/.

89 **what magazines people read** Elaborated from an example in Huff, op. cit., p. 16.

90 **Gleason scoring** This definition taken verbatim from http://www.cancer .gov/publications/dictionaries/cancer-terms?cdrid=45696. Accessed March 20, 2016.

91   **they had made an error in measurement** Jordans, F. (2012, Feb. 23). CERN researchers find flaw in faster-than-light measurement. *Christian Science Monitor.* http://www.csmonitor.com/Science/2012/0223/CERN-researchers -find-flaw-in-faster-than-light-measurement.

91   **1960 U.S. Census study recorded** This is from De Veaux, R. D., & Hand, D. J. (2005). How to lie with bad data. *Statistical Science, 20*(3), 231–238, p. 232. They cite Kruskal, W. (1981). Statistics in society: problems unsolved and unformulated. *Journal of the American Statistical Association, 76*(375), 505–515, and Coale, A. J., & Stephan, F. F. (1962). The case of the Indians and the teen-age widows. *Journal of the American Statistical Association, 57*, 338–347.

92   **claimed measurement error as part of their defense** Kryk, J. Patriots strike back with compelling explanations to refute deflate-gate chargers. *Ottowa Sun*, May 15, 2015. http://www.ottawasun.com/2015/05/14/pat riots-strike-back-with-compelling-explanations-to-refute-deflate-gate -chargers.

94   **statistic you encounter may not have defined homelessness** This example from Spirer, H., Spirer, L., & Jaffe, A. J. (1998). *Misused Statistics,* 2nd ed., revised and expanded. New York: Marcel Dekker, p. 16.

95   **Imagine that you've been hired by a political candidate** This example based on one in Huff, op. cit., p. 80.

95   **A newspaper reports the proportion of suicides** From Best (2005), op. cit.

97   **I'm not going to wear my seat belt because** This example comes from Best, J. (2012), and my childhood friend Kevin.

98   **the idea of symmetry and equal likelihood** The principle of symmetry can be broadly construed to include instances where outcomes are not equally likely but still prescribed, such as a trick coin that is weighted to come up heads two-thirds of the time, or a roulette wheel in which some of the troughs are wider than others.

98   **If we run the experiment on a large number of people** We could also conduct the experiment with a small number of people many times, in which case we would expect to obtain different numbers. In this case, the true probability of the drug working is going to be somewhere close to the average (the mean) of the numbers obtained in all the experiments, but it's an axiom of statistics that larger samples lead to more accurate results.

98   **Both classic and frequentist probabilities deal with** Classic probability can be thought of in two different ways: empirical and theoretical. If you're going to toss a coin or draw cards from a shuffled deck, each time you do this is like a trial in an experiment that could go on indefinitely. In theory, you could get thousands of people to toss coins and pick cards for several years and tally up the results to obtain the proportion of time that different

outcomes occur, such as "getting heads" or "getting heads three times in a row." This is an *empirically derived* probability. If you believe the coin is fair (that is, there's no manufacturing defect that causes it to come up on one side more than the other), you don't need to do the experiment, because it should come up heads half of the time (probability = .5) in the long run, and we arrive at this *theoretically*, based on the understanding that there are two equally likely outcomes. We could run a similar experiment with cards and determine empirically and theoretically that the chances of drawing a heart are one in four (probability = .25) and that the chances of drawing the four of clubs is one in fifty-two (probability ≅ .02).

99 **When a court witness testifies about the probability** Aitken, C. G. G., & Taroni, F. (2004). *Statistics and the Evaluation of Evidence for Forensic Scientists*, 2nd ed. Chicester, UK: John Wiley & Sons.

100 **In Tversky and Kahneman's experiments** Tversky, A., & Kahneman, D. (1974). Judgment under uncertainty: heuristics and biases. *Science, 185*(4157), 1124–1131.

101 **A telltale piece of evidence that this is subjective** For further discussion, and more formal treatment, see Iversen, G. R. (1984). *Bayesian Statistical Inference*. Thousand Oaks, CA: Sage, and references cited therein.

106 **the case of Sally Clark** I thank my student Alexandra Ghelerter for this example. See also Nobles, R., & Schiff, D. (2007). Misleading statistics within criminal trials. *Medicine, Science and the Law, 47*(1), 7–10.

108 **relative incidence of pneumonia** http://www.nytimes.com/health/guides /disease/pneumonia/prognosis.html.

108 **Bayes's rule to calculate a conditional probability** Bayes's rule is:

$$P(A \mid B) = \frac{P(B \mid A) \times P(A)}{P(B)}$$

111 **The probability that a woman has breast cancer** This paragraph, and this discussion, quotes nearly verbatim from Krämer, W., & Gigerenzer, G. (2005). How to confuse with statistics or: the use and misuse of conditional probabilities. *Statistical Science, 20*(3), 223–230.

112 **To make the numbers work out easily** How do you know what number to choose? Sometimes it takes trial and error. But it's also possible to figure it out. Because the probability is .8 percent, or eight people per thousand, if you chose to build a table for 1,000 women you'd end up with eight in one of the squares, and that's okay, but later on we're going to be multiplying that by 90 percent, which will give us a decimal. There's nothing wrong with that, it's just less convenient for most people to work with decimals. Increasing our population by an order of magnitude to 100 gives us all

whole numbers, but then we're looking at larger numbers than we need. It doesn't really matter because all we're looking for is probabilities and we'll be dividing one number by another anyway for the result.

116 **If you read that more automobile accidents occur at seven p.m.** Still confused? If there were eight times as many cars on the road at seven p.m. than at seven a.m., the *raw* number of accidents could be higher at seven p.m., but that does not necessarily mean that the *proportion* of accidents to cars is greater. And *that* is the relevant statistic to you: not how many accidents happen at seven p.m., but how many accidents occur per thousand cars on the road. This latter formulation quantifies your risk. This example is modified from one in Huff, op. cit., p. 78, and discussed by Krämer & Gigerenzer (2005).

118 **90 percent of doctors treated the two** Cited in Spirer, Spirer, & Jaffe, op. cit., p. 197: Thompson, W. C., & Schumann, E. L. (1987). Interpretation of statistical evidence in criminal trials, *Law and Human Behavior, 11*(167).

118 **One surgeon persuaded ninety women** From Spirer, Spirer, & Jaffe, op. cit., first reported in Hastie, R., & Dawes, R. M. (1988). *Rational Choice in an Uncertain World.* New York: Harcourt Brace Jovanovich.

The original report of the surgeon's work appeared in McGee, G. (1979, Feb. 6). Breast surgery before cancer. *Ann Arbor News,* p. B1 (reprinted from the *Bay City News*).

119 **As sociologist Joel Best says** Best, op. cit., p. 184.

PART TWO: EVALUATING WORDS

125 **Steve Jobs delayed treatment for his pancreatic cancer** Swaine, J. (2011, Oct. 21). Steve Jobs "regretted trying to beat cancer with alternative medicine for so long." http://www.telegraph.co.uk/technology/apple/8841347 /Steve-Jobs-regretted-trying-to-beat-cancer-with-alternative-medicine-for -so-long.html.

125 **an article in *Forbes* that claims** Rees, N. (2009, Aug. 13). Policing word abuse. *Forbes.* http://www.forbes.com/2009/08/12/nigel-rees-misquotes -opinions-rees.html.

125 ***Respectfully Quoted,* a dictionary of quotations** Platt, S., ed. (1989). *Respectfully Quoted.* Washington, D.C.: Library of Congress. For sale by the Supt. of Docs., USGPO.

125 **That book reports various formulations** Billings, J. (1874). *Everybody's Friend, or Josh Billing's Encyclopedia and Proverbial Philosophy of Wit and Humor.* Hartford, CT: American Publishing Company.

129 **humans had twenty-four pairs of chromosomes instead of twenty-three** Gartler, S. M. (2006). The chromosome number in humans: a brief history.

Nature Reviews Genetics, *7*, 655–660. http://www.nature.com/scitable/con tent/The-chromosome-number-in-humans-a-brief-15575. Glass, B. (1990). *Theophilus Shickel Painter*. Washington, D.C.: National Academy of Sciences. http://www.nasonline.org/publications/biographical-memoirs /memoir-pdfs/painter-theophilus-shickel.pdf. Retrieved November 6, 2015.

133 **If people in the arts and humanities have won a prize** Paul Simon, Stevie Wonder, and Joni Mitchell can be considered experts in songwriting. Although they do not hold university positions, university scholars have written books and articles about them, and Mr. Simon and Mr. Wonder were recognized by the president of the United States with Kennedy Center Honors, reserved for individuals who have made great contributions to performing arts. Ms. Mitchell received an honorary doctorate of music and won the Polaris Music Prize.

136 **Some people, including Noam Chomsky, have argued** Chomsky, N. (2015, May 25). The *New York Times* is pure propaganda. *Salon*. http://www.salon .com/2015/05/25/noam_chomsky_the_new_york_times_is_pure_proganda _partner/.

Achbar, M., Symansky, A., & Wintonick, P. (Producers), and Achbar, M., & Wintonick, P. (Directors). (1992). *Manufacturing Consent: Noam Chomsky and the Media* (Motion picture). USA: BuyIndies.com Inc. and Zeitgeist Films. https://www.youtube.com/watch?v=BsiBl2CaDFg.

136 **A 2011 fake tweet** Melendez, E. D. (2013, Feb. 1). Twitter stock market hoax draws attention of regulators. http://www.huffingtonpost.com/2013/02/01 /twitter-stock-market-hoax_n_2601753.html; http://www.forbes.com /forbes/welcome/.

136 **"The use of false rumors and news reports"** Farrell, M. (2015, July 14). Twitter shares hit by takeover hoax. *Wall Street Journal*. http://www.wsj .com/articles/twitter-shares-hit-by-takeover-hoax-1436918683.

137 **Jonathan Capehart wrote a story** (2010, Sept. 7). *Washington Post* writer falls for fake congressman Twitter account. *Huffington Post*, updated Sept. 7, 2010. http://www.huffingtonpost.com/2010/09/07/washington-post -writer-fa_n_707132.html; http://voices.washingtonpost.com/postpartisan /2010/09/obama_deficits_and_the_ditch.html.

138 **Who is behind it?** This is taken verbatim from *The Organized Mind*. Levitin, D. J. (2014). *The Organized Mind*. New York: Dutton.

139 **2014 congressional race for Florida's thirteenth district** Leary, A. (2014, Feb. 4). Misleading GOP website took donation meant for Alex Sink. *Tampa Bay Times*. http://www.tampabay.com/news/politics/stateroundup /misleading-gop-website-took-donation-meant-for-alex-sink/2164138.

141 **A court case ruled that Degil** Pink, D. (2013). Deceiving domain names not allowed. Wickwire Holm. http://www.wickwireholm.com/Portals/0 /newsletter/BLU%20Newsletter%20-%20January%202013%20-%20Deceiving

%20Domain%20Names%20Not%20Allowed.pdf; Bonni, S. (2014, June 24). The tort of domain name passing off. *Charity Law Bulletin* 342, Carters Professional Corporation. http://www.carters.ca/pub/bulletin/charity /2014/chylb342.htm.

141 **vendor operated the website GetCanadaDrugs.com** https://www.canada drugs.com/; https://www.getcanadadrugs.com/ (no longer available); Naud, M. (n.d.). Registered trade-mark canadadrugs.com found deceptively misdescriptive. ROBIC. http://www.robic.ca/admin/pdf/682/293 .045E-MNA2007.pdf.

141 **MartinLutherKing.org contains is a shameful assortment** The inflammatory quote from the website comes from the book by Taylor Branch, *Pillar of Fire*, but the author notes that he did not hear the tapes himself, he took them from three FBI agents who reported them to him.

141 **Stormfront, a white-supremacy, neo-Nazi hate group.** Sources which identify Stormfront as the Internet's "first hate site" include:

Levin, B. (2003). "Cyberhate: A legal and historical analysis of extremists' use of computer networks in America," in Perry, B., ed., *Hate and Bias Crime: A Reader*. New York: Routledge, p. 363.

Ryan, N. (2004). *Into a World of Hate: A Journey Among the Extreme Right*. New York: Routledge, p. 80.

Samuels, S. (1997). "Is the Holocaust unique?," in Rosenbaum, Alan S., ed., *Is the Holocaust Unique?: Perspectives on Comparative Genocide*. Boulder, CO: Westview Press, p. 218.

Bolaffi, G.; et al., eds (2002). *Dictionary of Race, Ethnicity and Culture*. Thousand Oaks, CA: Sage Publications, p. 254.

145 **Energy-drink company Red Bull paid** O'Reilly, L. (2014, Oct. 8). Red Bull will pay $10 to customers disappointed the drink didn't actually give them "wings." http://www.businessinsider.com/red-bull-settles-false-advertising -lawsuit-for-13-million-2014-10.

145 **Target agreed to pay $3.9 million** Associated Press. (2015, Feb. 11). Target agrees to pay $3.9 million in false-advertising lawsuit. http://journal record.com/2015/02/11/target-agrees-to-pay-3-9-million-in-false-advertising -lawsuit-law/.

146 **Kellogg's paid $4 million to settle** Federal Trade Commission. (2009, April 20). Kellogg settles FTC charges that ads for Frosted Mini-Wheats were false [Press release]. https://www.ftc.gov/news-events/press-releases /2009/04/kellogg-settles-ftc-charges-ads-frosted-mini-wheats-were-false.

147 **The *Washington Post* also runs a fact-checking site** https://www.washing tonpost.com/news/fact-checker/.

148 **Politifact summarized its findings** Carroll, L. (2015, Nov. 22). Fact-checking Trump's claim that thousands in New Jersey cheered when

World Trade Center tumbled. http://www.politifact.com/truth-o-meter/statements/2015/nov/22/donald-trump/fact-checking-trumps-claim-thousands-new-jersey-ch/.

148 **only one grandparent was born abroad** Sanders, K. (2015, April 16). In Iowa, Hillary Clinton claims "all my grandparents" came to the U.S. from foreign countries. http://www.politifact.com/truth-o-meter/statements/2015/apr/16/hillary-clinton/hillary-clinton-flubs-familys-immigration-history-/.

150 **322,000,000** The population of the United States, as of this writing. http://www.census.gov/popclock/.

150 **For coronary heart disease** American Heart Association (2015). AHA Statistical Update. *Circulation,* 131, p. 434–441. I thank McGill University Librarians Robin Canuel and Genevieve Gore for help in finding these statistics.

154 **In 1968, Will and Ariel Durant wrote** Durant, W., & Durant, A. (1968). *The Lessons of History.* New York: Simon & Schuster.

155 **The FBI announced in 2015** Federal Bureau of Investigation (2015, April 20). FBI testimony on microscopic hair analysis contained errors in at least 90 percent of cases in ongoing review [Press release]. https://www.fbi.gov/news/pressrel/press-releases/fbi-testimony-on-microscopic-hair-analysis-contained-errors-in-at-least-90-percent-of-cases-in-ongoing-review.

155 **Without these pieces of information** Aitken, C. G. G., & Taroni, F. (2004). *Statistics and the Evaluation of Evidence for Forensic Scientists,* 2nd ed. Chicester, UK: John Wiley & Sons, p. 95, citing Friedman, R. D. (1996). Assessing Evidence. *Michigan Law Review, 94*(6), 1810–1838.

156 **In one case in the U.K.** R v. Dennis John Adams, (1996) 2 Cr App R, 467; And Aitken, C. (2003). Statistical techniques and their role in evidence interpretation. In Payne-James, J., Busuttil, A., & Smock, W., eds., *Forensic Medicine: Clinical and Pathological Aspects.* Cambridge, UK: Cambridge University Press.

156 **the *New York Times* described a mysterious formation** Blumenthal, R. (2015, Nov. 3). Built by the ancients, seen from space. *New York Times,* p. D2.

159 **How much *more* productive and creative might she have been** I thank Stephen Kosslyn for sharing a version of this example with me.

159 **Two twins were separated at birth** Grimes, W. (2015, Nov. 13). Jack Yufe, a Jew whose twin was a nazi, dies at 82. *New York Times,* p. B8.

159 **They were reunited twenty-one years later** Much of this is taken verbatim from Grimes (2015), op. cit.

160 **A statistician or behavioral geneticist would say** Dr. Jeffrey Mogil, personal communication.

163   **if you ask a hundred people in a room** The formula is $1 - (1 - 1/2^5)^{100}$.

164   **Paul McCartney and Dick Clark** I thank Ron Mann for this observation.

165   *Larger samples more accurately reflect* Note that in a large sample, you are more likely to find an anomalous (outlier) observation than in a small sample, but when looking at the *mean*, the mean of the large sample is far more likely to reflect the true state of the world (because there are so many more observations that can swamp the anomalous one).

166   **if the study was on the incidence of preterm births** Krämer, W., & Gigerenzer, G. (2005). How to confuse with statistics or: the use and misuse of conditional probabilities. *Statistical Science, 20*(3), 223–230. See also Centers for Disease Control and Prevention. Preterm birth. http://www.cdc.gov/reproductivehealth/maternalinfanthealth/pretermbirth.htm.

166   **Consider a street game in which a hat** Krämer, W., & Gigerenzer, G. (2005). Technically, they note, this is an incorrect enumeration of simple events in a Laplacian experiment in the subpopulation composed of the remaining possibilities.

167   **similar mistakes were made by mathematical philosopher Gottfried Wilhelm Leibniz** Ibid.

168   **Counterknowledge, a term coined by** Thompson, D. (2008). *Counterknowledge: How We Surrendered to Conspiracy Theories, Quack Medicine, Bogus Science, and Fake History.* New York: W. W. Norton, p. 1.

168   **Damian Thompson tells the story** Thompson, D. (2008), op. cit.

170   **Shot on a consumer-grade camera** Trask, R. B. (1996). *Photographic Memory: The Kennedy Assassination, November 22, 1963.* Dallas: Sixth Floor Museum, p. 5.

170   **A *handful* of unexplained anomalies** Thanks to Michael Shermer for this.

170   **The difference between a false theory** This is a direct quote from Thompson, D. (2008), op. cit.

173   **As Damian Thompson notes** Thompson, D. (2008), op. cit., p. 17. The previous two sentences are from pp. 16–17.

173   **die each year of stomach cancer** National Cancer Institute. SEER stat fact sheets: stomach cancer. http://seer.cancer.gov/statfacts/html/stomach.html.

173   **than of unintentional drowning** Centers for Disease Control and Prevention. Unintentional drowning: get the facts. http://www.cdc.gov/HomeandRecreationalSafety/Water-Safety/waterinjuries-factsheet.html.

174   **A front-page headline in the *Times* (U.K.)** Smyth, C. (2015, Feb. 4). "Half of all Britons will get cancer during their lifetime." *Times.* www.thetimes.co.uk/tto/health/news/article4343681.ece.

175   **Cancer Research UK (CRUK) reports that** Boseley, S. (2015, Feb. 3). Half of people in Britain born after 1960 will get cancer, study shows. *Guardian.*

175 **Heart disease is better controlled** Griffiths, C., & Brock, A. (2003). Twentieth century mortality trends in England and Wales. *Health Statistics Quarterly, 18*(2), 5–17.

177 **This is based on reports by a variety** http://www.nrdc.org/water/drinking /qbw.asp; http://www.mayoclinic.org/healthy-lifestyle/nutrition-and -healthy-eating/expert-answers/tap-vs-bottled-water/faq-20058017; http:// www.consumerreports.org/cro/news/2009/07/is-tap-water-safer-than-bottled /index.htm; http://news.nationalgeographic.com/news/2010/03/100310 /why-tap-water-is-better/; http://abcnews.go.com/Business/study-bottled -water-safer-tap-water/story?id=87558; http://www.telegraph.co.uk/news /health/news/9775158/Bottled-water-not-as-safe-as-tap-variety.html.

177 **In New York City; Montreal; Flint, Michigan; and many other older cities** Stockton, N. (2016, Jan. 29). Here's how hard it will be to unpoison Flint's water. *Wired.* http://www.wired.com/2016/01/heres-how-hard-it -will-be-to-unpoison-flints-water/.

PART THREE: EVALUATING THE WORLD

179 **Nature permits us to calculate only probabilities** Feynman, R. P. (1985). *QED: The Strange Theory of Light and Matter.* Princeton, NJ: Princeton University Press.

182 **A case of fraud occurred in 2015** Reardon, S. (2015, July 1). US vaccine researcher sentenced to prison for fraud. *Nature, 523*, p. 138.

182 **controversy about whether the measles, mumps, and rubella MMR vaccine causes autism** Wakefield, A. J., et al. (1998, Feb. 28). RETRACTED: Ileal-lymphoid-nodular hyperplasia, non-specific colitis, and pervasive developmental disorder in children. *Lancet, 351*(9103), 637–641. http://www .thelancet.com/journals/lancet/article/PIIS0140-6736(97)11096-0/abstract.

Burns, J. F. (2010, May 25). British medical council bars doctor who linked vaccine with autism. *New York Times*, p. A4. http://www.nytimes.com/2010 /05/25/health/policy/25autism.html.

Associated Press (2011, Jan. 6). Study linking vaccine to autism is called fraud. *New York Times.* http://query.nytimes.com/gst/fullpage.html?res =9C02E7DC1E3BF935A35752C0A9679D8B63.

Rao, T. S., & Andrade, C. (2011). The MMR vaccine and autism: sensation, refutation, retraction, and fraud. *Indian Journal of Psychiatry, 53*(2), 95–96.

192 **For example, Holmes concludes that** From Thompson, S. (2010). The blind banker. *Sherlock* (TV series, first aired October 31, 2010).

194 **The germ theory of disease** I first learned about this story from Hempel, C. (1966). *Philosophy of Natural Science.* Englewood Cliffs, NJ: Prentice-Hall.

199 **To fill out the rest of the table** I'm using the 168-hour week (7 days × 24 hours a day) to account for thoughts you might have while dreaming, and people who might call and wake you from a sound sleep. Of course, one could subtract out eight hours of sleep per night (or whatever) and then use only the 112 hours of wakefulness to come up with a different probability, but it doesn't change the conclusion.

202 **less safe mode of travel** In retrospect this switch was foolish, at the time, but it may have been the rational thing to do. Four hijacked, suicide planes at once was unprecedented in aviation history. When confronted with a big change in the world, often the best thing to do is to think Bayesian: update your understanding, stop relying on the old statistics, and seek alternatives.

202 **conclude that air travel** Based on Huff, op. cit., p. 79.

202 **There were not nearly as many flights in 1960** See, for example, Iolan, C., Patterson, T., & Johnson, A. (2014, July 28). Is 2014 the deadliest year for flights? Not even close. CNN. http://www.cnn.com/interactive/2014/07 /travel/aviation-data/; and Evershed, N. (2015, March 24). Aircraft accident rates at historic low despite high-profile plane crashes. *Guardian*. http:// www.theguardian.com/world/datablog/2014/dec/29/aircraft-accident-rates -at-historic-low-despite-high-profile-plane-crashes.

203 **An FBI page reports that** http://www.fbi.gov/about-us/cjis/ucr/crime-in -the.u.s/2011/crime-in-the.u.s.-2011/clearances.

204 **All home robberies in a neighborhood** Image from http://www.clipart of.com/portfolio/bestvector/illustration/retro-black-and-white-criminal- carrying-a-flashlight-and-box-210461.html.

206 **In a famous psychology experiment** Nisbett, R. E., & Valins, S. (1972). Perceiving the causes of one's own behavior. In Kanouse, D. E., et al., eds. *Attribution: perceiving the causes of behavior.* Morristown, NJ: General Learning Press, pp. 63–78.

And, Valins, S. (2007). Persistent effects of information about internal reactions: ineffectiveness of debriefing. In London, H., & Nisbett, R. E., eds. *Thought and Feeling: the cognitive alteration of feeling states.* Chicago, IL: Aldine Transaction.

207 **Between 1990 and 2010, six times as many** What is causing the increase in autism prevalence. *Autism Speaks Official Blog*, Oct. 22, 2010. http:// blog.autismspeaks.org/2010/10/22/got-questions-answers-to-your-questions -from-the-autism-speaks%E2%80%99-science-staff-2/.

208 **The majority of the rise** Ibid.

208 **the Internet to guide your thinking on why autism** Suresh, A. (2015, Oct. 13). Autism increase mystery solved: no, it's not vaccines, GMOs, glyphosate— or organic foods. Genetic Literacy Project. http://www.geneticliteracyproject

.org/2015/10/13/autism-increase-mystery-solved-no-its-not-vaccines-gmos
-glyphosate-or-organic-foods/.

208 **She also couches her argument** Kase, A. (2015, May 11). MIT scientist
uncovers link between glyphosate, GMOs and the autism epidemic. *Reset
.me*. http://reset.me/story/mit-scientist-uncovers-link-between-glyphosate
-gmos-and-the-autism-epidemic/.

209 **no evidence that thimerosal was linked to autism** Honda, H., Shimizu,
Y., & Rutter, M. (2005). No effect of MMR withdrawal on the incidence of
autism: a total population study. *Journal of Child Psychology and Psychiatry*, 46(6), 572–579. http://1796kotok.com/pdfs/MMR_withdrawal.pdf, and
many other sources.

Reardon, S. (2015). US vaccine researcher sentenced to prison for fraud.
Nature, 523(7559), p. 138.

211 **as we know, there are known knowns** Defense.gov News Transcript: DoD
News Briefing—Secretary Rumsfeld and Gen. Myers, United States
Department of Defense (defense.gov).

214 **We can clarify Secretary Rumsfeld's four possibilities with a fourfold
table** I thank Morris Olitsky for this.

217 **One of the cornerstone principles of forensic science** Inman, K., & Rudin,
N. (2002). The origin of evidence. *Forensic Science International, 126*(1),
11–16.

Inman, K., & Rudin, N. (2000). *Principles and Practice of Criminalistics:
the profession of forensic science*. Boca Raton, FL: CRC Press.

218 **Suppose a criminal breaks into the stables** I'm basing this section on the
discussion found in Aitken, C. G. G., & Taroni, F. (2004). *Statistics and the
Evaluation of Evidence for Forensic Scientists,* 2nd ed. Chicester, UK: John
Wiley & Sons, pp. 1–2, and using their setup and terminology.

218 **take literally the assumption in the American legal system** Aitken,
C. G. G., & Taroni, F. (2004), op. cit.

221 *the prosecutor's fallacy* Thompson, W. C.; Shumann, E. L. (1987). Interpretation of statistical evidence in criminal trials: the prosecutor's fallacy and
the defense attorney's fallacy. *Law and Human Behavior 2*(3), 167–187.

230 **The quality of the photographs is high** Hasselblad.com. https://www
.hq.nasa.gov/alsj/a11/a11-hass.html; http://www.wired.com/2013/07
/apollo-hasselblad/.

231 **It's been claimed that the chances of life forming on Earth** Estimates
include 1×10^{390}. http://evolutionfaq.com/articles/probability-life. See also
Dreamer, D. (2009, April 30). Calculating the odds that life could begin by
chance. *Science 2.0.* http://www.science20.com/stars_planets_life
/calculating_odds_life_could_begin_chance.

231 **In a TED talk with more than 10 million views** https://www.ted.com /talks/david_blaine_how_i_held_my_breath_for_17_min?language=en.

232 **there are more than 5,000 TED-branded events** Bruno Guissani, Curator of TEDGlobal Conference, personal communication, September 28, 2015.

232 **Fox television reported his ice-block demonstration** https://www.you tube.com/watch?v=U6Em2OhvEJY.

233 **Out of a sense of ethics** Glenn Falkenstein, personal communication, October 25, 2007.

234 **There was even a peer-reviewed paper** Korbonits, M., Blaine, D., Elia, M., & Powell-Tuck, J. (2005). Refeeding David Blaine—studies after a 44-day fast. *New England Journal of Medicine, 353*(21), 2306–2307.

234 **The current editor of the journal searched** J. Drazen, MD, email communication, December 20, 2015.

234 **The lead author on the article told me in an email** M. Korbonits, MD, email communication, December 25, 2015.

234 **a physician *did* monitor Blaine throughout the fast** Jackson, J. M., et al. (2006). Macro- and micronutrient losses and nutritional status resulting from 44 days of total fasting in a non-obese man. *Nutrition, 22*(9), 889–897.

237 **Blaine's record was broken in 2012** http://www.huffingtonpost. com/2012/11/16/breath-world-record-stig-severinsen_n_2144734.html; Grenoble, R. (2012, Nov. 16). Breath-holding world record: Stig Severinsen stays under water for 22 minutes (Video), *Huffington Post*. http://www.huffing tonpost.com/2012/11/16/breath-world-record-stig-severinsen_n_2144734. html.

240 **An article in the *Dallas Observer*** Liner, E. (2012, Jan. 13). Want to know how David Blaine does that stuff? (Don't hold your breath). http://www .dallasobserver.com/arts/want-to-know-how-david-blaine-does-that-stuff -dont-hold-your-breath-7083351.

240 **preparation for the breath holding for an article in the *New York Times*** Tierney, J. (2008, April 22). This time, he'll be left breathless. *New York Times*, p. F1.

240 **the *Oprah* appearance in his blog** Tierney, J. (2008). David Blaine sets breath-holding record. http://tierneylab.blogs.nytimes.com/2008/04/30 /david-blaine-sets-breath-holding-record.

241 **Tierney writes, "I was there"** John Tierney, email correspondence, January 13 and 18, 2016.

245 **Eventually all of the elements between 1 and 118** Netburn, D. (2016, Jan. 4). It's official: four super-heavy elements to be added to the periodic table. http://www.latimes.com/science/sciencenow/la-sci-sn-new-elements-20160104 -story.html.

247 **Here's Harrison Prosper, describing this plot** Prosper, H. B. (2012, July 10). International Society for Bayesian Analysis. http://bayesian.org/forums /news/3648.

248 **Louis Lyons explains "The Higgs"** Lyons, L. (2012, July 11). http://bayesian .org/forums/news/3648.

248 **Although CERN officials announced in 2012** In articles on the Higgs boson, you may encounter reference to the 5-sigma standard of proof. Five-sigma refers to the level of probability that the scientists agreed upon before conducting the experiments—the chance of their misinterpreting the experiments had to have a confidence interval within five standard deviations (5-sigma), or 0.0000005 (recall earlier we talked about 95 and 99 percent confidence intervals—this is a confidence interval of 99.99995 percent). See http://blogs.scientificamerican.com/observations/five -sigmawhats-that/.

248 **Prosper says, "Given that the search"** Prosper, H. B. (2012, July 10). http:// bayesian.org/forums/news/3648.

248 **Physicist Mads Toudal Frandsen adds** (2014, Nov. 7). Maybe it wasn't the Higgs particle after all. Phys.org. http://phys.org/news/2014-11-wasnt-higgs -particle.html.

249 **Joseph Lykken, a physicist and director of the Fermi National Accelerator Laboratory** http://phys.org/news/2014-11-wasnt-higgs-particle.html.

249 **as *Wired* writer Signe Brewster says** http://www.wired.com/2015/11 /physicists-are-desperate-to-be-wrong-about-the-higgs-boson/.

250 **physicist Nima Arkani-Hamed told the *New York Times*** Overbye, D. (2015, Dec. 16). Physicists in Europe find tantalizing hints of a mysterious new particle. *New York Times*, p. A16.

CONCLUSION: DISCOVERING YOUR OWN

252 **A lot of things that should be scientific** These ideas and their phrasing come from: Frum, D. (2015). Talk delivered at the Colleges Ontario Higher Education Summit, November 16, 2015, Toronto, ON.

APPENDIX: APPLICATION OF BAYES'S RULE

255 **To compute Bayes's rule** Iversen, G. R. (1984). *Bayesian Statistical Inference*. Quantitative Applications in the Social Sciences, vol. 43. Thousand Oaks, CA: Sage.

ACKNOWLEDGMENTS

The inspiration for this book came from Darrell Huff's *How to Lie with Statistics*, a book that I've read several times and appreciate more with each reading. I was also a huge fan of Joel Best's *Damned Lies and Statistics*, and Charles Wheelan's *Naked Statistics*. I owe all three authors a great debt for their humor, wisdom, and insight, and I hope this book will take its place alongside theirs for anyone who wants to improve their understanding of critical thinking.

My agent, Sarah Chalfant at the Wylie Agency, is a dream: warm, attentive, supportive, and indefatigable. I feel privileged to work with her and her colleagues at TWA, Rebecca Nagel, Stephanie Derbyshire, Alba Ziegler-Bailey, and Celia Kokoris.

I am thankful to everyone at Dutton/Penguin Random House. Stephen Morrow has been my editor through four books and has made each of them incalculably better ($P < .01$). His guidance and support have been valuable. Thanks to Adam O'Brien, LeeAnn Pemberton, and Susan Schwartz. Hats off to Ben Sevier, Amanda Walker, and Christine Ball for the numerous things they do to help books reach readers who want to read them. Becky Maines was a wonderful more-than-a-copy-editor, whose breadth and depth of knowledge and clarifying additions I very much enjoyed.

ACKNOWLEDGMENTS

I'm grateful to the following for helpful discussions and comments on drafts of this manuscript: Joe Austerweil, Heather Bortfeld, Lew Goldberg, and Jeffrey Mogil. For help with specific passages, I'm indebted to David Eidelman, Charles Fuller, Charles Gale, Scott Grafton, Prabhat Jha, Jeffrey Kimball, Howie Klein, Joseph Lawrence, Gretchen Lieb, Mike McGuire, Regina Nuzzo, Jim O'Donnell, James Randi, Jasper Rine, John Tierney, and the many colleagues of mine of the American Statistical Association who helped proof the book and review the examples, especially Timothy Armistead, Edward K. Cheng, Gregg Gascon, Edward Gracely, Crystal S. Langlais, Stan Lazic, Dominic Lusinchi, Wendy Miervaldis, David P. Nichols, Morris Olitsky, and Carla Zijlemaker. My students in McGill's honors and independent research seminar helped provide some of the examples and clarified my thinking. Karle-Philip Zamor, as he has through four books, helped me enormously to prepare figures and solve all manner of technical problems, always with good cheer and great skill. Lindsay Fleming, my office assistant, has helped me to structure my time and keep my focus and assisted with the end notes, index, proofreading, fact-checking, and many other details of the book (and thanks to Eliot, Grace, Lua, and Kennis Fleming for sharing her time with me).

INDEX

Note: Page numbers in *italics* refer to illustrations.

Every Word is a
Bird We Teach to Sing

DANIEL TAMMET

Every Word is a Bird
We Teach to Sing

ENCOUNTERS WITH THE MYSTERIES
AND MEANINGS OF LANGUAGE

HODDER &
STOUGHTON

First published in Great Britain in 2017 by Hodder & Stoughton
An Hachette UK company

1

Copyright © Daniel Tammet 2017
All quotations from the works of Les Murray have been reproduced
from *New Selected Poems* by Les Murray, with the permission
of Carcanet Press Limited

The right of Daniel Tammet to be identified as the
Author of the Work has been asserted by him in accordance
with the Copyright, Designs and Patents Act 1988.

A CIP catalogue record for this title is available from the British Library

ISBN 978 0 340 96130 8
eBook ISBN 978 1 848 94688 0

Typeset in Sabon MT Std by
Palimpsest Book Production Ltd, Falkirk, Stirlingshire

Printed and bound in Great Britain by Clays Ltd, St Ives plc

Hodder & Stoughton policy is to use papers that are natural, renewable
and recyclable products and made from wood grown in sustainable forests.
The logging and manufacturing processes are expected to conform to
the environmental regulations of the country of origin.

Hodder & Stoughton Ltd
Carmelite House
50 Victoria Embankment
London EC4Y 0DZ

www.hodder.co.uk

In loving memory of my father

CONTENTS

1

FINDING MY VOICE

Though English was the language of my parents, the language in which I was raised and schooled, I have never felt I belonged to it. I learned my mother tongue self-consciously, quite often confusedly, as if my mother were a foreigner to me, and her sole language my second. Always, in some corner of my child mind, a running translation was struggling to keep up. To say this word or that word in other words. To recompose the words of a sentence like so many pieces of a jigsaw puzzle. Years before doctors informed me of my high-functioning autism and the disconnect it causes between man and language, I had to figure out the world as best I could. I was a misfit. The world was made up of words. But I thought and felt and sometimes dreamed in a private language of numbers.

In my mind each number had a shape – complete with colour and texture and occasionally motion (a neurological phenomenon that scientists call synaesthesia) – and each shape a meaning. The meaning could be pictographic: eighty-nine, for instance, was dark blue, the colour of a sky threatening storm; a beaded texture; and a fluttering, whirling, downward motion I understood as 'snow' or, more broadly, 'winter'. I remember, one winter, seeing snow fall outside my bedroom window for the first time. I was seven. The snow, pure white and thick-flaked, piled many inches high upon the ground, transforming the grey concrete of the neighbourhood into a virgin, opalescent tundra. 'Snow,' I gasped to my parents. 'Eighty-nine,' I thought. The thought had hardly crossed my mind when I had another: nine-hundred and seventy-nine. The view from my window resembled nine-hundred and seventy-nine – the shimmer and beauty of eleven expanding, literally multiplying eighty-nine's wintry swirl. I felt moved. My parents' firstborn, I had been delivered at the end of a particularly cold and snowy January in 1979. The coincidence did not escape me. Everywhere I looked, it seemed, there were private meanings writ large.

Was it from that moment – the sudden sense that my meanings corresponded to the wider world – that I first had the urge to communicate? Until that moment, I had never felt the need to open up to another person: not to my parents or siblings, let alone to any of the other children at my school. Now, suddenly, a feeling lived in me, for which I had neither name nor number (it was a little like the sadness of six, but different). I eventually learned the feeling was what we call loneliness. I had no friends. But how could I make myself understood to children from whom I felt so estranged? We spoke differently, thought differently. The other children hadn't the faintest idea (how could they?) that the relationship between eighty-nine and nine hundred and seventy-nine was like the relationship between, say, *diamond* and *adamant*. And with what words might I have explained that eleven and forty-nine, my mental logo-grams, rhymed? A visual rhyme. I would have liked nothing better than to share with my classmates some of my poems made of numbers:

Sixty-one two two two two eleven
One hundred and thirty-one forty-nine

But I kept the poems to myself. The children at school intimidated me. In the playground every mouth was a shout, a snort, an insult. And the more the children roared, the more they laughed and joked in my direction, the less I dared approach them and attempt to strike up a conversation. Besides, I did not know what a conversation sounded like.

I renounced the idea of making friends. I had to admit that I wasn't ready. I retreated into myself, into the certainties of my numerical language. Alone with my thoughts in the relative calm and quiet of my bedroom, I dwelled on my number shapes, on their grammar. One hundred and eighty-one, a prime number, was a tall shiny symmetrical shape like a spoon. When I doubled it – modified its shape with that of two, which was a sort of 'doing' number – it equated to a verb. So that three hundred and sixty-two had the meaning of 'to eat' or 'to consume' (more literally, 'to move a spoon'). It was the mental picture that always announced that I was hungry. Other pictures that rose up in me could morph in a similar way, depending on the action they described and whether it was external or internal to me: thirteen (a rhythmic descending motion) if a raindrop on the windowpane caught my gaze, twenty-six if I tired and sensed myself drifting off to sleep.

My understanding of language as something visual carried over to my relationship with books once I became a library-goer and regularly tugged large, slender, brightly coloured covers down from the shelves. Even before I could make out the words, I fell under the spell of *The Adventures of Tintin*. The boy with the blond quiff and his little sidekick dog, Snowy. Speech in bubbles; emotions in bold characters and exclamation marks; the story smoothly unfolding from picture to altered picture. Each frame was fit to pore over, so finely and minutely detailed: a mini-story in itself. Stories within stories, like numbers within numbers: I was mesmerised.

The same understanding, the same excitement, also helped me learn to read. This was my luck, since reading had not initially come easily to me. Except for the occasional word of comfort the night after a nightmare, my parents never read me bedtime stories, and because the antiepileptic medicine I was prescribed at a young age made me drowsy in class, I was never precocious. I have memories of constantly falling pages behind the other children, of intense bouts of concentration in order to catch up. My delight in the shapes of the words in my schoolbooks, their visual impression on me, made the difference. One of the books, I remember, contained an illustration of a black-cloaked witch, all sharp angles, astride her broom. To my six-year-old imagination, the letter W was a pair of witches' hats, side by side and hanging upside down, as from a nail.

Back in those days, the mid-eighties, it was possible for a teacher to give her young charge a repurposed tobacco tin (mine was dark-green and gold) in which new words, written in clear letters on small rectangular cards, were to be brought home for learning. From that time on I kept a list of words according to their shape and texture: words round as a three (*gobble, cupboard, cabbage*); pointy as a four (*jacket, wife, quick*); shimmering as a five (*kingdom, shoemaker, surrounded*). One day, intent on my reading, I happened on *lollipop* and a shock of joy coursed through me. I read it as *1011ipop*. One thousand and eleven, divisible by three, was a fittingly round number-shape, and I thought it the most beautiful thing I had yet read: half number and half word.

I grew; my vocabulary grew. Curt sentences in my schoolbooks' prim typeset; lessons the teacher chalked up on the blackboard; breathless adjectives on crinkly flyers that intruded via the letterbox; pixelated headlines in the pages of Ceefax ('See Facts'), the BBC's teletext service. All these and many more besides I could read and write, and spell backward as well as forward, but not always pronounce. Only rarely did words reach me airborne, via a radio or a stranger's mouth. (I watched television for the pictures – I was forever lowering the sound.) If I surprised my father talking to the milkman at the door, or my mother sharing gossip with a neighbour over the garden fence, I would try to listen in – and abruptly tune out. As sounds and social currency, words could not yet hold me. Instead, I lavished my attention on arranging and rearranging them into sentences, playing with them as I played with the number shapes in my head, measuring the visual effect of, for instance, interlacing round three-y words with pointy four-y ones, or of placing several five-y words, all agleam, in a row.

A classmate called Babak was the first person to whom I showed my sentences. He was his parents' image. They were thin, gentle people who had fled the Ayatollah's Iran several years before for the anonymity of a London suburb; they had recently enrolled their son at my school. Babak was reassuringly unlike the other children, with his thick black hair and crisp English and a head both for words and numbers. In his backyard one warm weekend, sitting opposite me on the grass, he looked up from the Scrabble board to read the crumpled sheet of lined notepaper I was nervously holding out to him.

'Interesting. Is it a poem?'

I sat still, my head down, staring at a spot between the numbered tiles. I could feel his inquisitive brown eyes on me. Finally, I shrugged and said, 'I don't know.'

'Doesn't matter. It's interesting.'

This was also the opinion of my headmaster. How exactly my writing reached him remains, to this day, something of a mystery to me. I was ten. The class had been reading H. G. Wells's *The War of the Worlds;* and, in a state of high excitement induced by the graphic prose, I had been rushing home every day after class to the solitude of my bedroom to write –

cautiously, to begin with, then compulsively. Of this story, my first sustained piece of writing, my mind has retained only fragments: winding descriptions of labyrinthine tunnels; outlines of sleek spaceships that blot out the sky; laser guns spending laser bullets, turning the air electric. No dialogue. The story inhabited me, overpowered me. It quickly exceeded every line of every page of every pad of notepaper in the house. So that the first my teacher heard of it was the afternoon, after class, when I blushed crimson and asked whether I might help myself to a roll of the school's computer printout paper. I could, but in exchange I had to confide in her the purpose. The following week, softly, she asked me how the story was coming along. She wanted to see it. I went away and brought back, with difficulty, the many pages filled with my tiny, neat hand to her desk. She said to leave them with her. I hesitated, then agreed. Did she, upon reading the story, decide to urge it upon the headmaster? Or did the headmaster, visiting the teacher or simply passing by, happen on it? However it came to him, one morning during the school assembly, breaking from his usual headmaster patter, he announced that he was going to read an extract from my story to the hall. I hadn't expected that. Not without so much as a word of warning from my teacher! I had never seen the headmaster read aloud a pupil's work. I couldn't bring myself to listen along with the other students. Out of nerves and embarrassment, I put my palms to my ears – it was one of my habits – and fixed my eyes on the whorls of dust on the floor. But after the assembly, children who had never so much as given me the time of day came up and greeted me smilingly, tapping me on the shoulders, saying 'Great story' or words to that effect. The head-master would have awarded my story a prize, he made a point of telling me later in his office, if only he had had such a prize to give. His encouragement was a fine enough substitute, which I treasured. So I was crestfallen when I had to move on to high school soon afterward and, in lieu of deploying my imagination to compose new stories, was made to regurgitate umpteen examination-friendly facts. The talent peeping out from under my shyness and social bewilderment I would have to nourish more or less on my own, I realised, foraging for whatever extracurricular susten-ance I might find.

It was among the bookcases of the municipal library that I spent most of my adolescence, as fluent by then in the deciphering of texts as I remained inept in conver-sation. These years of reading, I see now, were a way of apprenticing myself with voices of wisdom, the multitudinous accents of human experience, listening sedulously to each with my bespectacled eyes. Growing in empathy book by book, from puberty onward I increasingly set aside the illustrated encyclopaedias and dictionaries in favour of history books, biographies, and memoirs. I pushed myself to go further still, intel-lectually and emotionally, into the fatter novels of Adult Fiction.

I was afraid of this kind of fiction. Afraid of feeling lost in the intricacies of a social language I had not mastered (and feared I never might). Afraid that the experience would shake whatever small self-confidence I had. A good part of the fault for this lay on my high school English classes and their 'required reading.' If Shakespeare – his outlandish characters and strange diction (which we read in a side-by-side translation to contemporary English) – had fascinated me, Dickens had seemed interminable and Hardy's *Jude the Obscure* very obscure indeed.

But in the municipal library I had the freedom of the shelves. I could browse at my ease. The works I looked at were not the thematic or didactic stories told by wordy, know-it-all narrators that examiners use for their set questions. They were shorter novels by living writers: artfully concise personal reflections on modern life (ranging from the 1950s to the year just past) written by and for a socioeconomic class that was not my own. But for all that, they were approachable. Partly I went to them for the past readers' marginalia – for the crabbed, scribbled words of agreement or annoyance or wonder, which imparted unintended clues to the meaning of a particular sentence or paragraph. Also for the creases and thumbprints and coffee stains on the pages, reminders that books are also social objects – gateways between our internal and external worlds. And partly for the characters' dialogue, their verbal back-and-forth clearly set out and punctuated, integral to the story. So this is how people talk, I would think, as I read. This is what conversation looks like.

And some nights, I dreamed I watched the dialogic patterns converted into my number shapes:

'Twelve seventy one nine two hundred and fifty-seven.'

'Two hundred and fifty-seven?'

'Two!'

'Four. Sixteen.'

'Seventeen.'

When I was nearing matriculation from high school, Frau Corkhill, who had been my German teacher for several years, began inviting me over to her house for late-afternoon lessons in conversation – more in English than German.

I sorely needed such practice. Outside of the family, where so much can be meant and understood without even needing to utter a word, I was able to say little that didn't come out sounding clunky, off-topic, or plain odd. For templates I relied mostly on the dialogues I had studied in the library novels; but such schemas, however many I studied, however well performed, would only ever, I came to realise, get me so far. I was almost a young man: the urge to communicate had begun to take on a new charge. One day, in history class, the sight of a new boy brought my thumping heart into my throat, and my attraction compelled me try to converse with him. I talked and talked, happy to be anxious, but what had looked so good and persuasive on the page of a novel fell flat in my strangled voice, which was unpractised and monotonous. The courage I'd mustered vanished into mortification. More than mortification. Seven hundred and fifty-seven (a shape which I can only compare to a ginger root): an acute feeling – arising from immense desire to communicate, aligned with a commensurate incapacity to do so – for which English has no precise equivalent.

Frau Corkhill was a short and stout and red-haired woman, at retirement age or thereabouts, and the object of much sniggering from some of the pupil population for her various eccentricities. She ate raw garlic cloves by the bulbful. She wore flower print dresses and fluorescent socks. She merely smiled a bright red lipstick smile and gazed up wistfully at the ceiling when any other teacher would have bawled an undisciplined student out of the room. Such behaviour was, in my view, neither here nor there. She doted on me. She was like a grandmother to me. She seemed to intuit the invisible difficulties against which I had fought all through my childhood. I remember the day she gave me her telephone number, an attractive medley of fours and sevens, shortly before I was to change classes. The first three digits after the area code became my nickname for her. Before long, I called and accepted her invitation to the house. Every week for the next year, I rode the red double-decker the twenty or so minutes to her door.

These lessons-slash-discussions with Frau Corkhill were the highlight of my week. She was a woman of infinite patience, a professional at making light of others' mistakes, at correcting by example rather than by admonishment. Her home was a space in which I could talk and exchange without fear of being taken for a conversational klutz. We sat in the living room next to a bay window overlooking the rose garden, on high-backed chairs at a table dressed in a frilly white cloth, a tray and china tea set in its centre, like a scene out of a library novel.

We talked about the school, about whatever was in the news. Sometimes we changed language, English to German and back again. Frau Corkhill's English was unique, her accent part German and part Geordie (*Corkhill*, her married name, is a common surname in northern England). Strange to think, I had not noticed people's accents before. Strange to remember my surprise when a classmate informed me that my pronunciation of *th* was off (my Cockney father's fault). I had not known to notice.

But now, talking with Frau Corkhill, I understood how many Englishes must exist. Hers, mine: two among countless others.

In writing the story of my formative years in the words I had back in 2005 (I was twenty-six), with feeling but without confidence or high finish, I found my voice. The international success of *Born on a Blue Day* began a conversation with readers from around the world. Where some British and American critics saw only a one-off 'disability genre' memoir, the account of a 'numbers wiz,' German and Spanish and Brazilian and Japanese readers saw something else, and sent letters urging me to continue writing. Many referred to a closing chapter in which I recounted a public reading I had given at the Museum of the History of Science, in Oxford, in 2004. The subject of my reading was not a book, not the work of any published name, but a number: pi. Over the three preceding wintry months, like an actor analysing his script, I had rehearsed the number from home, assimilating its unstinting digits by the hundreds of hundreds, until I knew the first 22,514, a European record's worth, by heart. On the fourteenth of March, I narrated this most beautiful of epic poems, an *Odyssey* or an *Iliad* composed of numbers, in a performance spanning five hours, to the hall. For the first time in my life I spoke aloud in my numerical language (albeit, necessarily, in English words), at length, passionately, fluently. And if, in the early minutes of my recitation, I worried that the small crowd of curious listeners might comprehend about as much as if I were performing in Chinese, shake their heads, turn their backs on me and leave, all my fears quickly evaporated. As I gathered momentum, acquired rhythm, I sensed the men and women lean forward, alert and rapt. With each pronounced digit their concentration redoubled and silenced competing thoughts. Meditative smiles broadened faces. Some in the audience were even moved to tears. In those numbers I had found the words to express my deepest emotions. In my person, through my breath and body, the numbers spoke to the motley attendees on that bright March morning and afternoon.

The numbers also spoke through the printed page to my far-flung readers, came alive in their minds, regardless of the translation that conveyed them. My lifelong struggle to find my voice, my obsession with language, appeared to them, as it did to me, like a vocation.

I'd written a book and had it published. But it

remained unclear whether a young man on the autistic spectrum could have other books in him. No tradition of autistic writing existed (indeed, some thought *autistic author* a contradiction in terms). I had no models (though, later, I made the discovery that Lewis Carroll – possibly – and Les Murray, the Australian poet and Nobel Prize candidate, to name only two, shared my condition), no material. I was on my own.

But then another reader's letter arrived. It was in French, a language I had studied in high school, from a young Frenchman named Jérôme, who would, in time, become my husband. Through months of thoughtful and playful correspondence, Jérôme and I fell in love. For him, for his country and language, I chose willingly to leave the country and the language I had never felt were mine. We moved to Avignon, then north to Paris, settling among the bistros and *bouquinistes* of Saint-Germain-des-Prés.

Before Jérôme, I had largely given up on literature. Novels and I had long since parted company. Now, though, in our apartment, surrounded by our books (Jérôme owned many books), we sat together at a brown teak table and, taking turns, read aloud from the French translation of Dostoyevsky's *L'Idiot*. My voice when I read, as when I had recited the number pi, seemed at once intimate and distant: another voice in mine, enlarging and enriching it. And, as with pi, I understood and became enthralled.

Reading a Russian work in French, I was not invaded by the feeling of foreignness that the pages of English novels had roused in me. On the contrary, I felt at home. I could, at last, read unencumbered by my self-consciousness, solely for the pleasure of learning new words and discovering new worlds. I could read for the sake of reading.

Dostoyevsky's reputation, a powerful intermediary between his work and modern readers, would once have daunted and kept me away. But Dostoyevsky's language proved to be picture perfect. A case in point is the character General Ivolgin, the smell of whose cigar provokes a haughty English lady travelling with her lapdog in the same compartment to pluck the cigar from between his fingers and toss it out the train. Yet the general just sits there, seemingly unfazed by the lady's behaviour. Quick as a flash and ever so smoothly, he leans over and chucks her little dog out after his cigar. I remember my voice, in the telling, interrupted by my own shocked laughter, and how my merriment communicated itself to Jérôme and had him in stitches.

It wasn't only Dostoyevsky who could so affect us. In the following months we laughed and gasped over Isaac Babel's short stories. Kawabata's *Le Grondement de la Montagne* (*The Sound of the Mountain*) – the tale of an old man's ailing memory – brought tears to my voice. The visual music of *Paroles* by Jacques Prévert reverberated in my head long after I closed its covers.

Then, one day, as if removing the stabilisers from a child's bicycle, Jérôme ceased to accompany my literary reading. I did not wobble. And, after devouring both tomes of Tolstoy's *Guerre et Paix,* I tried the Russian master's *Anna Karenina* in English, and the heroine's passions, Levin's and Kitty's foibles, Vronsky's contradictions all affected me so greatly that I clean forgot the apprehension of my former reading life. Something had worked itself in my head. All literature, I finally realised with a jolt, amounted to an act of translation: a condensing, a sifting, a realignment of the author's thought-world into words. The reassuring corollary – reassuring to a novice writer like myself, just starting out: the translatorese of bad prose could be avoided, provided the words were faithful to the mental pictures the author saw. I had more than one book in me. And each of my subsequent books – a survey of popular neuroscience, a collection of essays inspired by mathematical ideas, a translation/adaptation into French of Les Murray's poetry – was different. Each taught me what my limits weren't. I could do this. And this. And this as well. All the time that I was writing, I was also studying in my after-hours with the UK's distance learning higher education institution, the Open University. In 2016, at the age of thirty-seven, I graduated with a first-class honours bachelor of arts degree in the humanities. I published my first novel that same spring in France.

I have not yet written my last English sentence, despite ten years spent on the continent and despite the increasing distillation of my words from French. That choice, renewed here, is an homage to my British parents and teachers. A recognition, too, of the debt

I owe to a language commodious enough even for a voice like mine. English made a foreigner of me, but also a writer. It has become the faithful chronicler of my metamorphosis.

2

THE LANGUAGE TEACHER

Everything I know about teaching a foreign language, I learned in Lithuania.

It was 1998. I was nineteen, unready for university, full of wanderlust and good intentions. I enrolled in a government-run volunteer programme that sent young men and women overseas. I could have been sent to Poland to nanny little Mateuszes and Weronikas or to a clinic in Russia short of file finders or to wash dishes in a hotel who knows where in the Czech Republic or to the British embassy in Slovenia, whose front desk needed manning.

Instead I was dispatched to Lithuania, to the city of Kaunas. I couldn't speak a word of Lithuanian. My innocence of the language didn't seem to matter, though. A young Englishman with passable French and German (Lithuanian bore no relation to either) was apparently sufficient for instructing the job-seeking inhabitants eager to learn English.

I remember taking the aeroplane from London to the capital city of Vilnius. The thrill of take-off. To feel airborne! No one in my family had ever flown before. 'Head in the clouds,' my father had sometimes said of me. And now his words, once a mere expression, had come literally true.

The nations of the former Soviet Union were shown to us in Western newscasts as uniformly grey, dilapidated, Russified. But the Lithuania I arrived in, only a few years after Moscow's tanks had slunk away, had reason for optimism. The population was youngish, new shiny buildings were sprouting up here and there, and, despite fifty years of foreign occupation, Lithuanian habits and customs had lived on.

It took time to adjust. Little shocks of unfamiliarity had to be absorbed. October in London was autumnal; in Kaunas, the cold reminded me of a British winter. Snow was already in the offing. And then there was the funny money, the *litas,* in which my volunteer stipends were paid. But strangest, in those first days, was the language, so unlike the sounds and rhythms of any other language I had heard. An old man in my apartment block stops me in the stairwell to tell me something keen and musical – what is it? Children in the street sing a song – what is it about? Unintelligible, too, were the headlines and captions the inky newspapers carried. They looked like a secret code. How I wished to work out the cipher!

A codebreaker. But the Lithuanian learning kit the programme's staff had given me was small. In less experienced hands, the kit – really a pocket dictionary and phrasebook – would have seemed futile; there was nothing an imagination could fasten onto. I knew better. I sat at my apartment desk, opened the dictionary, about the size of a deck of cards, and flicked the wispy, nearly transparent pages to the word for 'language': *kalba.* As words went, it struck me as beautiful.

Beautiful and fitting. Suddenly other words, in other languages, swam in my head: the English *gulp,* the Finnish *kello* ('bell'). Less the words than the various meanings behind them: *gulp,* a mouthful of air; *bell,* a metal tongue. In this way, *kalba* I understood intuitively as something of the mouth, of the tongue. (Like *language,* whose Latin ancestor, *lingua,* means 'tongue.')

Fingering the pages again, hearing them crinkle, I turned them at random and read *puodelis,* cup. If *kalba* was a word to savour, *puodelis,* I felt, belonged between the palms. I closed my eyes and rubbed my hands together as though palping the syllables: *puo-de-lis, puo-de-lis.*

I roved five pages, ten, as many as I could soak up in a sitting. My eyes went from entry to entry. I was looking for the kind of wonderful juxtaposition you otherwise see in the fairy tale and the surrealist poem, the kind at which the unwitting lexicographer excels. *Cat hair* and *cathedral. Mushroom* and *music-hall. Umbrella* and *umbilical cord.* Lithuanian, in this respect, can be just as Grimm or Dada, let me tell you. A short way into the *D*'s, I hit on the Lithuanian for *thistle* and *combustible –* in the words of my dictionary, *dagis* and *dagus –* two ordinarily distinct ideas only a vowel apart. They recalled Exodus, put a Baltic twist on the story of Moses and the burning bush. Musing over this, I could not help asking myself, what sermon would a desert thistle have spoken?

What surprises this pungent little dictionary contained! What pleasure! And the more pages I turned, the more my pleasure in its company grew. In the excitement and anxiety of those first days in Kaunas we became inseparable.

A week after my arrival I was already on the job, teaching for two hours Monday to Friday at a women's centre in the city's downtown, a trolleybus ride away from my apartment. In the classroom, the dozen women in front of me didn't look at all like the squat, kerchiefed babushkas I scrimmaged with each morning on the bus. They wore smart skirts and makeup and their hair in varying degrees of stylishness. And when, during our first lesson, I introduced myself and tried out some words of Lithuanian, they chuckled good-naturedly at my accent – the women had never heard their language in a British voice before. I asked them their reasons for coming. One student, Birutė (a common name, I learned), turned into something of a spokeswoman for the others. She stood up and said, in excellent English, 'We want to improve our English. Because English has become the language of skilled employment here. If you speak Lithuanian and Russian and Polish but no English, you are worse than illiterate. Look at the job advertisements in the newspapers! *Anglų kalba reikalinga*, "English required."'

Birutė was by far the strongest student. She was in her forties, very slim and elegant, her dyed black hair cut boyishly short. 'I studied English at university. But that was long ago.' Her confidence in her English sometimes wavered.

Aida, Birutė's friend, wanted to say something as well. She was younger than Birutė, and shyer. Her voice was soft and hesitant. Understandably so. All she had for English were a few scattered phrases. Birutė intervened. 'She says she hopes you will do better than the last teacher. An American. She says she could not follow a word he said.'

At Aida's comments the other women in the class began shouting all at the same time, apparently eager to join in the criticism of my predecessor. Their shouts held months of accumulated frustration, annoyance, despair. Down with boring textbooks! Down with paedagogic jargon! We want to learn English, not a bunch of useless rules!

I was taken aback. I had expected the calm of a classroom; I hadn't expected this brouhaha. To tell the truth, I started to feel a bit afraid. And I was embarrassed by my fear. I thought, I'm nineteen. I don't know what to tell them. I've only just arrived here. I was of two minds whether or not to walk out.

Just then Birutė waved her arms and hollered something at the room. An awkward silence came over the women.

'*Atsiprašau* ["excuse me"],' she said. 'I should not [have] let Aida say what she said. She becomes excited and excites the others. We are very happy and we thank you for teaching us your language.'

We spent the rest of our first encounter looking at the textbooks provided by the centre and trying to get some sort of discussion going. But the women were right. The pages were soporific beyond any teacher's skill or enthusiasm. If I continued to work from them, as the volunteer before me had, whatever remained of the women's hopes of speaking serviceable English might have been crushed for good. I resolved to drop the book. To teach differently. How? I did not know. Even so, my attitude was that I would find another approach in time for the next lesson.

I racked my brain to find a more natural, more enjoyable method.

It came to me late that evening at my apartment while I sat in an armchair reading from the little Lithuanian–English dictionary as had become my habit. I was up to the letter O when the entry *obuolys* ('apple') made me stop and put the book down. I closed my eyes. Suddenly I recalled the moment, ten years ago, when I discovered the existence of non-English words, that is to say, other nations' languages.

Back in east London, exceedingly shy, almost house-bound, I had got to know one of my little sister's girlfriends, who lived a few doors down. The blond mother of this blond girl was Finnish (I had no idea what *Finnish* meant), and, to teach her daughter the language, one day she gave her a bright Finnish picture book. The gift, as it turned out, went unopened; the girl had no interest in words my sister and her other friends would never have understood. She left the picture book with us.

Cover-wise it looked like any other unthumbed picture book, but once inside I sat astonished. On every page, below the colourful illustration of an everyday object, a word that didn't quite look like a word. A word intended for another kind of child. Finnish!

Of all the impressions this book made on me, the red apple accompanied by the noun *omena* left the deepest. There was something about the distribution of the vowels, the roundness of the consonants, that fascinated. I felt that I was seeing double, for the picture seemed to mirror the word and vice versa. Both word and picture represented an apple by means of lines.

The next day, on my way to the centre, I stopped in at a grocery store and bought a bag of apples. When the women filed into the classroom and saw the pyramid of red and green apples on my table, I said, 'Yesterday some of you said you knew no English. That's not true. You know lots of English words. You know *bar*.'

'*Baras*,' Aida said.

'Right. And *restaurant*.'

One of the women at the back shouted, '*Restoranas*.'

'Yes. And *history, istorija*, and *philosophy, filosofija*.'

Birutė, sitting near the front, said, '*Telephone*.'

'*Telefonas*. You see? Lots of words.' I turned to the apples.

'*Taksi*,' someone said.

'Yes, well, the list is long. What about these on my table?'

The women replied as one, '*Obuoliai!*'

Apples.

I told my students about the picture book, and the story of the red apple. Birutė translated for me. I said, 'If you can draw an apple, you can learn the word *apple*.' After I had asked them to take out their pencils and paper, I went to lift and give out the pile of fruit; but I misjudged the gesture and heard the hapless apples slip from my grasp and roll off along the floor.

Women's laughter.

I bent down and picked up the apples and put one on each of the students' desks. I was laughing, too. But concentration quickly replaced the levity. Heads were lowered; brows were creased; pencils were plied. A quarter of an hour or so later I told the students to stop. Their drawings ranged from coloured-in circles to Birutė's delicate sketch, complete with shading.

'When you put pencil to paper you don't draw the apple as such, you draw its shape and texture and colour,' Birutė translated. 'Each aspect is proportional to the drawer's experiences. So one apple might be round like a tennis ball; a second, glossy as plastic; a third, baby-cheek red.' I said the word *apple* was another form of drawing. 'You draw *a-p-p-l-e*.' As I spoke, I wrote the letters in red on the whiteboard. 'An initial *A*, consecutive *P*'s, an *L*, and a final *E*. Your imagination can play with them as it plays with shape and colour. Mix them around. Subtract or add a letter. Tweak the sound of *P* to *B*.' In the way that an apple can make a sketcher think of a tennis ball, or plastic, or a baby's cheeks, an *apple* can bring to an English mind a *stable*, or a *cobbler*, or *pulp*, I explained.

Then I told the women to take out their dictionaries and find other *apple*-like words.

Birutė's face lit up: she understood. Her pen, busy with words, ran quickly across the sheet of paper. The others wrote more tentatively. Empty lines glared at the women with least English.

'Turn the pages of your dictionary in the direction of the letter *P*,' I encouraged. 'Look for possible words in combinations like *P* something *L*, or *PL* something, or *P* something something *L*, and so on. Or turn to the front and search for words that begin *BL*. Or think about how English words handle a pair of *P*'s or *B*'s, how they push them to the middle – *apple* and *cobble* – or out to the extremes – *pulp*. Birutė, can you translate that please?' Birutė repeated my words, but in Lithuanian.

When the students had finished writing down their findings, they took turns reading them aloud to the class. One lady came up with *bulb;* another, *appetite;* a third, *palpable.* A fourth in the corner, relishing the sudden attention of the room, shouted, 'Plop!' Just the sound conjured apples falling out of trees from ripeness.

'Apple pie,' Aida suggested suddenly.

I nodded. On the whiteboard appeared *apple pie*.

Out of her store of words, duly put to paper, Birutė joined in: 'Pips. Peel. Plate. Ate. Eat.'

I was delighted. She had let the language think for her.

We stayed with the exercise for the following lessons. We found *car* in *chair* and *wet* in *towel*, and *window* brought us, word by word, to *interview;* and as the students' vocabularies filled out, so did their confidence. The mood in the classroom lightened; betterment seemed only another lesson away. Even those with the least English found themselves writing and speaking more and more. Enthusiastic students don't make good dunces.

Some English words, my students and I decided, are diagrams. We looked at *look* – the *o*'s like eyes; and at how the letters in *dog* – the *d* like a left-looking head, the g like a tail – limned the animal. We admired the symmetry, so apt, of *level*. Other words are optical illusions: *moon*, after you have covered the first or third leg of the letter *m*, turns night into day: *noon*. *Desserts* is a mouth-waterer of a word, or a mouth-dryer, depending on which direction – left to right, or right to left – the reader takes it in. Still other words are like successive images in a flip book. See how the *T* advances:

Stain

Satin

Saint

I spent a whole lesson explaining a type of word I might have classified as impressionistic. They are the words that most sway the eye, tease the ear, intrigue the tongue. Those that give off a certain vibe just by their being seen and heard and repeated. Consider *slant*, I said. I wrote the word on the whiteboard. Did Birutė know it? No, Birutė didn't know it. None of the students had read or heard it before. That could have made them tetchily impatient, but it didn't. With my new teacher's nerves abating, and Birutė translating, I felt sure I wasn't in any danger of losing the room. I was in complete command. So I said, 'Let's stay a few moments with *slant*. What kind of a word picture is *slant*? Do its letters, their corresponding sounds, give the impression that the word refers to something light or heavy? Or to something opaque? Shiny? Smooth?' (Part of teaching a language is educating your students' guesses, taming them.) Opinion in the class was divided. A good many of the women, though, said the sight and sound made them think of something negative rather than positive, something on the heavy side. I went to the spot on the board beside *slant* and continued writing: *sleep, slide, slope,* and *slump*. What did they all have in common? Visually, and audibly, lots. The words were the same length; they had the same onset – *sl*; they closed on a *p, t,* or *d*. And their meanings? I raised my left hand to eye level and lowered it. *Sleep:* a stander or sitter lies down. *Slide* and *slope:* a descent. *Slump:* a company's stock plummets. The words formed a polyptych, a series of interrelated pictures. *Slant,* then? The women raised their hands and lowered them. 'Like this,' I said, raising my left hand again and lowering it diagonally: my hand, a translator of *slant.*

With the forefinger of my left hand I drew a circle around my nose and mouth. '*Smell,*' I said. '*Smile.*' I smiled. '*Smirk.*' I made a face. '*Smoke.*' I brought an air cigarette to my lips. '*Smother.*' I clapped a palm over my mouth. '*Sneeze.*' I pretended to sneeze. '*Snore.*' I pretended to snore. '*Sniff,*' I said, sniffing. '*Sneer,*' I said, sneering. Another polyptych in words.

'*Snail,*' Birutė said. 'What about *snail?*'

'Like tongues,' I said. 'Tongues with shells.' And added, once the laughter had subsided, 'Of course, not every word fits into a particular frame.'

But many did. Our imaginations, during part of the rest of the lesson, painted in *thump*s and *stomp*s and *bump*s and *whomp*s the colours of a bruise. Next, the broken, kinetic lines of *z*'s – like moving points which have lost their way – excited in us a sensation of perplexity: ears *buzzed* as though filled with *jazz*, eyes were *dazzled*, heads felt *fuzzy* and *dizzy*. At the end of class, the other side of *drizzle* and *blizzard*, the women left with their senses thoroughly soaked.

Always I situated the words the students would learn in sentences. Every sentence was an experiment in composition; I was less interested in realistic description. I wanted the women to view the words from different perspectives, to study the effects of layout on meaning, to understand grammar as the arranging and blending of sounds and letters.

'Wind down the window!'

'The second-hand watch's second hand has stopped.'

'Her teacher's smooth ink taught thought.'

Winter came, Lithuanian style. It snowed and snowed and snowed. (*Snow*. What does it have to do with the mouth or nose? Some memories of my childhood in winter are of tasting snowflakes melt on the tip of my tongue.) Ground-floor homes were up to their window-sills in snow. I shivered, one floor up, in my draughty apartment built before double glazing, on a street that is forgivably forgotten by tourist maps. Under a blanket I curled up with the pocket dictionary and watched the weather on TV. Compared to Britain's and the female forecaster's long trim outlines, Lithuania appeared somewhat bulky and flat. The temperatures in every part of the country were all preceded by minuses. I had never before seen the numbers go so low.

To take my mind off my shivering, I turned to a page near the end of the dictionary. My Lithuanian was coming along. I was finally able to understand the old man's neighbourly banter when he stopped me as

I carried my groceries up the stairs. The headlines on display at kiosks had become as mundane to me as any in London. And my feel for the language was improving; I noticed myself noticing more and more, making more and more connections – like the entries on this page under the letter *V*. They were loud words, those that began with *var-*. The sounds they depicted were predictable, repetitive. I looked at the words of this polyptych and heard the caw of *varna,* a crow; the ribbet of *varlė,* a frog; the dingdong of *varpas,* a bell; the chug of *variklis,* an engine; the drone of *vargonai,* an organ; the squeak of *vartai,* a gate. I heard someone shout my *vardas,* my name, over and over again. I listened to the verb *varvėti,* 'to drip,' its familiar tin drumbeat.

Jo vardas Valdas. ('His name is Valdas.')

Iš variklio varva benzinas. ('Petrol drips from the engine.')

Soon my Lithuanian outgrew the pocket dictionary. I was hungry for other books. But on the shelves in the apartment's living room stood only black-and-white photos: a thin man in a dark suit, a thinner woman in a pale dress. The landlord's family, I presumed. No novels. Zero stories. Years of communism had been hard on language. Schoolbooks singing the praises of Comrades Lenin and Stalin had given the printed word a bad name. I turned the apartment upside down. I pulled out drawers: buttons, discontinued stamps, a sprinkling of rusty coins. I opened the cupboard abutting the bedroom: a bottle of vodka, four-fifths bottle, one-fifth vodka. In the closet, beneath

a spare bedspread, I dredged up only a yellowing telephone directory. I thought, if I want to read, I'd better
join the city library.

At the library I had to fill out forms. '*Vardas*,' the
top of the form asked. I wrote my name and, under
it, the apartment's address. I gave the women's centre
as my employer. The muted man at the desk exchanged
my responses for a pass and watched the library's
newest recruit set out among the aisles. I roamed from
bookcase to bookcase, pausing here and there to sample
the pages. Quickly, however, my initial excitement
shrivelled. So dull, the Soviet-era books, so very dry.
So full of the words for work and happiness. Work,
work, work. Happiness, happiness, happiness. As the
Lithuanians say, thanks for the poppies, but I would
like bread. I almost gave up.

But chance intervened. I had wandered out to the
dust-collecting reaches of the library. I stumbled on a
slim volume – very old to judge by the worn, flaky
cover – by a poet named Kazys Binkis. Suddenly my
imagination woke up. Clouds that sauntered like calves
along fields of sky; forests in May colours; recipes in
which thoughts were measured out in grams – I
instantly decided not to hand in my pass. 'Was there
perhaps a bilingual edition?' I asked one of the librarians. I was thinking of using it in my class. The sallow,
grey-haired librarian (he didn't look like he had ever
tasted a snowflake) shook his head. He pointed at a
remote bookcase – foreign literature; *foreign* here
meaning English, mostly – where I found an anthology
of British and American poems and checked it out with

the Binkis. The library's poetry section henceforth kept my students and me in texts.

I was coming out of the classroom one afternoon when I heard the director's door open, heard my name called in her stentorian voice, and her jewellery jingle as she stepped back inside her office. It wasn't the first time that the director had asked a staff member into her office, but until now, if her voice resounded in the centre, it was never with the curiously accented syllables of 'DAN-i-el'. When I tapped at her door and went in, she was at her desk leafing through the English text-book. The impressive perm made her head look very big. She said, 'I hear strange things about your class. I don't understand. What a secretary of bees? What does it mean?' My students had steadily been working through the library's anthology, and in the past few lessons we had been looking at the poems of Sylvia Plath. '"Here is the secretary of bees" is from "The Bee Meeting",' I explained.

'But there no such thing. No such thing as secretary of bees.' Incomprehension aged her. She was suddenly all frowns and worry lines. 'It not correct English. The centre has textbooks to teach correct English. See?' Her ringed finger tapped a sentence on the page in front of her. 'Here. Like this.' She read aloud from the text-book: 'John's secretary makes coffee in the morning.' She read the sentence as crisply as a prosecutor putting her case to a court. 'Why not use this sentence instead?'

'John's secretary makes coffee in the morning.' It was a grammatical sentence. But then so was 'Here is the secretary of bees.' And without any of the textbook's blandness. Only the latter induced the students' attention. I spoke carefully.

'The textbook's sentence is, shall we say, factual. It contains a lot of facts. There is someone called John; John has a secretary; the secretary makes coffee; the coffee is made in the morning. One fact after another and another. They make no pictures. Everything is simply assumed. The world is the world. And in the world Johns have secretaries, and secretaries make coffee, and coffee is drunk in the morning.'

'What wrong with that?' the director demanded. Her accent was Russian.

'Memory – for one thing. Lots of facts go forgotten. No fact, no word. The student's language becomes full of gaps. Whereas the other kind of sentence is different; it doesn't assume anything. It's not a fact; it's a picture. The students can imagine what a secretary of bees would look like. And imagining, they understand and remember better.'

As I spoke, I sensed that the director and I had irreconcilable differences concerning how a language ought to be taught. Even so, she heard me out. I said that each word in a textbook, being a fact, could mean more or less only one thing. A word in a poem, on the other hand, could say ten different things. When Plath writes of hearing someone's speech 'thick as foreign coffee,' *coffee* here means so much more than it does in the hands of John's docile secretary. The line stimulates the reader's interest. *Thick* and *foreign* are lent an aura of unfamiliarity. Questions begin to multiply. How can a word be thick (or thin)? Why describe the coffee as foreign? Were the words Plath heard like coffee because they were bitter? Did they give her black thoughts? So that instead of repeating a word – *coffee* or *thick* or *foreign* – ten times over in ten different sentences ('Can I have a cup of coffee please?', 'My favourite drink is coffee,' 'His wife is foreign,' 'This cinema shows foreign films every other week'), hammering it into the memory, the same word can be understood in ten different ways in a single reading and absorbed instantly.

For the director, poetry was only a side effect of language, peripheral; for me it was essential. A student would learn phrases like 'arrange hair' or 'arrange an appointment' far more easily, I thought, after reading Plath's line 'Arranging my morning.' Not the other way around. Grammar and memory come from playing with words, rubbing them on the fingers and on the tongue, experiencing the various meanings they give off. Textbooks are no substitute.

The director relented. It wasn't as if she could fire me; labour cannot get any cheaper than free. But before I went, she had a job she needed doing. The European Union money on which the centre depended for its annual budget was fast disappearing; and in order to acquire new funds, the director had drafted in English a project proposal that she wanted me to look over. I read the printout. The old Soviet way with words! Bureaucratese. Syntax minus meaning – every second word replaceable, every third or fourth dispensable. An Amstrad's pride. What could I tell her? I thought of my students. I bit my lip and said it should be fine.

Wednesday evenings we were usually five or six at my apartment: me, Birutė, Aida, and two or three of the other students. Conversation practice. These no-pressure sessions over tea were my way of reciprocating the women's regular gifts of food and advice and hospitality, and the respect with which they dealt with me, a young man – young enough to be, for some of them, a son – living alone in the snowy depths of a post-Soviet country, a thousand miles from his family. To Lithuanians, Birutė confided one sleety evening, such informal gatherings were still rather new; they had only in recent years started to forget to close their curtains, to talk without fearing walls with ears, and to not jump when a drunken neighbour bangs at the wrong door. I discovered then why the apartment, which had been vacant for some time before my coming, contained no books.

During the decades of Soviet occupation, an army of censors had stalked Lithuanian for the tiniest inkling of protest, of satire, of ambiguity. Back then every word was a potential suspect, every misprint a potential crime. Up and down the country, millions of books were hauled away from homes and pulped. Book inspectors could turn up on your doorstep at all hours. Why is this Cyrillic-print paperback propping up a table? How come Comrade Stalin's *Dialectical and Historical Materialism* sits in your bathroom? Who spilt tea on the novel (by an author awarded a state prize) here? In this climate of trepidation many readers chose to take precautionary steps. Children's books, Russian-language bestsellers, popular stories and novellas, and even works signed Vladimir Lenin were all fed to stoves. It would take days, sometimes weeks, for the odour of vanished words to lift.

What a difference, then, for the women who freely fraternised every Wednesday evening, the five or six of us speaking English (and me, my best Lithuanian) to our hearts' content. Cradling the poetry anthology on her knees, one of the students, like an inexperienced reader, would turn the pages gingerly and read a few lines or verses aloud. Then the others would describe a memory that the picture made by the words had dislodged. Sometimes the women's memories were of nursery rhymes they had learned as a little girl from a mother or grandmother. Lithuanian words of course, but the same sorts of rhymes, the same rhythms. The iambic beat of the heart.

> Musė maišė, musė maišė, ['The fly mixed, the fly mixed,']
> uodas vandens nešė, ['The mosquito carried water,']
> saulė virė, saulė virė, ['The sun boiled, the sun boiled,']
> mėnesėlis kepė. ['The little moon baked.']

It was Aida, if I remember rightly, who sang. She might have been our mother, gently lulling her children to sleep. Her voice was warm and soft.

It was not unusual for my apartment to ring with the nation's songs and sayings, for the English the women conversed in was often supplemented by Lithuanian puns, asides, exclamations. They were like the compositions I taught in class. In a little notebook,

I entered my gleaned favourites, with the odd doodle of a comment:

> 'Buckle rhymes with night: sagtis/naktis. Think "black buckle".'
>
> 'Rankų darbo sidabro (handmade silver) – said when describing a wedding ring. Similar effect in handmade diamond.'
>
> 'Crust on bread is "pluta" – froth on beer is "puta" – "alus be putos, duona be plutos" (beer without froth, bread without crust) = "good as nothing".'
>
> 'Nettles are compared to wolves: they bite. "Wolf" = vilkas, which rhymes with šilkas (silk). Glossy leaves. Nettles – nets – threads.'

The flashes of native wit elated me; the words of a riddle or proverb could make my Wednesday. They were the words of Lithuanian's infancy, the forefathers around fires, faces lit up and coloured amber – words so musical and picturesque that I listened and blinked with pleasure. They made me wonder who was truly the language teacher – the women or I? They seemed to me to be troves of knowledge. Inestimable.

In April, the last of the snow turned to slush, then to water, then to an increasingly distant memory. *Snow* – and its Lithuanian counterpart, *sniegas* – returned to the dictionary to estivate. The Wednesdays were getting longer; the women arrived at my apartment looking less tired, and stayed later. They read the poems aloud and spoke in English more easily. Assurance rejuvenated them, made their skin shine. I had never seen the women look as beautiful as they did then.

Two months later my class broke up for the summer. After making my goodbyes at the centre, I had a little time on my hands before the flight home. Birutė, considerate Birutė, took me to the theatre. As luck would have it, a Lithuanian-language production of *My Fair Lady* (*Mano puikioji ledi*) was running. I hadn't read the original play; I hadn't seen the film or musical. Everything – the songs, the set, the curious clothes – was new to me. Or nearly. Halfway through the spectacle, my reserve of patience ebbing since the actors rushed their lines, or so I imagined, all of a sudden I found I understood:

'*Daug lietaus Ispanijoje!*'

'The rain in Spain stays mainly in the plain.'

Triumph was in the young actress's voice. She repeated the line, as if savouring every word.

'*Daug lietaus Ispanijoje!*'

I could hardly contain my joy.

3

YOU ARE WHAT YOU SAY

The first time I met with scientists eager to examine the goings-on of my mind, I was given a vocabulary test. Fifteen years on, I still remember how my heart sank. I had volunteered for their research with the intention of talking, explaining, recounting – of condensing my multicoloured thoughts, my unusual (as I had been told) creative processes, into words. I had never had that conversation before, could scarcely wait to have it (though, as it happened, the wait would ultimately grow by three more years). Before leaving for central London and the psychology department there, not knowing the sorts of test that lay in store for me, I had made a mental inventory of what I hoped to find: television-handsome doctors, open ears and open minds, answers to my zillion questions. I was prodigiously naive. My disappointment was almost immediate. No sooner had I arrived than I contemplated turning on my heels and taking the next underground train back. The whitecoats asked only for my age, my school grades, whether I was left- or right-handed. Then, they gestured toward a thin, impassive lady who led me down a narrow corridor. The woman introduced where we were going and rattled off a list of rote instructions to follow. I was told to speak clearly because her pen would record my every error. A taker of tests, that was all the people here saw in me. But I was too intimidated to back out. And so, once I had been shown into a low-ceilinged, white-walled rabbit-cage of a room, the vocabulary test began.

Fifty words, printed on a foolscap sheet, which I was to read aloud one by one. Clearing my throat gave me enough time to take in a surprising pattern in the list of words: quite a few of them – *aisle, psalm, debt* (as in 'forgive us our debts, as we forgive our debtors'), *catacomb, zealot, leviathan, beatify, prelate, campanile* – had to do with the Church. And there were, it seemed to me, further clues to the identity of the person behind the list. Medical words – *ache, nausea, placebo, puerperal* – suggested a doctor. Still others – *bouquet, cellist, topiary* – hinted at a life far removed from the East End factories and pound shops of my own childhood.

'*Chord*,' I muttered to the invigilator. It was the first word on the list.

'Please go on,' she said. Her voice was glassy.

I wanted to say, 'It is a golden word. Gold, white, and red. Like the colours of the flag of Nunavut. And if you retype the word in small letters, reordering them to spell *dcorh*, and then trim the tops off the tall letters, the *d* and the *h*, do you see the anagram *acorn*?' But I didn't say any of this.

'Please go on,' she said again. My responses came reluctantly.

Some words later, when I reached *equivocal,* I stopped and let out a small gasp, which was like laughter. *Equivocal!* A green word, a shiny word. A word comprising each of the five vowels. So beautiful. I was beside myself with enthusiasm. But the woman with the glassy voice didn't seem to notice. *Equivocal!* A word cool to the touch. The greenness. The shininess. The coolness. They all came at me simultaneously. The word radiated the sea on a late British summer afternoon – the briny, occasionally garlicky, smell of the sea – and aroused a momentary nostalgia for the coast. *Equivocal,* lifting my mood, offering me colour and beauty in the ambient drabness – suddenly made my coming here seem worth the trouble. Even if I had to reduce it to a game of pronunciation, to four syllables with accent on the second.

Much of my learning had come from library books. If I could recognise *prelate* and *beatify* (though I was unchurched, and my parents nonbelievers), knew that *gaoled* was an olde worlde way of spelling *jailed,* and was able to place *quadruped* as the Latin for *four-footed,* it was because of all the years I'd spent around assorted dictionaries and encyclopaedias. But the one-sidedness of this instruction had its limits, something the test would now proceed to amply demonstrate. *'Aeon.'* It brought me to a halt. From this word forward, I apprehended the certainty of fumbles. While understandable enough in print, 'aeon' was a sound that I could only vaguely make. A-on? Air-on? E-on? It was like a shibboleth, like pronouncing Lord Cholmondeley as 'chumly', or Magdalen College, Oxford, as 'maud-lin'. You had to be in the know.

You are what you say. But this notion of 'verbal

intelligence' – so dear to the box-tick-happy psychologist – that language is something you can measure precisely and put an exact number on, struck me as false. And the list that had been compiled by this notion contained some kind of assessment. What was it? What action, I wondered, what social task might ever require me to say something like *drachm* (and pronounce it 'dram')? It was hard to dispel my doubts. The dubiety ran deep in me. I knew from my days at school that obscure and rare words, 'dictionary words' – spoiled dialogue more often than not and banished their user to solitude. The experience had not been easy on me.

And yet I didn't speak my mind. I was to regret that afterward. But looking back, I can understand. Just what could I have said, assuming that I would have even been granted a hearing? That simply knowing a word like *drachm,* its pronunciation, did not make you a better, more intelligent, speaker? That listing words on their own, denuded of context and meaning, was impoverishing? That language, that feeling and thinking and creating in language, bore no resemblance to so pointless and inane an exercise? But the men and women here were psychologists, with the brisk deskside manners and the degrees of psychologists. Who was I to say anything? I, a twenty-two-year-old, under-employed man. Parents, a former sheet-metal worker and a stay-at-home mum. Higher education, none (at that time) to speak of. Obediently, I kept my head down and stuck to reading the list through to the end.

I remember now that there was a moment of confusion over the final item on the list. *'Campanile.'* I said it in the French way, *à la française*. I had learned French at school; during my teens I'd had French penpals; I had holidayed one summer at the home of a family in Nantes. So I had a good accent. But the invigilator hesitated; she raised her free hand to her bell-bottomed black hair. And let it drop. 'No,' she said at last, since it was time for the debriefing, and added, '*Campanile* comes from Italy.' 'Cam-pa-nee-lee', she insisted, was the correct pronunciation. It was what her notes told her. I started to object, but checked myself and let it pass. Another concession. According to the online version of the *Oxford English Dictionary*, either way of saying the word is right.

After a few other tests (equally mind-numbing, like having a story read to me and being asked to recall it), I left and never went back to the department. I never heard from its staff again. But whenever, in the years that followed, I came across one of those newspaper or magazine advertisements for 'word power packs' – 'At the Crucial Moments of Your Life, Do Your Words Fail You? Give yourself the chance to become the articulate man who succeeds in business . . . the fluent man who enjoys a rich social life . . . the facile speaker who is esteemed by his community' – it would remind me of that wasted day in central London and the psychologist with the glassy voice. The smudgy shill in the ad, the whitecoats' picture of articulateness in his thick-framed glasses and three-piece suit, appeared inches across from the pitches of scalp specialists and Positive Thinkers. A Get-Glib-Quick scheme. I didn't buy it. But the message continues to seduce to this day. The salesmen understand their buyers, the complexes of working-class vocabulary, the desire of many to improve themselves. They charge a few cents for every ten-dollar word and make you believe you've got yourself a bargain.

I'm not divulging any great trade secrets if I tell you these 'word power packs' don't work. The associations they teach, like the vocabulary test's list, seem pointlessly artificial. Take, for instance, the cousin of *timid* and SAT favourite, *timorous* ('timorous, *adj.*: Full of or affected by fear'). Instead of picturing a Tim Burton movie, or the British tennis choker Tim Henman, better to read from the poem 'To a Mouse,' by Robert Burns:

> *Wee, sleekit, cowrin, tim'rous beastie,*
> *O, what a panic's in thy breastie!*

As a teenager, that was how I learned the adjective: I let the poet's imagery animate it for me. The learning did not cost me a penny; and I suppose if it had, I would have been sore as hell at the seller, because not once have I heard or said it since. Most native English speakers seem to get happily by with *afraid* or *frightened* or *fearful,* or even, in the north of the United Kingdom, *frit*. Few, if any, have ever found themselves short a *timorous*.

Even as a very young man, I had certain intuitions about language being irreducible to individual pieces of vocabulary. But for a long time that was all they were. My thoughts needed clarifying. But I couldn't rush them. After my diagnosis at 25 – high-functioning autistic savant syndrome ('with excellent adaptation in adulthood') and synaesthesia, big words in their own right – I travelled, I read, I wrote. I became a writer. Time and experience did their work: clarity gradually came to me. Two things helped my thinking in particular. The first and biggest, almost ten years after reading aloud *aeon* and *drachm* and *campanile,* was my decision, upon entering my thirties, to finally study at a university. Remembrance of that list, of not speaking my mind out of self-consciousness or lack of social skill or confidence, niggled and played its part in prompting me to put a portion of my author royalties into a bachelor of arts degree. Among my other courses in the humanities, I chose sociolinguistics and discovered in it, like a revelation, the work of Shirley Brice Heath.

A sociolinguist and ethnographer, too little known in her native United States, Shirley Brice Heath concluded as early as the 1970s that schoolchildren from different social classes and cultures were set apart academically, not by deficits in IQ or vocabulary size, but by the differing 'ways with words' in which they were raised. Heath, who has risen to the academic stratosphere at Stanford, is a former foster daughter of a milkman and a factory employee. She is passionate about her studies in language development and child literacy. Her painstaking fieldwork, in the Piedmont Carolinas, took close to ten years to complete. With the chameleon qualities of a born anthropologist, Heath lived, talked, and played with the children and their relatives and acquaintances in two neighbouring textile mill communities: one ('Roadville') white, the other ('Trackton') black. She compared her findings with the behaviours of young 'Maintown' families like her own, white and middle class, whose children had doctors and professors for parents. In all three groups, Heath showed, the children grew up speaking a rich, fully-formed American English; only the manner in which they attained that fluency and the uses to which they put their English differed.

Maintown parents, Heath noted, treated even toddling sons and daughters as their conversational equals. 'School-oriented', the parents behaved much like teachers, accompanying a bedtime story with bright-voiced explanations of the narrative, riffs on theme and character, and little quizzes, concocted on the spot, to assist (and assess) the child's comprehension. By the time the child began school, classroom decorum came more or less naturally. The pupil had appropriate responses to Mr Brown's or Mrs Cooper's questions down pat.

In Roadville, boys and girls learned to read whatever picture books their parents' dollars could stretch to buy; but, contrary to their peers in Maintown, come reading time, they fidgeted. The books, picked out from grocery checkout stands, quite often failed to hold the attention of either party for very long, making the experience a chore. Parental explanations of a narrative tended to be perfunctory. No riffs, then. No little quizzes. Instead, the children would mostly hear stories by listening to adults 'tell on' themselves for laughs among friends, family, or townsfolk. A Roadville story was typically a story that diligently recounted a past event and held some kind of moral: storytelling in the normal sense was considered tantamount to lying. If asked by the schoolteacher to write a story about coming down from outer space or levitating in the air, the Roadville-raised pupil would hardly even know where to start.

Trackton's preschoolers were hardly ever addressed directly by adults. As one grandmother explained it, 'Ain't no use me tellin' 'im: learn this, learn that, what's this, what's that? He just gotta learn, gotta know; he see one thing one place one time, he know how it go, see sump'n like it again, maybe it be the same, maybe it won't.' But there was no speech delay. The infants learned as if by osmosis. Many became uncanny mimics, able to reproduce the walk and talk of family, neighbours, and regular visitors down to the gas meter man. Trackton talk was more allusive than that heard in the other communities: listeners were expected to mentally fill in any gaps. Impromptu stories had no formulaic 'once upon a time' beginnings or verbal wrappings-up to signal the end – the narratives ran on analogy, leaping from one thought or incident to another, and lasted however long audiences would entertain them. Schoolteachers often found pupils from Trackton boisterous (or else withdrawn) in class and less ruly in their compositions, finding in a lesson parallels that might not have been intended.

In 1983, Heath published her findings. She urged teachers to ponder the background of every pupil and their Maintown biases carefully. Too often, red pens penalised Roadville and Trackton students' different ways with words, mistaking them for deficiencies; the bar of academic ambition was correspondingly lowered. Abandoned to their frustration, their incomprehension, class laggers found themselves relegated to the shock of remedial lessons from which their young confidence frequently never recovered.

Such a waste! And yet, more than thirty years after Heath's publication, the cruel circle remains intact. As an ideological shortcut, the equation of poverty with poverty of thought is as popular as ever. Media and lawmakers fret that the less well-off are 'word poor'. Nothing, though, could be further from the truth. As Curt Dudley Marling, a professor in the Lynch School of Education at Boston College writes, 'All children come to school with extraordinary linguistic, cultural, and intellectual resources, just not the *same* resources . . . Respect for students' knowledge, who they are, where they come from . . . is the key to successful teaching.' Indeed, in those schools where children's differences have been roundly welcomed, their particular strengths nurtured (and weaknesses thoughtfully attended to), poorer pupils demonstrate just as supple an attention and just as ravenous an appetite for words, stories, and puzzles as other students do.

Shirley Brice Heath's linguistic anthropology fed my understanding of language. So, too, did my encounter with the work of a young American lexicographer, Erin McKean. In her various media appearances, widely relayed on the Internet, her slogan – *if a word works, use it* – struck home. I liked what she had to say about modern dictionaries, that they map a language and

explore every nook and cranny of its well- and lesser-
trod acres; that they avoid the snobbery of their older
counterparts: high culture's doormen charged with
excluding all words they judge 'bad', 'uncouth', or
'incorrect'. And I liked how, in keeping with her
colourful published guide to the semiotics of dresses
(her other passion), McKean's sole editorial criterion
is whether this or that word 'fits', whether its form and
the speaker's (or author's) purpose sufficiently 'match'.

McKean says she always wanted to be a lexicographer, ever since she read about one in a newspaper
as a kid. She started out doing manuscript annotation
('a fancy word for "underlining with coloured pencils"')
for the Chicago Assyrian Dictionary, before landing
positions consecutively at Thorndike-Barnhart
Children's Dictionaries, and the American Dictionaries
division for Oxford University Press. In 2004, only
thirty-three, she was tapped to become *New Oxford
American Dictionary*'s editor-in-chief. But lexicography has evolved so rapidly since then that she now
works exclusively online, having come to the conclusion that 'books are the wrong containers for
dictionaries.' The nonprofit *Wordnik*, which McKean
launched in 2009, is billed as the world's biggest online
dictionary. It forages for words within meaningful
phrases and sentences on the Internet and in the
contents of millions of digitised books going back
centuries: words that, for want of space or knowledge
of their existence, appear in no print dictionary, no
vocabulary list, no SAT test. *Wordnik* includes never-
before documented words such as *slenthem* (a Javanese

musical instrument) and *deletable*. The resulting map of English, its boundaries expanding by the day, is like no other; it is changing how we define *English*, since – on the conservative estimations of data scientists – the lexicon is more than twice as big as anyone ever previously imagined.

'My biggest frustration is not having enough hours in the day to do all the work that I would like to do,' writes McKean in an email. 'Although (going by the actuarial tables) I'm barely halfway through my expected lifespan, it's more likely than not that I will die before *Wordnik* is "finished" – the nature of the lexicography is that most lexicographers don't live to see the culmination of their projects, because English never stops and so the projects never really end.'

'English never stops.' It could be McKean's motto. Every minute of every hour of every day, someone in the English-speaking world plays a new combination of sounds, letters, and meanings on a listener or reader. Who, understanding, nods. English! Only tweaked, stretched, renewed. Words – but also what we talk about when we talk about words – are constantly shifting. When I clicked on *Wordnik*'s 'random search' option, 'to-dos,' the eminently logical plural of 'to-do,' popped up. Intrigued, I performed a quick side search on the Internet and turned up a 2001 *New York Times* article featuring 'not-to-dos.' Back on *Wordnik,* I clicked again:

'Nonfraud, *adj.* Not of or pertaining to fraud: "The SEC alleges that Assurant violated certain nonfraud-related sections of the Securities Exchange Act of 1934 with the accounting of this reinsurance contract." (from a 2010 article in *Insurance Journal*)'

And once more:

'Goaltend, *v*. To engage in goaltending: "Saturday afternoon, we were certain that . . . NBA commissioner David Stern would plop atop the Orlando rim and goaltend all of Rashard Lewis's three-point attempts." (from a 2009 article in *The Wall Street Journal*)'

Predictably enough, some of McKean's correspondents, particularly those fondest of the Caps Lock key, will complain that a *Wordnik* entry 'isn't a word' because it isn't a word that they like. The pedants' *grr*ing is a smaller frustration, since she knows she cannot reason with her critics. Better to save her attention for mail like the enthusiastic message that came last Christmas from a class of Australian schoolchildren, proposing their own (and who knows, perhaps impending) words, such as *insaniparty, kerbobble,* and *melopink*.

McKean's own English, her own 'way with words', doesn't seem all that much like the English of a lexicographer (though, apparently, her son's is that of a lexicographer's child: at the age of six he wrote – and without a letter out of place – *discombobulated* in his first-grade class journal). She can tweet, 'just fyi, if I ever "snap" it will be 100% caused by someone loudly popping their chewing gum on public transit.' She is not above typing *gonna* and *gotta*; *lookupable* (as in 'every word should be lookupable'), *madeupical,* and *undictionaried* are all words of her invention, and in her sentences they read just fine. ('Words only have meaning in context,' she notes. 'If I just say "toast" to you, you don't know if you're getting strawberry jam or champagne.')

'Chord'. 'Drachm'. 'Aeon'. 'Campanile'. What, I still wonder, did the psychologist listening believe she understood? You are what you say – well, maybe, up to a point. Every voice carries certain personality traits – the tongue-tiedness of one; of another, the overreaching vowels. Every voice, in preferring *dinner* to *supper,* or in pronouncing *this* as *dis,* betrays traces of its past. But vocabulary is not destiny. Words, regardless of their pedigree, make only as much sense as we choose to give them. We are the teachers, not they. To possess fluency, or 'verbal intelligence', is to animate words with our imagination.

Every word is a bird we teach to sing.

4

A POET SAVANT

The Australian poet Les Murray makes life hard for those who wish to describe him. It isn't only his work, which has been put out in some thirty books over a period of fifty years, and which, having won major literary prize after major literary prize, has made its author a perennial candidate for a Nobel. It is the man. In PR terms, Murray seems the antipode of Updikean dapperness, cold Coetzee intensity, Zadie Smith's glamour. His author photographs, which appear to be snapshots, can best be described as ordinary. The bald man's hat, the double chin, the plain t-shirt. A recent photograph, accompanying his *New Selected Poems*, shows him at a kitchen table, grandfatherly in his glasses. The artlessness is that of an autodidact. Murray has always written as his own man. Fashions, schools, even the occasional dictionary definition, he serenely flouts. To read him is to know him.

A high hill of photographed sun-shadow
coming up from reverie, the big head
has its eyes on a mid-line, the mouth
slightly open, to breathe or interrupt.

The face's gentle skew to the left
is abetted, or caused, beneath the nose

by a Heidelberg scar, got in an accident.
The hair no longer meets across the head

and the back and sides are clipped ancestrally
Puritan-short. The chins are firm and deep
respectively. In point of freckling
and bare and shaven skin is just over

halfway between childhood ginger
And the nutmeg and plastic death-mottle
of great age. The large ears suggest more
of the soul than the other features:

(These lines are from his 'Self-Portrait from a Photograph', in *Collected Poems*.)

His is a singular way with words; Murray admits to interviewers that he is something of a 'word freak'. His linguistic curiosity is immense, obsessive. *Gnamma* (an Aboriginal word for desert rock-holes wet with rainwater), the Anglo-Hindi *kubberdaur* (from the Hindi *khabardaar* meaning 'look out!'), *toradh* (Irish Gaelic for 'fruit' or 'produce'), *neb* (Scots for 'nose'), *sadaka* ('alms' in Turkish), the German *Leutseligkeit* ('affability'), *rzeczpospolita* (what Poland, in its red-tape documents, calls itself), *halevai* (a Yiddish exclamation): these are just a few of the many exotic words which Murray, over the years, has lavished on his poetry. 'His genius,' reported *The Australian,* 'is an ability to process the 1,025,000 words in his brain and select precisely the right one to follow the one that came before it.' The estimate is the reporter's – an unfortunate piece of silliness in an otherwise serious article. The estimate – not something to consider, something only to gawp at – comes from old attitudes, still publishable in 2014, toward 'freaks' like Murray.

Murray, as the same article notes, lives with high-functioning autism.

I discovered Les Murray in the early 2000s (a short while before my own high-functioning autism diagnosis). It happened in an English bookshop. The bookshop was in Kent, my home at the time. It was a large shop, bright with snazzy book covers, and superintended by a discreet staff. It welcomed browsers. Every now and then I came into the shop and thumbed its wares and pretended to have money. The pretence did me good. It taught me the prospective book buyer's graces. In the poetry section one day, I was looking at the stock – short books written by long names like Annette von Droste-Hülshoff, Guillaume de Saluste Du Bartas, Keorapetse Kgositsile, Wisława Szymborska – when a cover that bore the surprisingly modest name of Les Murray caught my eye. The familiarity of that 'Les', its ungentlemanliness, so seemingly out of place in a poetry section, appealed to me. And the book's title was as intriguing as the author's name was likeable: *Poems the Size of Photographs*. I picked it up and read. And read. It was the first time I read a book from cover to cover while standing in the shop. It helped, of course, that the pages, compared to a novel's or a biography's, were fairly few, and many of the hundred-odd poems rather short. Short, occasionally, to the point of evoking haiku-like riddles:

This is the big arrival.
The zipper of your luggage
Growls valise round three sides
And you lift out the tin clothes.

Short or long, each of the poems had something to pique my interest. The word pictures were vivid and felt right: magpies wore 'tailcoats', which, at the faintest noise, broke 'into wings'; headstones complete with crosses were the 'marble chess of the dead'; in a house, 'air [had] sides.' And Murray's patent delight in language, his fascination with it, matched mine. He wrote like a man for whom language was something strange, and strangely beautiful. 'Globe globe globe globe' is Murray mimicking a jellyfish. A soda bottle a little girl taps against her head, Murray informs the reader with onomatopoeic precision, sounded 'boinc'. 'Bocc' was the bottle's response when it struck the side of a station wagon. A woman could simultaneously describe a cheese in Australian English and, with copious gesturing, in 'body Italian'. And in the modern landscape of ever-present signs – airport signs, road signs, door signs, computer signs – the poet saw a 'World language' of pictographs, a language that could be 'written and read, even painted but not spoken.' He imagined its vocabulary:

Good is thumbs up, thumb and finger zipping lips
is *confidential. Evil* is three-cornered snake eyes.
. . .
Two animals in a book read *Nature,* two books

Inside an animal, *instinct*. Rice in bowl with chop-
sticks

denotes *food*. Figure 1 lying prone equals *other*.

Every pictograph would be findable, definable, in a
'square-equals-diamond book', a dictionary. I wanted
very much to buy Murray's book. But I had no money.
I had to wait. At my next birthday I came into a small
sum, my age in pounds sterling, and spent it on the
poems.

Without Murray's poems, I might never have become
a writer. At the time, so early in my adulthood, their
oddness reassured. I could see myself in them. In
London, my English had never been the English of my
parents, siblings, or schoolmates; my sentences –
oblique, wordy, allusive – had led to mockery, and the
mockery had caused my voice to shrink. But Murray's
was sprawling; it held a hard-won ease with how its
words behaved. If I had known then what I finally
confirmed years later, that this poet's voice, so beautiful
and so skilful, was autistic, and that Murray was an
autistic savant, I might have seen myself differently. I
might have written my essays and my first novel and
my own poetry years before they finally made their
way into print. But nobody told me; and what was only
a hunch, after being diagnosed myself, that Murray too
had autism, remained only that. Happily, this hunch
was enough to restore confidence in my own voice, to
nourish it back up to size. My voice eventually grew
as big as a book: a memoir of my autistic childhood.

Only while writing my second book – a survey of

scientific ideas about the brain – did I learn why no doctor or researcher would have told me that this famous poet (and more than a poet, one of the English language's great men of letters) was autistic. It wasn't due to lack of candour or self-awareness on the poet's part. As early as the 1970s Murray wrote, in a poem entitled 'Portrait of the Autist as a New World Driver', of belonging to the 'loners, chart-freaks, bush encyclo-paedists . . . we meet gravely as stiff princes, and swap fact' (a pity it was only collected thirty years later, and brought to my attention quite recently); in several more recent newspaper profiles, he described himself as a 'high-performing Asperger' (a pity, too, I didn't see these at the time). The reason, I discovered, was that most scientists thought autism inhospitable to cre-ativity, especially the literary kind. Their attitudes were coloured by having studied a very particular type of savant: those with a low or lowish IQ, nonverbal or mumbling. I pressed the scientists I met or wrote to. Had they really never heard of any autistic person with a gift for words? They had, I was told, and his name was Christopher. Christopher, in the photographs I saw, was coy and pasty, moustached and middle-aged, a Briton with brain damage, who spent his days turning the pages of his impressive collection of Teach Yourself books. Reading nearly around the clock he had taught himself, to varying extents, some twenty languages. And yet Christopher couldn't write.

In the end, it was Murray's following release, *The Biplane Houses*, that turned my hunch to certainty. By that time I had left Kent and its bookshop and was living in the south of France. I ordered my copy in a click of Internet: it arrived in a many-stamped airmail envelope – a small, slender, red and white book looking none the worse for the cross-Channel journey. As I flipped through the poems, I read, in the form of a sonnet, not a declaration of love, but the author's contemplation of his mind:

Asperges me hyssopo
the snatch of plainsong went,
Thou sprinklest me with hyssop
was the clerical intent,
not Asparagus with hiccups
and never autistic savant.

Asperger, mais. Asperg is me.
The coin took years to drop:

Lectures instead of chat. The want
of people skills. The need for Rules.
Never towing a line from the Ship of Fools.
The avoided eyes. Great memory.
Horror not seeming to perturb –
Hyssop can be a bitter herb.

I had read in who knows how many scientific papers that people with autism took up language – even twenty varieties – merely to release it as echoes. So I was thrilled to see Murray's lines, full of wit and feeling and prowess, prove their theories wrong. They made me want to learn all about the man behind the pen. I went to my computer; I punched in the poet's name and autism; the resulting webpages were many, well stocked, enlightening. The information had been there all along. But the reporters had pushed it to the margins. Each had written about Murray's autism only briefly, sketchily, as something that didn't fit. They had not thought to look any further, to look at the poet's oeuvre or his life in another light. To learn this side of Murray's story I had to piece it together by myself, click by click and link by link, following up every byte of every source that I could verify. Slowly, scattered facts took on the vivid colours of anecdote; interview digressions lost their blur. The developed picture of the poet's early life – the many years his mind took to adjust to language – was as revealing as it was compelling.

Leslie Allan Murray was born in 1938 in Nabiac, a small bush town and inland port in dairy farming country, 175 miles northeast of Sydney. His ancestors, agricultural labourers, had sailed from the Southern Uplands of Scotland in the mid-nineteenth century, bringing with them their Presbyterian faith and Scots dialect: *fraid* for a 'ghost', *elder* for 'udder'.

Les was an only child. From his first shriek, the baby's senses overworked: his indigent parents' weatherboard house frequently trembled with tantrums. Nothing but a warm bath in water heated on the boxwood-burning stove and then poured, crackling, into a galvanised washtub could soothe him.

The house was poky. There wasn't room for homebodies. Life was in the main outdoors – on the outskirts of the house, the white of sheep, the black of hens, the green of paddocks. The little boy imitated the calls of willie wagtails and played with cows and crows. He

rambled by himself the hilly hectares around the settlement, returning at dusk soaked as if in a cloudburst
of perspiration.

Very early, he learned the alphabet from canned-food
labels. He took at once to reading the few books around
the home: the Aberdeen-Angus studbook, the *Yates
Seed Catalogue,* the Alfa Laval cream separator manual,
and all eight volumes of the 1924 edition of *Cassell's
Book of Knowledge.* The books and he became the
best of friends. He slept with them every night out on
the veranda.

There were no educational establishments for miles,
so when Les was seven, school started going to him.
The postman was the intermediary. The lessons by
correspondence – in handwriting, grammar, and arithmetic – came every week, postmarked from Sydney, and
the boy delighted in tracing pothooks, building phrases,
and solving sums at a kitchen table surrounded by
buckets and chips of wood.

Two years later, and three and a half miles away, a
fifteen-pupil school opened. Les decided he could not
go there empty-handed; he asked his mother for several
sheaves of white kraft paper, and with a regularly
resharpened pencil wrote a long essay about the Vikings
(the information courtesy of volume eight of his encyclopaedia). The helmeted berserkers, the swishing axes,
the longboats: no historical detail was too small. He
lost himself in the writing. After trekking off at dawn
and arriving hours later, he read his work aloud to the
class: both short- and long-winded. Les was lucky in
his teacher. Just out of graduate school, the teacher

hadn't had enough experience to classify the boy as odd. Les's essay was patiently listened to; its precocious author thanked. Then Les and the fourteen other pupils were told to open their textbooks to page one.

He was more or less left to his own devices and soon drifted away from class and sequestered himself in the school's bookroom. With the exception of a few cousins, amiable and of similar age, he much preferred the company of books to other children. But even with the cousins Les wasn't much of a talker. If he wasn't talked to, he wouldn't start. When the boys induced him to play with them, it was only by going along with his obsessions. His greatest was the war: he had been a child of the war, and though incomprehensible, *war,* he knew, meant something excitingly big and far away, a silencer of grownups, about which the wireless had gone on and on in quick, breathy voices. So the cousins agreed to play Germans. Along the rabbit-ridden creek they raced and leapt and hid. Les spoke to them in orders. He tried sounding German; it made sense to him that he should try. Compared to the other boys, he had always felt and seemed somewhat foreign.

His mother died when he was twelve. Her widower's sobbing was incessant. It could be heard by the son, lying beneath several bedcovers, on the veranda. Les hadn't known his father could make such sounds. He interrupted his learning to attempt to grieve too. But what came so naturally to his father mystified him: crying was something he couldn't get the hang of. He grieved with his feet instead of his eyes, tramping them red, crisscrossing the valley until every fencepost hole, every patch of gravel, every spot, to a blade of grass, was part of him. So that whenever distress tightened his stomach and threatened tears, he closed his eyes and called up the texture of a pebble here, the colour of the earth there, and journeyed mile after mile in his head until composure settled on him.

A year taller, he returned to school. It was a bigger school, farther away (he rode the town's milk truck to its gates) with a bigger bookroom. The books soothed. And it helped that the room was not frequented by many children. His desire to keep to himself was as strong as at primary school. When the bookroom closed for lunch, he waited out the break, impatiently. With his back to a playground wall, shyness and anxiety pinning him there, he surveiled the kicking, skipping shadows as they passed.

Taree High, where Les boarded during the week, was rougher. Tall and large for sixteen, he was an easy target of mockery. To every fashion, every norm of appearance, the country boy was oblivious: he stood out from the students' well-laundered clothes and up-to-the-minute hairdos as a bumpkin. Class toughs and teases regularly turned on him, harassed him, pelted him with taunts. He could hardly open his mouth without being laughed at; he spoke like a walking encyclopaedia, pedantic. He used – sometimes, misused – long, obscure words: once threatening to 'transubstantiate' a taunter through a wall.

Most of the teachers didn't turn a hair to help. But they were dazzled by the sweep and precision of Les's knowledge, by the nonchalance with which a mention of, say, the pre-Columbians in an art lesson, might cause him to hold forth on the Aztecs, Zapotecs, Toltecs, and Totonacs. And Mr McLaughlin, the English teacher, a mild school counsellor of limitless patience, was understanding. To teach Les was to be assured of an eager hearing. He was a quick learner. It was Mr McLaughlin who introduced Les to poetry. Eliot. Hopkins. Les was entranced by Gerard Manley Hopkins's words, tried them out on his tongue. *For skies of couple-colour as a brinded cow.* He read on. Repeated. *With swift, slow; sweet, sour; adazzle, dim.* He knew this and, before long, each of his teacher's other set poems by heart. *Praise Him.* He dreamed of one day composing his own verse.

Les spent the rest of the 1950s in an undergraduate's navy-blue bell bottoms, not the green Dacron overalls he had half-hoped to climb aboard and pilot planes in (the Air Force medical officer failed him on sight). His initial disappointment faded quickly, however, because at Sydney University he discovered a remarkable library – a Gothic, gargoyled sandstone building that seated close to a million books. A million! So he wouldn't feel out of place at the university after all. And then he thought, where will I find the time? He was being serious. Every free hour he had, and quite a few taken ones, went into his reading. He skipped classes and dodged lecturers to skim, scan, browse, peruse. Encyclopaedias, poems, novels (baiting boredom – he couldn't enter the domestic scenes or believe the plots), short stories, hymnals, plays. He returned again and again to the *Encyclopaedia Britannica*; he pored over

Cicero, Kenneth Slessor, the King James Bible. And then, one day, running the length of a bookcase shelf, a yellow row of Teach Yourself language primers attracted his attention. The exotic letters, words, and sentences enticed. Danish by Hans Anton Koefoed. Russian by Maximilian Fourman. Norwegian by Alf Sommerfelt and Ingvald Marm. French by John Adams. Italian by Kathleen Speight. One after another, in quick succession, Les read them all from end to end. He found them easy. He had the knack of recalling the words, the phrases. And of playing with them, locating rhymes and other patterns, inventing new sentences, freer and more imaginative, out of the authors' dry, stiff examples. To round out his enjoyment, he enrolled for courses in German and preliminary Chinese. In next to no time, a couple of semesters, the languages in which Les could read and write numbered over ten.

Les the reader and linguist flourished; but Les the student floundered. The astringency of the curriculum – read this, not that; write your thesis in this manner, not that – demoralised him. He dropped out, age twenty-one, to see the country; he hitched long rides in trucks. He overnighted on building sites, in patches of long grass, wherever he could find a dry spot. It was during these vagabond months, in beige, hole-punched notepads, that he wrote his first significant poetry.

He hitchhiked back to Sydney, returned to student-ship unsuccessfully, dropping out a second time (he would earn his missing credits only a decade later in 1969), but not before he had met – backstage at a play put on by the German department – his future wife and the future mother of their five children. Usually, with strangers, he couldn't do small talk; conversation would fail him. But with Valerie, to his surprise, he had no such trouble. Valerie, who was born in Budapest and came to Australia after World War II by way of Switzerland, spoke English with a slight Mitteleuropean accent. She was a young woman of Hungarian, Swiss-German, and English words. Murray instantly felt at ease with her, and the attraction was mutual. (In 2012 they celebrated their golden wedding anniversary.)

Les's great luck continued. He got a job as a translator for the Australian National University in Canberra. He worked for several departments, rewriting in English one day an Italian paper on 'nodular cutaneous diseases in Po Valley hares,' the next a Dutch study of the trading history of Makassar. 'Languages,' in the words of a 1964 newspaper article about the young translator, had become his 'bread and butter'. For four years he and his growing family lived on Italian and Dutch, German and Afrikaans, French and Spanish and Portuguese. But it was in English that Murray's poems began to circulate in magazines. They were collected, in 1965, in *The Ilex Tree*. His first literary prize followed. His reputation and his confidence quickly soared. On the strength of the book and prize, he was invited to give readings as far away as Europe. Soon he lost his taste for translating; he left what he would later call his 'respectable cover' occupation to write full time. It was the beginning of one of poetry's most extraordinary careers.

* * *

Words have been knots of beauty and mystery as long as I can remember. But Murray's oeuvre was the leaven for my own literary beginnings. So when my debut collection of essays was published several years ago, I dedicated a copy to the poet as a token of gratitude and admiration. But I hesitated before posting it to the bush of New South Wales. What if the book never reached him, or arrived but didn't please him? For a while I couldn't make up my mind. Finally, though, I let my timidity pass. I wrote to the address I had found for Murray in a literary journal, and several weeks later – restless weeks, since each seemed to me to last a month – I had a letter from him.

Time and care had been expended on the letter. Murray's penmanship was firm and fluent (I especially liked his *a*'s, neat little pigtails); the letter was quite long. Far more than a simple thanks then, or even some compliments: the tone throughout was expansive, even confiding. I was touched. What touched me most was Murray's request that I write again to keep him abreast of my news and my work. An invitation to correspond.

I took him at his word. I wrote. He wrote back. And so it was that letters later, in the course of our now regular correspondence, I proposed to translate a selection of his poems into my second language, French.

I had a seen a gap in the international literary market: Murray's poetry, in translation, was already on sale in Berlin bookshops, Moscow libraries, Delhi bazaars. But not, for some strange reason, in the city, a cultural capital, that I had begun to call my home: Paris.

The poet's agent agreed; my editor in Paris agreed.

And Murray, in his expansive Australian way, gave me a free hand, carte blanche. I reread the poems, collection after collection after collection; for days, weeks, I immersed myself in them; I ended up selecting forty of my favourites to translate.

Julian Barnes, another Francophile Englishman, has written well of the challenge of translating a literary text. There are, Barnes sighs, as many ways to translate one sentence of a classic as there are translators. In 'Translating *Madame Bovary*' he offers, by way of illustration, a straightforward enough line from Flaubert: 'aussi poussat-il comme un chêne. Il acquit de fortes mains, de belles couleurs,' followed by a half-dozen attempts over the years to render the words in convincing English:

'Meanwhile he grew like an oak; he was strong of hand, fresh of colour.'

'And so he grew like an oak-tree, and acquired a strong pair of hands and a fresh colour.'

'He grew like a young oak-tree. He acquired strong hands and a good colour.'

'He throve like an oak. His hands grew strong and his complexion ruddy.'

'And so he grew up like an oak. He had strong hands, a good colour.'

'And so he grew like an oak. He acquired strong hands, good colour.'

And Flaubert, for all his similes, remains prose, albeit prose of the very highest order. Poetry, by virtue of its nature, is widely thought to be more or less untranslatable (consider Robert Frost's aphorism 'poetry is what gets lost in translation').

So translating Murray, I knew, wasn't something to be taken lightly. But I didn't share Frost's limited conception of translation (or of poetry). I could never have rejoiced in Szymborska's ode to the number pi, nor gasped at Mayakovsky's 'An Extraordinary Adventure which Befell Vladimir Mayakovsky in a Summer Cottage', had it not been for the ingenuity of their respective translators. All those Polish accents would have scratched my eyes; the black iron gates of Cyrillic would have barred my way. Without translation, some of my favourite works, Bashō's haikus, the Bible's Song of Songs, would have remained beyond reach of my comprehension, forever the preserve of Japanese's and Hebrew's right-to-left readers.

Murray's English is vivid, inventive, jaunty. He is fond of the striking word choice. In recent years the poet has done double duty as an occasional contributor to Australia's *Macquarie Dictionary*. *Pobblebonk* (an informal name for a kind of Australian frog), *doctoring* (the regular seeking of medical assistance), and *Archie* (a World War I anti-aircraft gun) are just a few of his recent submissions. If you read 'The Dream of Wearing Shorts Forever', Murray's paean to pants that reach to or almost to the knees, you encounter several amusing stanzas on legwear history and culture, and then these lines:

To moderate grim vigour
With the knobble of bare knees

Knobble! I love this poem; it was one of the forty I chose to put into French. I did everything translatorly possible to stay faithful to the text. 'Knobble' is Murray's reference to the British (but otherwise little-known) expression 'knobbly knees', and I had a hard time coming up with a French equivalent. Knobble. Knobble. I scratched my head. *Bosse?* Bump? *Nœud?* Knot? Neither of these would do. Too abstract. Too neutral. Knobbly knees are funny-looking knees; *knobble* is a funny word. After a lot more head-scratching, I finally hit upon an idea. In place of bump or knot or whatnot could go *tronche*. To French eyes and ears, *tronche* looks and sounds just as comic as *knobble* does to English ones. *Tronche* can mean face (as in 'to make a funny face'), expression, the (dodgy, weird, or amusing) look or demeanour of an object or person. '*La tronche!*' is a typical French exclamation, meaning roughly 'look at the state of him!' or 'get a load of her!' It was as close a parallel as you could get in French. Readers of my translation imagined the stern shirt and tie softened, *par la* (by the) *tronche* (funny face) *des genoux nus* (of bare knees).

Then there is assonance and alliteration, poetry's nuts and bolts. Murray is a master of verse mechanics. 'The Trainee, 1914', which was another of my selections, tells of an Australian lured from his hovel to fight in a foreigners' war:

> *Till the bump of your drum, the fit of your*
> > *turned-up hat*
> *Drew me to eat your stew, salute your flag*

Bump, drum; drew, stew; and the end rhyme: *hat,*

flag. The poem runs on assonance (there is also simi-
larity of sounds between *fit* and *hat*). All the poetry
of the lines resides in these patterns, so in my translation I worked to retain them. I made adjustments. *Bump*
became *tempo* (alliterating with *tambour,* the French
for 'drum'); *fit* turned to *chic* (alliterating with *chapeau,*
hat). *Stew* was trickier. What suitable French verbs
rhyme with *ragout?* So I had the poem's character eat
not stew but *soupe;* and, for the sake of rhyme, tweaked
the tense:

> *Jusqu'à ce que le tempo de votre tambour, le chic*
> *   de votre chapeau*
> *Me poussent à manger votre soupe, à saluer votre*
> *   drapeau*

The hardest thing, of course, was to keep the translations Murray in structure while French in language.
Perhaps the most ambitious of my selections, 'The
Warm Rain' is pure Murray – controlled lines, graphic
words, surprising rhymes. Consider the beginning:

> *Against the darker trees or an open car shed*
> *is where we first see rain, on a cumulous day,*
> *a subtle slant locating the light in air*
> *in front of a Forties still of tubs and bike-frames.*
>
> *Next sign, the dust that was white pepper bared*
> *starts pitting and re-knotting into peppercorns.*
> *It stops being a raceway of rocket smoke behind cars*

By themselves, these seven lines abound in complexity.
If the first two slip fairly smoothly into French,

Sur un fond d'arbres sombres ou un carport ouvert
La pluie nous apparaît, un jour nuageux

The others resist. It is necessary to visualise the succession of word pictures as a film. In the third line, the play of raindrops in the air; in the fourth, the shed's exposed interior ('tubs and bike-frames'), by dint of the rainfall, looks like something in a grainy 'Forties still'.

Des fils subtils [subtle threads] qui rayent et
   floutent [which scratch and blur] l'air
Comme un retour [like a flashback] sur les années
   quarante [to the Forties] en images.

The beginning of the second stanza, too, shifting focus, poses difficulties. The imagery is fast-moving, complex. Dust particles, 'white pepper bared', become engorged with rain and turn to 'peppercorns'. Ordinarily thrown up by passing drivers, the puffs of dust 'behind cars' no longer resemble smoke. The translator's task of preserving the alliterations, consonance, and assonance from line to line – 'pepper bared', 'peppercorns', 'behind cars' – is formidable. Something, in order for my translation to be viable, had to give. It was the white pepper. I used *sel,* salt, instead.

Puis la terre jusqu'alors fine comme le sel fin
Se mouille et ses particules grossissent en fleur de
   sel.
Les voitures qui passent n'engendrent plus leur
   fumée de fusil

[Then the earth until then fine-grained as table salt
Grows moist and its particles fatten to fleur de sel.
The cars that pass no longer produce their rifle
smoke]

Translation takes away, but it also gives. Gone, in French, are the car shed's tubs and bike frames; gone, also, is the rocket's smoke (replaced by a rifle's, to reproduce the pattern of repeating sounds: '<u>sel</u> <u>fin</u>', '<u>fleur</u> de <u>sel</u>', '<u>leur</u> <u>fumée</u> de <u>fusil</u>'). Two gones, modest losses, but there is a counterbalancing gain: the French has *fleur de sel;* it more than compensates with the startling image of a rain that causes the salty earth to flower.

My translation, *C'est une chose sérieuse que d'être parmi les hommes* (*It is Serious to be with Humans*), was published in the autumn of 2014. On October twelfth, radios tuned in to *France Culture* spoke for thirty minutes in my voice. They told listeners of Murray, his life and work, the repayment of a young writer's debt to his mentor. The broadcast went around the French-speaking world.

Some months later, Murray wrote with pride of a Polynesian neighbour, a native French speaker, who admired the translation and was using it in her French classes. And though his own French was rusty, he said, he could nevertheless understand and enjoy quite a lot. In other letters and postcards – their news ten days older for being sent by snail mail (Murray dislikes computers) – he affectionately recounted long-past trips to Britain, Italy, Germany, and France. But there was nothing in these messages suggesting he might one day return; he hadn't been back in years. He was well into his seventies; his health was unreliable; his farm in Bunyah was over ten thousand miles away.

So the news that he would be in Paris in September 2015 came out of the blue. He had accepted a festival's invitation to speak at la Maison de la Poésie. I was thrilled. Honoured, too, when the organiser asked me to share the stage with the poet. Copies of the translation would be on sale after the talk.

Murray, in a multicoloured woolly pullover – a dependable 'soup-catcher' – braved the twenty-plus hours on an aeroplane and met me in a bistro. To be able to put a voice, so rich and twangy, to the poems after all this time! At la Maison, the following evening, our conversation had an audience and spotlights and was interspersed with readings in both languages.

After the event, a book signing. The queue was gratifyingly long. I was backstage when someone from the festival came to fetch me; Murray wanted me to sign the book with him.

Murray said, 'It's only fair. We're co-authors.'

5

CUECUEIUCA

Mexico. It means, some say, 'centre of the world'. It was there, in Puebla, the centre of the centre of the world, that I met an indigene who spoke lightning.

Our meeting took place soon after Día de los Muertos. It wasn't my idea. The day before, jetlagged, I had had my nose firmly in my notes when one of the conference organisers chaperoning me spotted the book I had taken with me for the plane: *The Nahuas after the Conquest: A Social and Cultural History of the Indians of Central Mexico, Sixteenth through Eighteenth Centuries* by James Lockhart.

'The *Nawas*,' she said, giving me the correct pronunciation of Mexico's largest indigenous group, descendants of the Aztecs, 'you are interested in the *Nawas?*'

Who could not be? I thought. I was particularly interested in the Nahuatl ('clear as a bell') language, which has given the Western lexicon *avocado* (*ahuacatl*), *guacamole* (*ahuacamolli* – literally 'avocado sauce'), *chocolate* (*xocolatl*), *cacao* (*cacahuatl*), *tomato* (*tomatl*), *chili*, *chia*, *peyote* (*peyotl*), *ocelot* (*ocelotl*), and *axolotl* (a type of salamander). 'No culture ever took more joy in words,' Lockhart writes. No culture ever revered more the power and the magic of sounds.

Mexico City (then Tenochtitlan) was always booming, ringing, resounding in the days of Montezuma's glory. The wind whistled, the Aztecs – with flutes and ocarinas – whistled; to the tinkling of a rain shower they added the tinkling of their bracelets, anklets, ceramic pendants and beads; after a night ablare with thunder, a morning of horns and conches, copper gongs and tortoiseshell drums. Singers in iridescent feathers roared like jaguars, squawked like eagles, cooed like quetzals. Mellifluous orations, 'flower songs,' offered the listener colour and beauty, and could inspire and pacify.

My chaperone said, 'I know a Nawa man. Quite a talker. He's in town now. I'll arrange a meeting for tomorrow morning, I'll tell him you're only passing through. There's a café close by. I'll write the address down for you. You can meet there.'

'Does he speak English?'

'He speaks Spanish. I'll go and telephone him now. Primero Dios, he'll be free.'

He was.

His name was Francisco (also his father's name, I learned). He arrived at the café bang on time. His white shirt was spotless and ironed, his brown face was clean-shaven and lively. I bought him coffee and introduced myself as best I could; my Spanish, for want of practice, was a little rough. I said I wanted to learn more about the Nahuatl language.

Nahuatl, he told me (pronouncing it 'Na-wat'; the *l* in *tl* is only for the eyes), is what scholars call it, a word for their books. He called his language *Mexicano*, pronouncing it 'Me-shi-ka-no'.

A couple of little sugar skulls, *calaveras* left over from the previous days' festivities, grinned sweetly at me from over behind the glass display cases.

'My mother spoke beautiful Mexicano.' His face softened. I found it hard to keep up with that face. The more I talked with him the more I felt he was a man of many moods, a man who had lived many lives. Now, smiling, his brown eyes shining, he looked fifty; now, frowning, his brow darkening as if shaded by an imaginary sombrero, he looked seventy. Frowning, he said, 'I am not my mother. I mix my words.' He meant that he used many Spanish words in his Mexicano sentences. It was a natural consequence of living in a country in which Spanish predominates.

Francisco explained that some Nahuas heap opprobrium on the practice. 'They are few but they are very touchy about our language. Sometimes, if I say something like *hasta moztla,* you know, like *hasta mañana,* someone will say, "Why are you talking to us in Spanish?" But that is just the way it is now. I'm too old to change. *Hasta moztla* feels natural to me. It is part of my language. Or I give someone the day of the week, and the person asking replies, "Say it in Mexicano!" To which I say, "Igual, it's the same, it's the same day in Spanish or in Mexicano." And another thing. Let me tell you, the most extreme among them, they are funny. According to them, you shouldn't say you drive a *coche*. You should say you drive a *tepozyoyoli*. But no one says that. It means "metal creature". Imagine! "I came here by metal creature."'

And he let out a curt laugh.

Lockhart writes that some Aztecs went to similar lengths to express in their own words what the Spaniards talked and gestured to them about. A dagger, for example, they tried calling *tepozteixilihuaniton* (literally, 'little metal instrument for stabbing people'); *nequacehualhuiloni* ('thing for shading one's head') was their neologism for a hat. *Tlamelahuacachihualiztli* ('doing things straight') was what they came up with for the concept of law or justice. But none of these circumlocutions ever found many speakers – understandably enough. Like Francisco, most preferred to adopt – with minor tweaks – Spanish terms.

What mattered to the great majority of Nahuas was not the origin but the sound of a word. The word could come from more or less anywhere, Spanish, American English, Portuguese, you name it, provided that it sounded good. Francisco said, 'When our mouths fall in love with a sound, we take it and speak with it.' Nahuatl grammar was no object. The Nahuas' notion of a word was, it seems, far more subtle and inclusive than the Western one.

Reduplication, the repetition of a syllable within a word to alter or enhance its meaning, is the underpinning of the distinctive Mexicano sound. (Reduplication is rare in English. Something like it occurs in *reread* – a more consuming, emphatic action than reading is – or the much decried *pre-prepared*). *Kochi* means 'sleep.' Repeat the first syllable and you get *koh-kochi*, to be sound asleep. *Xotla* is to burn, *xoxotla*, to burn intensely. And these are only two from among the many dozens, scores, perhaps hundreds of possible examples the language offers.

It makes, as Francisco said, for a language that is most welcoming to its environment's many shades of sound and nonhuman voices. A speaker can say *cocotl, cacalotl, papalotl* and instantly conjure up the white of a dove, the black of a crow, the fluttering of a butterfly, simply by naming each creature. Utter *huitzitzilin* and a hummingbird zips and thrums straight from your lips to the listener's imagination. It is why a Mexicano telephone does not merely 'ring'. To the Nahuas, its *tzitzilica* is reminiscent of the hummingbird's condensed, ear-catching frenzy. Rain here, too, is eloquent. A moderate shower, the kind that gradually empties markets and causes poplars to dribble, will *chichipica;* the same rain, turning white and fluffy in winter, will *pipixahui*. Indoors, on the stove, water heated in a pan starts to *huahualca* when it boils; corn cooked in a pan, on the other hand, is all *cuacualaca*. *Tzetzeloa* is another common house-hold sound. It describes a shirt, a skirt, a blouse shaken out with any vigour.

And occasionally, to Mexicano ears, things can speak.

'My mother, she would say to me, "You hear the clock? Listen to what the clock is telling you." Always, work this and work that. That is what it was saying. Work. You have to work. *Tequiti* – it means "to work."'

Certain things, his mother would say, her son was not to pay any attention to. Strong gusts. Wells. Various types of bird. They had dirty mouths, she warned.

And once, Francisco remembered, his mother had told him about the words of lightning. Had he never noticed what lightning said?

I stopped him. *'Relámpago?'* I repeated the Spanish word doubtfully. 'Yes,' he said, *'relámpago*, lightning.' I waited for him to continue.

He took a glug of his coffee first.

'She told me that lightning says *"cuecueiuca"*.'

I thought, *yes, that is the sort of thing lightning would say*. Something bright and jagged. It was as if the lightning were, from what I could gather, narrating its own effect on the night sky. The word means 'it is flaring' or 'it is glinting'.

The voice of lightning: *cuecueiuca!*

It is, I realised, an illustration of synaesthesia. Lightning makes a deep impression on the viewer's optic nerve. The Mexicano mind sees a pattern, and in the pattern discerns a snap meaning. Sense and sound, in a flash of intuition, become complementary. Atmospheric electricity addresses the Nahua in labio-velars ([kʷ]), and it simply feels right. Francisco said he couldn't explain any better why, to impersonate lightning, he had to round his lips.

A centuries-old culture, laden with an extraordinary sonic know-how, but it is on the wane. A reform of Mexico's constitution in 2003 granted Nahuas the belated right to education and government services in their language for the first time. But Francisco was worried. All around him, the evidence in his ears suggested ongoing decline. There are still more than a million up and down the country who speak Mexicano, but many resemble Francisco: men and women of a certain age. The young increasingly discard the language for fear of being thought *indios* by their Spanish-speaking compatriots and, unpersuaded by the arguments for bilingualism, disinherit the generations to come.

Outside the café, Francisco, full of coffee beans, whistled a bright little tune. He lived alone, he said. He missed talking. He didn't talk enough. Talking always did him a world of good.

The day after my conference – another warm, blue-sky day – I walked the streets of Puebla, all ears. The honks of passing cars. The sweet percussion of fountain water. Suddenly, not far from me, a squat pink church's bell began to toll. What did it say? I wondered. I would have liked to have had Francisco to ask.

6

A CLOCKWORK LANGUAGE

Even in the biggest cities of Europe and North America, you can go a lifetime without hearing or seeing a word of the international auxiliary language Esperanto. No *saluton* ('hello'), no *dankon* ('thank you'). No *ne* ('no'). The scarcity of speakers has always put Esperantists at a disadvantage. The irony of their situation is plain. They are left defending the principle of a universal idiom in the very languages Esperanto was meant to supplant. So they talk up their literature in the language of Proust, Rimbaud, and Sartre; resort to German to remark how efficient and precise Esperanto is; proclaim its internationalism and exult in its flexibility – all in English. They sing the praises of its euphony in Italian. *Il faut lire Baghy! Sehr Logische! Speak the World's Language! Una lingua bellissima! Esperanto!* But their enthusiasm is not persuasive. It seems to suffer in translation. Difficult, then, for them to press their case with composure. Incomprehension is the least of their worries; what they really fear is indifference. Indifference tinged with mockery. Esperanto? Didn't that go out with the penny farthing? They are frequently mistaken for crackpots.

During the spring, summer, and autumn of 2015, I got to know several Esperantists. *Denaskaj Esperantistoj* (native Esperanto speakers), they had been brought up speaking the invented language from birth. The speakers' names and email addresses came courtesy of various groups to whom I had sent brief enquiries in my own (quite elementary) Esperanto – a vestige of my years as an adolescent swot. There was Peter, a sexagenarian teacher born in Aarhus to a German father and a Danish mother, who proved to be a diligent correspondent. Another was Stela, a twenty-something whose other mother tongue was Hungarian. A few more would reply sporadically, half-answer a couple of my questions, and volunteer the odd detail, only to vanish behind their pseudonyms.

They were Esperantists replying to messages written in Esperanto, but Peter and Stela initially employed a strange English that omitted grammar and made their sentences gappy and curt. It was as though they wanted to keep the language to themselves. Each, I realised, was sizing me up. Disappointed in the treatment they had received from the media, they were wary of anyone unaffiliated with the cause. My credentials needed parsing. Who was I exactly? What did I want? My Esperanto, such as it was, went a long way toward reassuring them. Before long, their reluctance yielded to eagerness; they were suddenly keen to talk to me, someone who saw them as much more than fodder for copy. And Stela remembered her boyfriend, a Frenchman, reading my first book, *Born on a Blue Day*. *La blua libro*, 'the Blue Book', she called it.

Peter and Stela took me into their confidence. They switched to their mother tongue. I looked on their

Esperanto sentences with something approaching astonishment. It fascinated me how the two of them thought and conversed so easily and naturally, just as if their words had evolved, accreted, from head to head, from mouth to mouth, over millennia. 'To describe my Esperanto life is no easy task. So many experiences and memories. Which, among them, stand out? My mind resists. All my life the language has been a constant companion,' Peter wrote (in my translation – my Esperanto improved rapidly during our correspondence). Similar passages appeared in Stela's emails. They reminded me of a twenty-year-old feeling – a thought until now unexpressed – produced by the library book in which I first read 'iam estis eta knabo' ('once upon a time, a little boy'): this isn't some funny made-up lingo, but rather human language in one of its most recent and intriguing forms.

Ludwik Lejzer Zamenhof, who dreamed up Peter and Stela's mother tongue, was born in 1859 in the Polish city of Białystok, then a part of the Russian Empire. The son of a Jewish girls' tutor whose works included *A Textbook of the German Language for Russian Pupils*, young Ludwik grew up in a bookish, multilingual atmosphere. At home, he spoke Russian and Litvak Yiddish; at school, he acquired some French and German, conjugated Latin verbs, and deciphered texts in Greek. Polish, though frowned upon by the Tsarist authorities, he picked up from some of his neighbours. And since God spoke Hebrew, the boy learned to read its characters from right to left and to say his prayers with a good accent. Yet, despite this studiousness, he could not always follow what was quoted, laughed at, gossiped over on a street corner, nor purchase the family's bread from just any market stall; on the contrary, in a city divided along language lines, it required nothing more than taking a wrong turn or mistaking one face for another for Ludwik to suddenly feel himself a foreigner in the eyes of a Belarusian seller, a Lithuanian loudmouth, a Ukrainian passerby.

Zamenhof knew from experience that in Białystok, a foreigner was often an object of suspicion and anger. There was the time his shortsighted eyes were drawn past a window to a group of long beards trudging up the wintry road that ran beside his parents' house. Suddenly, snowballs coming from all directions pelted the stunned men a powdery white; the thud, thud, thud in the boy's head was like gunshot. Through the windowpane he watched as the snowballs continued to fly. He made out the shouts of 'Jewish swine!' that accompanied them, and his chest tightened. The shocked thought of the stones concealed within the snowballs made him wince. And when the throwers, out of ammunition, turned on their heels and fled, the boy heard them imitate the men's Yiddish in contemptuous snorts: 'Hra – hre – hri – hro – hru.'

'Hra – hre – hri – hro – hru.' The sounds galvanised Zamenhof's imagination. A clever and high-strung student, he felt set apart, touched by specialness, and now he became obsessed with the dream of a unifying language that would make such taunts, and the prejudices behind them, disappear. It was the first stirrings of a lifelong labour.

Much of what we know about the beginnings of Esperanto comes from the early Esperantist Edmond Privat, whose readable 1920 Esperanto-language biography of Zamenhof – Peter in one of his emails had recommended it – slips only occasionally into hagiography. 'Zamenhof had a head for outlandish ideas,' Peter admitted. 'The brotherhood of humanity, all that mystical stuff. Talking our way to world peace. Absurd! Of course, a shared language is no guarantee of mutual understanding. No panacea. Think of the former East and West Germany. Look at North and South Korea. Even we Esperantists are quick to quarrel. But he was on to something: an international language could at least carry ideas across seas, alleviate prejudice, broaden horizons. And he hit on just the right way to devise one that was both as simple as possible, and as complex as necessary, to put into words every human thought.'

Zamenhof reached this 'right way' through trial and error. He began by experimenting with a method first proposed by the Anglican clergyman John Wilkins in the seventeenth century and later written about in a little Borges essay: randomly assigning meanings to syllables. Each word arose from the steady concatenation of syllables and letters: the longer the word, the narrower the sense. Borges, in *The Analytical Language of John Wilkins,* informs us that *de* referred to an element in general; *deb,* 'fire'; *deba,* 'flame'. *Zana,* 'salmon', was a qualification of *zan,* 'river fish', which itself qualified *za,* 'any fish'. In a similar 'langue universelle', invented by Charles L. A. Letellier in 1850, *a* means 'animal'; *ab*, 'mammal'; *abo*, 'carnivore'; *aboj*, 'feline'; *aboje*, 'cat'. Zamenhof's own version of this scheme progressed no further than the sheets of school paper on which he jotted down line after line of unmemorable and barely distinguishable words.

Having failed at concocting an original vocabulary, Zamenhof busied himself with the idea of reviving his schoolmaster's Latin. He imagined populations once more orating as in the time of ancient Rome. But the language had those annoying declensions, which got his homework low marks; he would first have to simplify away all its endings, trim a noun like *domus* ('house') and its cohort of variants – *domuum*, *domōrum*, *domibus* – to a succinct *domo*. And then there were the cigars and the sugar cubes, the steam trains and the sewing machines, the bureaucrats and the rag-and-bone men for whom and for which no Latin word exists. The world had long since outgrown the Romans. And so, little by little, Zamenhof came to see his idea for what it was: another dead end. This real-isation, like his last, was a long time coming.

Finally, he settled on gleaning and blending words from all the languages that he heard at home, learned in school, or read in books. Each item of vocabulary would then be pressed into lexical moulds of the teen-ager's devising – the nouns cookie cut into an -*o* ending, the adjectives into an -*a* ending, infinitives cut to end in -*i* – so that the resulting coinages possessed their own look and logic: *kolbaso* (from the Russian колбаса meaning 'sausage'), *frua* (from the German *früh* meaning 'early'), *legi* (from the Latin *legere* meaning 'to read'). There remained, though, the danger that borrowing from many disparate sources might produce a linguistic mishmash. Zamenhof was missing a means to generate additional words from his coinages. The breakthrough came during one of his walks home from school, when a Cyrillic sign hanging outside a sweet-shop snagged the student's attention. *Konditerskaya*, literally 'a confectioner's place', bore a resemblance to the sign he'd once seen announcing a porter's lodge, *svejcarskaya* – literally, 'a porter's place'. The signs' resemblance, echoing the comparable function of the things they described, struck him as highly useful. Productive affixes like the Russian -*skaya* that marked various kinds of place, he realised, could give his nascent language further structure and help it cohere.

Words made out of such affixes dotted Peter's and Stela's messages. Esperanto contains a great many. In one email Peter wrote that he had been helping his *bofilino* ('daughter-in-law') move her things to a new address. A prefix, *bo-*, (from the French *beau,* indicating a relation by marriage) and a feminising suffix, -*ino, (filo* means 'son', *filino,* 'daughter') compose the noun. In another, he apologised for sending only a *mallonga* ('short') note. *Mal-* means 'the opposite of'. *Cerbumadi* ('to think about constantly'), an activity Stela indulges in – and a verb that I had never seen before – comprises *cerb-* ('brain'), -*um* (a suffix denoting a vague action), and -*adi* ('to do constantly'). Three composite words, and throughout my correspondents' pages there were dozens, scores, hundreds more.

Words that, in another country, another century, had been put together consciously, one by one by one, from scratch. To almost anyone else, the scale of such a task might have seemed impossibly discouraging. But once Zamenhof started, he found he could not stop. From time to time he must have paused to wonder whether he would ever tire of it, whether this mania would ever wear off, but that never happened. A schoolboy's hobby pushed to extreme lengths, the project consumed every minute of every hour of his free time. Day after week after month went into it. Alone though he was, the student was resourceful. He could be found flipping through the thickest dictionaries in the library, or compiling page after page of notes at his bedroom table, or asking his teacher unanswerable questions about which concepts might stick as well in a Malay's memory as in a Frenchman's. Every bout of progress in his tinkering, however small, however unstable, every newly conceived word or rule, gave him pleasure. *Anno*, his first pick for 'year,' he ultimately dropped in favour of *jaro*, swapping a Latin influence for a German – or perhaps Yiddish – one. His attempt at saying *and* – for a while *e* – turned to *kaj*, a rare Greek input, and made him shiver with delight. He was overjoyed to discover, single-mouthedly, that keeping the accent on a word's penultimate syllable helped with melody and ease of speech.

Mordechai Zamenhof did not share his son's enthusiasm. He hated the gibberish the boy came out with; he hated the many wasted hours diverted from studies – he had him down as a future doctor. Always on the edge of anger, he frowned and sighed and more than once gave Ludwik a stern talking-to. But Ludwik would not listen to reason. His refusal induced panic in his father. Would his eldest son shirk respectability for such a hare-brained scheme?

Mordechai spoke to his son's schoolmaster. In his long service of children's education, he would have seen many a boyish eccentricity, many a hobby gone wrong. The schoolmaster confirmed the father's fears. The lad hadn't a sane nerve in his brain. Only madmen behaved likewise, wasting their lives on pointless pursuits. A serious scholar knew better than to mix and muddle languages. Soon rumours flew around the school that the Zamenhof boy was definitely cracked.

Ludwik could not stop – but for two years he had to. The hiatus was occasioned by his being sent away to Moscow to read medicine. In his absence, bales of notes – every trace of his years of solitary labour – went to cinders in his father's fireplace. Mordechai was determined not to have his family made into a laughing stock. But what he destroyed, Ludwik, on his return, restored word by word from memory. He packed his folders and bags for Warsaw, where he set up a modest ophthalmological practice, and became engaged to a soap manufacturer's daughter. At last he was free. He divided his time between eyeglasses-fitting, fiancée-squeezing, and putting the finishing touches to his 'universala lingvo'.

In ten years, Zamenhof's creation had gone from a few lines to a language, complete with nouns, pronouns, verbs, proverbs, adjectives, synonyms, rhymes. The only thing it was now lacking were speakers. So he drew up a forty-page pamphlet – a guide like the one his father had written for Jewish girls wanting to talk like Fräulein – and published it in 1887 under the title (in Cyrillic) *International Language. Foreword and Complete Textbook (for Russian Speakers)*. He was only twenty-eight. His nom de plume was Doktoro Esperanto ('Doctor Hoper').

Peter and Stela, to my surprise, had never looked at Zamenhof's pamphlet, the original Esperanto primer. Not that they did not read in Esperanto – Peter mentioned owning or having owned *La Grafo de Monte-Kristo,* a translation of Dumas's *The Count of Monte Cristo;* a short story collection, *La Bato (The Boat)*, by Lena Karpunina; early Esperanto novels, such as *Mr Tot Aĉetas Mil Okulojn (Mr Tot Buys a Thousand Eyes)*, by Jean Forge, and *Sur Sanga Tero (On Bloody Soil)*, by Julio Baghy; and the poet William Auld's epic *La Infana Raso (The Infant Race)*, among others. Stela had long ago read her Esperanto edition of *Winnie the Pooh – Winnie la Pu –* to pieces. But like most contemporary Esperantists, the two were unaware of the primer's contents and unmoved by its history. And indeed, when I read a digitised version of the pamphlet, I saw that it had hardly aged well. Its opening sentence – the first published in the language – is the stiff and baffling: 'Mi ne scias kie mi lasis la bastonon; ĉu vi ĝin ne vidis?' ('I do not know where I left the stick; have you not seen it?'). And then there is the odd tone throughout: earnest, touchy, glib. Zamenhof, full of his setbacks and frustrations, had infused the pamphlet's list of vocabulary with them: fighting words like *bati* ('to flog'), *batali* ('to fight'), *bruli* ('to burn'), *ĉagreni* ('to chagrin'), *detri* ('to destroy'), *disputi* ('to quarrel'), *insulti* ('to insult'), *militi* ('to struggle'), *ofendi* ('to wrong'), *puni* ('to punish'), *ŝanceli* ('to stagger'), *trompi* ('to deceive'), *turmenti* ('to torment'), *venki* ('to vanquish'), abounded. A few years had sufficed to put many others out of date: *ekbruligu la kandelon* ('light the candle'), *kaleŝo* ('horse-drawn carriage'), *ŝtrumpo* ('stocking'), *telegrafe* ('telegraphically'), *lavistino* ('washerwoman').

But many saw the pamphlet differently at the time of its publication, having never seen anything of its like before. In the 1880s, voices the world over were crying out to be heard and understood. Geographic distance muffled them. Incomprehension among nations garbled them. So when the upper classes read the brand-new project in translation, they hailed 'Doctor Hoper' as a pioneer on the level of Bell and Edison. The pamphlet made them utopian. Esperanto quickly became a household word among the wealthy. Within a year or two, thousands were relearning how to say their name (*mia nomo estas* . . .), ask questions (*Ĉu vi* . . . ?), and count to ten: *unu, du, tri, kvar, kvin* . . .

Presently, letters addressed to the Doktoro in Warsaw started arriving. One correspondent complained that *'en mia urbo ĝis nun neniu ankoraŭ ion scias pri la lingvo Esperanto'* ('in my town to this day not one person knows anything about the Esperanto language'). Another wrote to ask whether it was correct to say *la malfeliĉo faris lin prudenta* ('misfortune made him prudent') or *la malfeliĉo faris lin prudentan* (Zamenhof replied 'both'), and went on to sniff that *'la signetoj superleteraj estas maloportunaj en la skribado'* ('the superscript characters are inconvenient when writing'). But most of the letters were laudatory. A gentleman from Birkenhead, England, asked to receive more Esperantist publications, a request echoed in mail from Philadelphia and Paris. A novice in Kiev wrote a single line to commend the project as *tre interesa* ('very interesting'). A dentist in Saratov proposed a translation of Gogol. Zamenhof must have been delighted by the sight of his words in other hands. And not long after he began receiving the letters, he heard his creation spoken back to him. Antoni Grabowski was the speaker's name. He was a chemical engineer, turning thirty, who had Zamenhof's moustache and goatee and also his passion for languages. He had examined the pamphlet with the near-sighted attention he usually reserved for blueprints. He was smitten. He became an instant convert to Esperanto. He took a train to Warsaw, sought out the author's house, and knocked on the door. A thin, little, prematurely bald man opened. Zamenhof listened with dazzled pleasure to the stranger's stuttering yet familiar sentences and welcomed him inside. Alas for history, their exchange went unrecorded – neither man thought to commit it to paper; for the following reconstruction I have leaned on a near-contemporary textbook by Grabowski on conversational manners:

'Ĉu vi estus Sinjoro Zamenhof?' ['Are you Mr
   Zamenhof?']

'Jes!' ['Yes!']

'Sinjoro, mi havas la honoron deziri al vi bonan
   tagon.' ['Sir, I have the honour of wishing you
   good day.']

'Envenu, mi petas. Sidiĝu. Kiel vi fartas?' ['Come
   in, please. Take a seat. How are you?']

'Tre bone, sinjoro, mi dankas. Kaj vi?' ['Very well,
   sir, thank you. And you?']

'Mi fartas tre bone.' ['I am very well.']

'Mi ĝojas vin renkonti.' ['I am delighted to meet
   you.']

'Vi estas tre ĝentila.' ['You are very kind.']

'Volu preni mian karton de vizito.' ['Please take my
   business card.']

'Ĉu mi povas proponi al vi kafon aŭ teon?' ['May I
   offer you coffee or tea?']

I can imagine Zamenhof saying something like this
out of habit, without reflecting, since outside books,
tea would still have been something of a rarity in
those parts.

'Kafon.' ['Coffee.']

'Ĉu vi deziras kremon?' ['Do you want cream?']

'Ne, mi trinkas nur kafon nigran.' ['No, I only
   drink black coffee.']

And after the niceties, the two men presumably got
down to business. Probably they talked about the
pamphlet, discussed ways to further the cause, or went

over some of the finer points of Esperanto grammar. They would have refrained from straying into politics, let alone matters of the heart. The conversation would have talked itself out after half an hour or so.

The letters! The conversation over coffee! Fervently, Zamenhof believed that his was an idea whose time had come. He pictured mouths by the thousand, million, rushing to pronounce his every sentence, foresaw the irresistible triumph of his utopia. He conceived it to be a revolution. Instead, it became merely a fashion. British publishers, angling for column inches, put out tiny print runs of *Teach Yourself Esperanto* phrasebooks. Continental hosts who wanted to impress their guests recited a few lines over wine and canapés. American dabblers in table-turning and simplified spelling switched fads. Every effort failed to turn these fair-weather speakers into dues-paying members. Most would not put their money where their mouths were. Most soon forgot the little they had once so enthusiastically learned. Months passed, years, and still no utopia. Zamenhof found himself with hardly a kopeck in his pocket; Esperanto had eaten all five thousand rubles of his wife's dowry.

His health was equally poor. He had to think about his heart. His heart was bad. People were surprised when he gave his age. Too much tobacco. Too little sleep. He had next to no appetite. He was always correcting some error-strewn letter, pulling together an exhorting sermon, thinking up another coinage. His life was one long language lesson.

Yet, increasingly, it must have seemed to Zamenhof

that the teaching was being taken out of his hands. Ten years after his pamphlet, the centre of the Esperanto-speaking world had moved westward to France. France became an economic lifeline: its intelligentsia subscribed many francs for the publication of periodicals. But the intellectuals also proposed to edit them as they saw fit. Strikingly ugly they thought the diacritics, those circumflexed *h*'s and *s*'s and the other supersigned characters, which continue to give today's keyboards hiccups; they wished to get rid of them. Zamenhof said he was willing to entertain the possibility. And there were other things the patrons wanted changed. They wanted the language to look more like Italian: don't say *du libroj* ('two books') but *du libri*. Again, Zamenhof said he could live with the idea. Adjectives, they added, should no longer have to agree with the nouns they described: don't say *grandaj libroj* ('big books') but *granda libri*. Nor, for that matter, should the object of a sentence require an *n*: not *mi legas grandan libron* ('I'm reading a big book') but *mi legas granda libro*. Zamenhof demurred. Dropping the diacritics and modifying words he could accept in principle; altering the way sentences worked, however, he could not.

The squabble continued on and off for years. In the end, the movement split into a mainstream of conservatives, led by Zamenhof, and a minority of reformers – tinkerers, to the conservatives – determined to do away with what they thought of as the creator's mistakes. The conservatives did away with them instead. Esperanto survived, but Zamenhof's optimism was shaken. In 1914, World War I broke out. Three years later, he died.

Peter: 'One day, not long after the Great War, the village chimney sweep told my father all about an "easy-to-learn international language". My father had little schooling, but he could read, so he looked up and found an Esperanto course in the pages of the popular encyclopaedia *Die Neue Volkshochschule*.' The four-volume work also taught its readers stenography, graphology, hygiene, sport, and art, among other topics. 'Later, he moved to the city, Hamburg, and became a policeman; he was active on the Esperanto scene there.'

Weimar Germany, in the 1920s, was a land of strikes and uprisings, anger and hunger. The country was rife with anti-Semitic feeling. For German speakers of Esperanto, whose creator had been a Jewish Russo-Pole, it was a hard place to live. Racist slurs needed ignoring whenever they exchanged greetings in a busy street, shared news and anxieties, or gathered in someone's apartment or a public hall to commemorate Zamenhof's birthday. As early as 1926, Peter's father could have read in the national weekly *Der Reichswart* that Esperanto was a 'freak of a language, without roots in the life of a people . . . part of Zionist plans to dominate the world, and aid Zion's slaves to destroy the Fatherland'. During a 1928 debate in the Bavarian state assembly about the possible introduction of Esperanto courses into German schools, the National Socialist deputy Rudolf Buttman put on record his contention that Esperanto was 'a Jew's cobbled-together language, a thorn in the side of German culture'. The

National Socialists' own newspaper, the *Völkische Beobachter*, complained in 1930 that some Germans were 'babbling' the language of 'the bloodsucking internationalists'.

Internationalist tendencies. It was for these that Peter's father was thrown out of the police force in 1932. Even among his fellow Esperantists he could find no refuge. Many, trying to accommodate themselves to the new regime, aping its bigotry, expelled 'Jews, pacifists, profiteers' from their clubs. The clubs did not last long: in 1935 the National Socialist Minister Bernhard Rust closed them all down. Around the same time, a ban was enforced on letters written in either Esperanto or Hebrew. By then, Peter's father had fled for Holland, then for Norway, then for Denmark, where he taught beginning Esperanto for a living. 'One of his students was my future mother. They married in 1937. After the German occupation, my father became an army chauffeur. He drove the military brass around and kept his head down. When the war started going wrong for the Nazis, he got called up, but he managed to wriggle out of it. His speaking Danish saved him: an interpreter was far more valuable to them than simply another pair of boots.'

Peter was born in his mother's hometown in 1947. In the following year, his father moved back with his wife and son to Hamburg. In Hamburg, the father returned to his policeman's green uniform, his daily rounds, and his Esperanto club, where he taught evening classes. The family home was like a continuation of those classes: Peter would sit with Eliza, his little sister, eating not *Brot*, *Liverwurst*, and *Marmalade* at the kitchen table but *pano, hepatokolbaso,* and *marmelado*. Brother and sister learned to speak their Esperanto with the father's earthy, *Plattdeutsch* accent. 'His Esperanto skimped on any finery. No wonder, since my father hailed from a little burg inhospitable to education. His speech lacked the nuances of the well-schooled man.' From their early childhood, Peter and his sister were regular clubgoers. He recalls smiling inwardly at the adults' blunders. 'On rare occasions, I would even catch my father – a very active instructor – teaching some point of grammar inaccurately (I didn't dare say anything).'

In one of his messages, Peter attached a colour photograph of himself as a small, blond, rosy-cheeked boy. The boy is dressed in a tie and beige shorts, his white socks pulled right up to the knees. To his left, in the centre, stands his father, rosy-cheeked and dark-suited, displaying proudly a child's drawing of the literary character Struwwelpeter. Behind Peter, a Christmas tree glitters. On the far left, his public: girls and women who sit or stand and beam. 'I was five. I had had to learn by heart the story of Struwwelpeter in Esperanto.' At his parents' club, he recited poems, sang songs, performed sketches. Sometimes foreign Esperantists came to visit. 'If my memory serves me right, I spoke fluently with them – as fluently as any small child could.'

Outside the home and club he was shyer. 'Once, during a tram ride, my parents and I were talking together in Danish (my mother always spoke to me in her native tongue) when a Danish passenger joined in. I did not like the sound of him. I turned my back on the interrupter and carried on in Esperanto.'

He began attending international Esperantist gatherings at the age of nine. 'In 1956, I took part in the first Children's Congress. It was held in Denmark.' Peter mixed with some thirty other children. One of the boys, about his age, had flown in from Texas. The Texan wore a cowboy suit, with a red sombrero on which someone had embroidered an eagle. The two started talking. 'That we chatted in Esperanto seemed normal – in no way remarkable.' The boys shared a group visit to the zoological garden in Copenhagen. 'But all that time I could not take my mind off that hat!' After returning to Hamburg, Peter's parents took him as usual on their weekly visit to the club. 'Turned out the boy's family had dropped in there on their way home to the States; they had left the cowboy hat there for me as a gift. I was very proud.'

Before his teens he had never given a thought to the language. 'For the first time I became aware of its rules. I began speaking more slowly, asking myself whether I had to cap this or that word with an -*n*.' More and more, he found himself needled by grammatical scruples: should he say such-and-such sentence like this or like that? How could he know for certain whether what he was saying made complete sense? It was a phase he seems to have quickly grown out of. 'I began to travel by myself. My confidence increased. I spent six weeks holidaying in Finland. I had no Finnish. My Danish and German were useless. And yet I never felt like a tourist. Unlike campers who have little contact with the country, I ate and slept in the homes of Finnish Esperantist families. My Esperanto was a passport to the world.'

Peter has spent the greater part of his life travelling. 'I have spoken Esperanto in Germany, Denmark, France, Finland, Sweden, the Netherlands, Poland, the Czech Republic, Slovakia, Slovenia, Croatia, Serbia, Bosnia and Herzegovina, Austria, Australia, Britain, Spain, China, Nepal, Iceland, and the United States.' In Croatia, he met his first wife. 'It was during an Esperantist seminar. She spoke no German, and I no Croatian.' Her Esperanto charmed him. 'The Serbs and Croats are the best pronouncers.' They whispered sweet nothings, gossiped with each other, disputed memories, all in the language. Their marriage was childless; Peter later adopted a boy and a girl. 'My son lives only for football. As for my daughter, she was already an adult; I only managed to teach her a smattering of words. I suppose, from the movement's viewpoint, I haven't been a very successful Esperanto parent.'

His sister has had more success. He told me, 'My niece and nephew are both native speakers. But, from what I can gather, the boy has let his Esperanto go. My niece married a native speaker but they aren't active in the movement.' He and his sister are no longer on speaking terms.

In all our exchanges Peter struck me as a man who lived for the cause. He had written many polite, rectifying letters to the papers. 'For many years I subscribed to *Die Zeit,* and would write in from time to time about Esperanto. An editor finally replied to inform me that the paper's policy forbade him to publish anything positive about the language. Times change, though, and gradually mention of Esperanto is creeping back into print.' He was not, he insisted, a proselytiser. 'I don't conceal my convictions. I often wear a lapel pin showing the movement's green flag. But I don't go in for preaching.'

At fifty, he decided to return to his birthplace; he now lives an hour's drive from Copenhagen. 'I can go for days without speaking a word of Esperanto. The nearest club is in the capital, and I only rarely venture over. The language is more for the telephone, and the computer.' He splits his life between his three native tongues. 'I do all my sums in German. How I jot down a grocery list depends on where I shop: here in Denmark, I use Danish. When I visit Germany, German. A note reminding me to buy such-and-such Esperanto book I make in Esperanto.' And some nights, he dreams in Esperanto. 'They seem to crop up after I've spent several days at one of the various get-togethers.'

Does he believe the language has a future? 'I hope so. It's like a legacy passed down through the gener-ations.'

Stela, a sociologist in her twenties, is part of the new generation of native speakers. Stela is her Esperanto name. Hungarians call her Eszter. Her father was a French schoolteacher; her mother, a Budapest-based commercial translator. Both were dedicated Esperantists. Her father, though, had another life back in France. 'He only visited my mother and me from time to time.

You might say he was a character. His Esperanto was good, not perfect; he was always dropping his *n*'s.' Stela's came from her mother, who learned to speak it faultlessly at university.

Stela, unlike Peter, remembers her first word. 'My mother was pushing me somewhere in the buggy when I saw an oncoming tram, raised my index finger, and shouted *Vidu!* ("Look!").' Hungarian she learned to speak at school. 'I never understood why my class-mates thought my speaking Esperanto with my mother was so curious. Some of the other children spoke foreign languages at home: Russian, for example, or German.' At times she had trouble telling her two languages apart: 'I would get my words mixed up. I would say *pampelegér* for "grapefruit": a combin-ation of the Esperanto *pampelmuso* and the Hungarian *egér*.'

Her bookish mother's daughter, Stela read and read and read. Her bedroom shelves were crammed with colourful Esperanto books. One of her favourite titles, she told me, had been *Кумеўаўа, la filo de la ĝangalo* (*Kumewawa, the Son of the Jungle*), by the Hungarian Esperantist author Tibor Sekelj. She told me this quite offhandedly, without preliminaries, as though the story were a classic. And, when I googled for more details, I discovered to my astonishment that it was. A children's adventure yarn, first published in 1979, and many times translated, it has enjoyed a long career throughout the globe's bookshops. In Tokyo it was apparently once as big as *The Little Prince*. I managed to track down the original summary:

On the river Aragvajo, an affluent of the Amazon, a group of tourists is in danger. Their ship has sunk. A good thing Kumewawa steps forward to help them. 'Fish we measure according to their length; men, according to their knowledge.' Kumewawa is only twelve years old, but belongs to the Karajxa tribe and knows everything there is to know about life in the jungle.

But despite Kumewawa's escapades, the sinking ships and the daring rescues, Stela, like many of her generation, slowly lost interest in reading. 'I wouldn't recognise one of Zamenhof's proverbs or expressions. I don't have that culture. Some learners might, but for me the culture resides in my family and friends.' Curiously, her mother ceased using the language with her once she turned twenty. 'She simply said to me: 'I've taught you all I know.' Before then the only time she spoke to me in Hungarian was when she lost her temper.' It is to her friends in the movement, her second family, that she feels closest. 'Growing up, I failed to understand why the other Hungarians considered their summer holidays as purely family time. I always spent my summers overseas with other Esperantists.'

Both Peter and Stela wrote warmly of the movement's various international meetings: in clubs, congresses, and other get-togethers. It was clear, from their differing accounts, that the meetings had varied over the years. In Peter's day, they had been formal, requiring ties and good manners. Today, bare feet and bottles of plonk are everywhere to be seen. Peter's generation had organised sparkly balls. Stela's slaps tables with playing cards.

Peter confided that the food, which students prepared for the attendees, was consistently bad: all grease and wilting vegetables; probably the badness of the food is one of the few things that hasn't changed.

Certainly, Esperanto has changed. Its evolution is one of the things that animated Peter and Stela most. The evolution betokens an adaptive, natural-like language. Some changes, the smallest, have been motivated by technological advances: Peter long ago stopped saying *kasedaparato* ('cassette player'); the same sounds, with few alterations, he recycles these days to talk about the *kafaparato* ('coffee machine'). Text messaging has abbreviated the most common words: Stela texts *k* for *kaj* ('and'), *bv* for *bonvolu* ('please'), *cx* for *ĉirkaŭ* ('about'). Other modifications are cultural, born out of a youngster's desire not to sound like Granny and Gramps. For Peter, something excellent is simply *bonega* (literally 'greatly good'), whereas for Stela it is *mojosa* ('cool'). 'I was at the teen meetup where it was first spoken,' Stela told me. The new word caught on fast. 'Now at every youth event it's mojosa this and mojosa that.'

The biggest and least reported shifts have been internal. Shades of meaning have progressively changed words. A hundred years ago, owing to the influence of Russian, boats and ships *naĝis* ('swam'); nowadays, they sail. The example is Jouko Lindstedt's, an Esperantist linguist. Another such case concerns the word *versaĵo* (literally 'a piece of verse'), which has long since been supplanted by *poemo,* a term that originally referred only to the epics. Also, the way words link up to convey more complex meanings has altered over the past hundred years, the phrases becoming progressively shorter. Zamenhof would have said something like *ĝi estus estinta ebla* ('it would have been possible'); Peter, to express the same thing, says *ĝi estus eblinta;* Stela says *ĝi eblintus.* And with this evolution, the idea of a good Esperanto sentence – one that feels more Esperanto than another – has gradually emerged. Claude Piron – an influential figure in the movement – discouraged a sentence like *en tiu epoko li praktikis sporton kun vigleco* ('at that time he practised sport with vigour') in favour of the swifter *tiuepoke li vigle sportis* (literally 'that-time-ly he vigorously sport-ed').

Swifter, perhaps, but Piron's sentence, like most in the language, remains confined to paper or a computer screen. Esperanto has always been an overwhelmingly written language. 'Quite often, I require a few minutes to get into my normal flow when I speak Esperanto,' Stela admitted. 'I'm forever writing in it, but speaking and writing aren't the same thing.' That is why the congresses were so important to her and Peter. They provided momentary spaces in which to converse, but such conversations could hardly be altogether satisfying. Or natural. Many of the attendees were eternal beginners, forever glancing down at their notes as if standing at a podium. Misunderstanding was only ever a fluffed vowel away.

It is one thing to flex syntax. It is another thing to understand what is being said or written. Ken Miner, another Esperantist linguist, has written several papers – in Esperanto – in which he highlights ambiguities.

Does the simple-looking sentence *mi iris en la ĝardenon* mean 'I went into the garden' or 'I was on my way into the garden'? Peter, when I asked him, thought the former, but Miner said quite a few other speakers, even the most experienced, disagree. In another paper, Miner discusses the suffix *-ad*. Textbooks tell learners that it indicates duration: *kuri* ('to run'), *kuradi* ('to run and run'). But Miner, burrowing into Esperanto literature, found plenty of sentences that contradicted the rule: *li atendadis dum horoj* ('he waited and waited for hours') but *pacientoj atendas dum monatoj* ('patients wait for months'), where the first uses *-ad* to describe a wait lasting hours, but not the second for one that goes on and on for months.

These ambiguities and inconsistencies, explains Miner, are the result of non-natives having shaped Esperanto. Without a native speaker's intuitions, Zamenhof and his first followers had had no choice but to rely on logic; languages, however, by their nature, are often illogical. The users' judgments, each under the sway of their respective grammars (Polish, Russian, English, French . . .), had clashed. To this day, they still do. Not even Peter and Stela can tell for sure whether certain sentences are right or wrong. Their intuitions, forged in a non-Esperanto-speaking society, are unreliable. Peter, at various points, told me as much. 'Don't think that native Esperanto speakers automatically speak flawlessly! There exist notable counterexamples.' Much depends on the parents' degree of fluency. Some touted as natives, it turns out, speak only a kitchen Esperanto.

Miner's research, Peter's remarks: they came as something of a surprise. They ran counter to the confidence with which the movement promotes its 'easy-to-learn' language. Peter, though, was unsurprised by my surprise: 'About the average level among learners, I can only speak from experience. Most have little grasp of the language. Esperanto isn't as easy as some of our adepts like to affirm.' Quirks and contradictions, no different from those that learners of any language encounter, are to blame. The *Plena Manlibro de Esperanta Gramatiko* (the complete handbook of Esperanto grammar) spans some seven hundred pages. And when I looked again at the assertions the movement had published – 'For a native English speaker, we may estimate that Esperanto is about five times as easy to learn as Spanish or French, ten times as easy to learn as Russian, twenty times as easy to learn as Arabic or spoken Chinese, and infinitely easier to learn than Japanese' – I felt that they were only salesman words, salesman numbers, without foundation.

One hundred and thirty years ago, Zamenhof's stated goal had been ten million speakers – ten million to start with. When asked by an Associated Press reporter in 1983 to estimate the global number of speakers, the movement's president, Grégoire Maertens, replied, 'I usually say 10 million people read and understand Esperanto. Speaking is another thing. But then, how many people speak their own language correctly?' I believe this is wishful thinking. I cannot take seriously the claims made in the papers, that more speak Esperanto than speak Welsh or Icelandic; that it is on a par with Hebrew and Lithuanian. And none of the promoters seem to have taken into account population inflation: ten million in 1887 would be fifty million in 2017. No article, though, has printed a figure of fifty million, not even in the most enthusiastic columns of the Esperantist press. In linguistics circles, the actual figure of active, competent Esperanto speakers is estimated much, much lower. Ken Miner and his colleagues put it at fifty thousand. Fifty thousand, give or take, is the number who speak the Greenlandic Kalaallisut.

Humans everywhere use language, but only rarely choose which. Each is some person's birthright. Peter and Stela were born into Esperanto, two of probably no more than a thousand such speakers. So why have the others, several tens of thousands, volunteered to master it? Not for any practical purpose: English, natural to hundreds of millions, has become the world's primary medium of communication. Nor is it because Esperanto learns itself: it is as complex and capricious as any other language. Deeper needs – to stand out or apart, to flatter the conscience, to integrate into a tight-knit community – lie behind the learner's decision. Esperanto simplifies the world, neatly divides it into two: on one side, all the men and women in the dark, those who have stopped up their ears, the trucklers to the powers-that-be; on the other, those who can truthfully say 'Mi parolas Esperanton.'

A comforting abstraction, but Peter, at least, is old enough to know better. Once, during our many exchanges, I broached the subject of favourite words. I had unearthed a survey conducted among Esperantist writers, poets, and other personalities, and sent it to him. I thought it a way to get him talking about Esperanto's emotional charge. I was wrong. Peter wrote back that words, by themselves, in whatever language, could never do justice to the world. It was what Stela had meant when she explained that her culture resided in her family and friends. 'These sorts of lists leave me cold,' he wrote. 'Why should *mielo* ["honey"] be considered a more beautiful word than *marmelado?* I have never seen or heard a *najtingalo* sing. But two winters ago, here in the Danish countryside, not far from my home, I taught a wild pheasant to eat out of my hand. For me, *fazano* has a much finer ring than *najtingalo.*'

7

THE MAN WHO WAS FRIDAY

Ngũgĩ wa Thiong'o, the Grand Old Man of African letters, is telling me how he wrote *Caitaani mũtharaba-Inĩ* (*Devil on the Cross*), the first Kikuyu-language novel, on prison toilet paper.

When did it all begin? The midnight before New Year's Eve, 1977. I was with my family in Limuru when armed policemen came and shoved me into a Land Rover. It was an abduction: there was no reason in law to drive me away. Next day, Saturday, they put me in chains. They took me to Nairobi, to the Kamiti Maximum Security Prison. I lost my name, Ngũgĩ. The prison officers called me only by the number on my file: K677. We, the other detainees and I, were considered political dissidents by the Kenyatta regime. For one year I was held without trial. My daughter Wamuingi was born in my absence. I received the news in a letter with a photograph enclosed. The first weeks were the hardest. I felt very lonely. But the other prisoners knew me and my work. They were encouraging. Some of them had books they lent me to read. Dickens. Aristotle. Some gave me biros and pencils as gifts. And there was another source of motivation to write. One day a warder complained to me that we educated Kenyans were guilty of looking down on our national languages. I could not believe my ears! That night

I sat at my desk in my cell and let the beginning of a story in Kikuyu pour out of me. For paper, of course, I had only that provided in bundles by the authorities to meet a prisoner's bodily needs. The toilet paper was thick and rough, and intended to be rough, but what was bad for the body was good for the pen. The bigger problem was the language. My Kikuyu looked funny on paper. I had no experience of writing at such length in Kikuyu. Once again, my friends in the prison were very kind, very helpful. They helped me to find the right word for this scene or that character. They taught me songs and proverbs I hadn't heard before. I wrote the story's final sentences only days before learning that I was to be released. The book was published in Nairobi two years later, in 1980.

Having bared his prison story to me, its trauma and its triumph, he smiles, looking younger than his seventy-seven years, and returns to his plate of *foie gras poêlé avec tartare de légumes*. We are talking over lunch in the French coastal town of Nantes. May in Nantes is already summery. We are on the busy terrace of a restaurant a short walk from Ngũgĩ's hotel. Below the terrace, literary festivalgoers, between events, traipse or cycle along the canal; a few look up when Ngũgĩ stands abruptly and steps out of the parasols' shade to the railing to take a call. His black shirt is embroidered with a bright golden-elephants design; his baggy, beige slacks finish in a pair of black, worn-out trainers. His Kikuyu (the indigenous language of over six million Kenyans) – fast and loud and emphatic – so unlike his English of the previous minutes, so unlike the French of the surrounding tables, fills the air. Nantes, as the town's statues and street names remind us, is the birthplace of Jules Verne. Is Ngũgĩ, to the festivalgoers on foot or bike who steal looks as they pass, 'Africa'? Verne's Africa?

Immense brambly palisades, impenetrable hedges of thorny jungle, separated the clearings dotted with numerous villages . . . Animals with huge humps were feeding in the luxuriant prairies, and were half hidden, sometimes, in the tall grass; spreading forests in bloom redolent of spicy perfumes presented themselves to the gaze like immense bouquets; but, in these bouquets, lions, leopards, hyenas and tigers, were crouching for shelter from the last hot rays of the setting sun. From time to time an elephant made the tall tops of the undergrowth sway to and fro, and you could hear the crackling of huge branches as his ponderous ivory tusks broke them in his way. (From *Five Weeks in a Balloon*.)

Ngũgĩ is on his way to Kenya. (Nantes, for its international writers' festival, is a brief stopover.) He has lived in exile in the West, mostly in the United States, where he has taught comparative literature at the University of California, Irvine, since 1982. Trips to his homeland are few and far between, and this one – his first in over ten years – might be his last. So it is a piece of luck, his coming here (a two-hour train ride from Paris), and my learning of it in time; and it is generous, or indulgent, of him to agree to meet with me to talk language and language politics when he must have many things on his mind.

In much of Kenya, where English, alongside Kiswahili,

is the official language, his politics are controversial. Ngũgĩ has long argued that African authors should write and publish in African languages ('Why should Danish, with five million people, be able to sustain a literature but not the Yoruba, who are forty million?'); that African intellectuals should reason and debate in African languages. English, he insists, is not an African language. It is an argument that has set him apart from many of his contemporaries. The Yoruba author Wole Soyinka wrote his plays and poems in English; Chinua Achebe, an Igbo, wrote his novels and essays in English. In his 1965 essay 'The African Writer and the English Language,' Achebe pointed out that his was 'a new English, still in full communion with its ancestral home but altered to suit its new African surroundings.' An Igbo English. But for Ngũgĩ, such defences fail to hold up. Soyinka and Achebe, he thinks, write as Africans estranged from their languages by the legacy of European colonialism. They write as white-cloaked surgeons whose African characters speak with transplanted tongues.

Consciousness of language politics came early to Ngũgĩ. In 1952, when thousands of the dispossessed Kikuyu began rising up against the colonial administration, he was a schoolboy of fourteen. The violent revolt by Kenya's largest ethnic group panicked the British into declaring a state of emergency. All nationalist-run school administrations across the land were fired and replaced with ones sympathetic to the Crown. The sudden changes deprived Ngũgĩ and the other rural children – pickers of tea leaves – of their indigenous-language curricula. Instead of traditional songs, the pupils read *Robinson Crusoe;* instead of tales learned around a village fire, Shakespeare. Ngũgĩ, whose Kikuyu compositions had once been his teacher's pride, saw his classmates punished for speaking the language. They were struck with a cane or made to wear metal plates around their necks bearing the words 'I am stupid' or 'I am a donkey.' Lucky, resolute, he himself was never a donkey. To his surprise and relief he took at once to the English books. English, in his skilful hands, would bring him various student prizes and, in 1964, to Britain on a scholarship to study at the University of Leeds.

I don't tell Ngũgĩ, when he returns from his call, my admiration for his first novels: *Weep Not, Child, The River Between,* and *A Grain of Wheat,* a trio originally published in London in the 1960s by the Heinemann African Writers Series. I don't tell him how the deceptively simple, evocative opening to *The River Between* gets me every time, a pleasure that no amount of rereading seems to dim.

The two ridges lay side by side. One was Kameno, the other was Makuyu. Between them was a valley. It was called the valley of life. Behind Kameno and Makuyu were many more valleys and ridges, lying without any discernible plan. They were like many sleeping lions which never woke. They just slept, the big deep sleep of their Creator.

I don't tell him, because Ngũgĩ would later call these works – which he wrote in English and published under his childhood name of James Ngũgĩ during his years in Leeds – his 'Afro-Saxon novels', the better to distance himself from what he considers to be their relative inauthenticity compared with his subsequent writings in Kikuyu. To write in Kikuyu while imprisoned was, for him, an act of autonomy, of self-determination. More than sounds and stories linking him to his formative years, the language, in Ngũgĩ's mind, became a repudiation of the country's English-speaking elite. The gesture was brave (he was always having to squirrel the toilet paper manuscript away from guards); it was eloquent. And yet, something in Ngũgĩ's disregard for his earlier work, his argument about English in Africa rankles with me. I voice my thoughts.

'You wrote your first, *the* first, novel in Kikuyu in the 1970s. But hasn't the relationship between Africa and the English language continued to change since then? What about the rise of post-independence Africans who feel completely comfortable in their English, authors like Chimamanda Ngozi Adichie and Chris Abani?'

(I mention Abani for a reason. We once met at a writers' event in the United States and discussed the 'language question'. Abani grew up in Nigeria. His father is Igbo. His mother is British. So English, in which Abani writes, is for him a native language. And it was for writing, in English, what were construed as dissident texts by the English-speaking Nigerian government, that he was put behind bars at the age of eighteen and then on death row. His release two years later was like a rebirth. But he never thought about writing in Igbo; it would have felt inauthentic, he said.)

Ngũgĩ says, 'These authors have an inheritance. They should contribute to it. They have a duty to protect it.'

And, listening between the lines, I think I understand the two ideas of inheritance that Ngũgĩ's word contains. The first, postcolonial, says African authors must take up pens to show that *third-world* doesn't mean 'third-rate', that African languages are the equals of their European counterparts: equally rich and complex and artful. The second, ancestral, says language is a natural resource of collective memory: a unique way of being in and knowing the world.

'There's now a whole generation of young people in Africa who, through no fault of their own, do not speak their African mother tongue. You might say, "English or French is their mother tongue." No. English isn't African. French isn't African. It isn't their fault, but that's how it is. What would I say to a native English-speaking African? If you are born in an English-speaking household, there's no reason for you not to learn an African language at school. Kiswahili. Kikuyu. Igbo. Yoruba. Whatever. And then use your English to translate works into African languages.'

There is nativist rhetoric in Ngũgĩ's references to 'African mother tongues' and 'African languages'. It reminded me, a little uncomfortably, of things I had heard in Britain about the decisions of several councils, with their large immigrant communities, to erect road and street signs in other orthographies: Polish, Punjabi. Those opposed to the signs had declared that neither Polish nor Punjabi was a British language. That, too, was intended to pass for an argument. But terms such as 'African languages', like 'British languages', make little, if any, sense, linguistically speaking. They are proxies for personal opinions, depending entirely for their meaning on whom you ask. I suppose every African draws the line somewhere, and Ngũgĩ draws it at English and French. Fair enough. But then, Ngũgĩ's position, I learned, is less personal than it is ideological, with the incongruities that ideology brings. For example, does he consider Kiswahili – a widely spoken Bantu lingua franca composed of many Arabic, Persian and Portuguese influences – African?

'Yes.'

'Arabic?' Language of the continent's first colonisers, ivory and slave traders.

'Yes.'

'Afrikaans?'

He hesitates. 'Yes.'

I try to get him to divulge his reasoning, but it isn't easy. He only says, by way of explanation, that many poor black children in South Africa speak Afrikaans. I do not press him. Already his knife and fork hover over the remains of the foie gras.

Ngũgĩ's thinking can be traced back to the mid- to late 1960s, his postgraduate student days in Leeds: the heady mix of Marx and Black Power. *Lumpenproletariat! Black Skin, White Masks!* No longer would he be the young author who could write, in a 1962 opinion piece for Kenya's *Sunday Nation,* 'I am now tired of the talk about "African culture": I am tired of the talk about "African Socialism" . . . Be it far from me to go about looking for an "Africanness" in everything before I can value it.' That was before Leeds. Now, on his return to Nairobi, a thirty-year-old professor at the capital's university, he proposed to abolish the English department in favour of a department of African literature and languages. He renounced his schoolboy Protestantism, and with it the name *James*, of his first books. His writing grew darker, more didactic.

In his 1985 essay *On Writing in Kikuyu*, Ngũgĩ describes the precursor to his imprisonment: the six months he spent in 1977 working with the villagers of Kamiriithu to stage the play *Ngaahika Ndeenda* ('I Shall Marry When I Want'), about greedy imperialists who exploit brave Kikuyu peasants. In keeping with his collectivist thinking, he intended the project to be a group effort. The men and women of the village were duly quick to correct his numerous Kikuyu mistakes: 'You university people, what kind of learning have you had?'

'I learned my language anew,' Ngũgĩ concedes. But he was proud, with much to feel proud about: The participatory theatre project did a lot of good. Some of the villagers learned to read and write. Others, feeling valued, having something for their minds to do, drank less, so that alcoholism, the scourge of many Kenyan villages, was alleviated. Several discovered in themselves an actor's bravado or other talents that might otherwise have stayed forever dormant. The popular enthusiasm was such that the play's rehearsals immediately drew large crowds. Hundreds, then thousands, from the surrounding villages, came and sat in the open air – and listened, shouted, clapped, laughed, booed. It was enough to make the authorities jumpy. Ngũgĩ became a man with enemies in high places.

The play was hastily banned; the village theatre group disbanded, its grounds razed. And on the night of December 30, 1977, armed policemen with an arrest warrant rode past maize fields, tethered goats, and chicken coops to the only house in Limuru with a telephone wire.

Hoping, by incarcerating Ngũgĩ's voice, to silence him, the regime unwittingly made him a cause célèbre. In London, members of the Pan African Association of Writers and Journalists massed before the Kenyan embassy and held up placards that read 'Free Ngũgĩ.' A letter to the editors in the June 1978 issue of the *New York Review of Books* urged governments to push for the author's release; its signatories included James Baldwin, Margaret Drabble, Harold Pinter, Philip Roth and C. P. Snow.

It is surprising, then, that after his liberation Ngũgĩ was allowed to publish his novel. Perhaps the regime intended to placate the international uproar the author's imprisonment had provoked. Perhaps, the Kikuyu was simply too much work for censors used only to reading in English and Kiswahili. Perhaps, going only on the book's original cover, a cartoonish drawing that showed a paunchy white man hanging from a dollar-studded cross, the regime thought the book nothing more than an anti-imperialist tract. (Indeed, the title, *Caitaani mũtharaba-inĩ*, can be understood as 'The Great Satan on the Cross').

Whatever its reason for allowing the novel's release,

the regime underestimated the Kikuyus' appetite for a publication in their own language. The story of a young village woman who confronts the seedy foreign-money corruption in her country proved popular. In his essay, Ngũgĩ relates how the book was read aloud by literate members of a family to their neighbours, all ears. In bars, he writes, a man would read the choicest pages to his fellow drinkers until his mouth or glass ran dry – at which point a listener would rush to offer him another beer in return for the cliffhanger's resolution.

Like most of Ngũgĩ's international audience, I don't know Kikuyu and had to content myself with the author's own English translation of the story. (Other translations have appeared in Kiswahili, German, Swedish – and Telugu, in a version created by the activist poet Varavara Rao during his own prison stint.) So I ask him now if he could teach me a few Kikuyu words from the book. He sees me readying my pen and notepad, and – so promptly and proprietorially that I'm taken aback – he reaches over and lifts them from my hands. He says, as my pen in his hand writes, '*kana,* infant.' It could also mean 'fourth' or 'to deny' or 'if' or 'or', depending on how you say it. *Turungi,* he says, is a very Kikuyu word. 'It means "tea".' *Kabiaru,* another 'very Kikuyu word', is 'coffee.' In Kenya, home of tea leaves and coffee beans, I can easily see how these words might have associations – social, cultural – that are particular to the Kikuyu imagination. But even if they do, both, as Ngũgĩ goes on to explain, happen to be English imports: *turungi,* from 'true tea' and 'strong tea'; *kabiaru* from 'coffee alone', meaning black, no milk.

(And, later, with the help of Kikuyu scholars, I will discover just how much vocabulary Ngũgĩ's Kikuyu works – plays, novels, children's stories – owe to English: *pawa* ['power'], *hithituri* ['history'], *thayathi* ['science'], *bani* ['funny'], *ngirini* ['green'], *túimanjini* ['let's imagine'], *athimairíte* ['while smiling'], *riyunioni ya bamiri* ['family reunion'], *bathi thibeco* ['special pass'], *manĩnja wa bengi* ['bank manager']. It also turns out that the extent of this borrowing is quite vexing to some Kikuyu critics. Why, for example, does Ngũgĩ write *handimbagi* ['handbag'] when Kikuyu has its own word, *kamuhuko?* Why does he write *ngiree* ['grey'] when Kikuyu already has *kibuu?*).

English has fed Kikuyu (as French long fed English, as Arabic fed Kiswahili). It is why history is too complex, too many-sided, it seems to me, for the binary readings – African versus European, indigenous versus imperialist, black versus white, poor versus rich – of much language politics. In the person of the Victorian settler, the British expelled many Kikuyu from their land and homes; in the person of the colonial policeman, they shouted and shot at African protestors demanding equal rights. Many colonisers committed atrocities. Nothing can justify the baseless stupidity of the colonialist worldview. But even in the grubbiest circumstances, men and women of all tongues can perform small acts of humanity. English was also the language in which the Kenyan-born paleontologist L. S. B. Leake – a fluent Kikuyu speaker – compiled his magnum opus *The Southern Kikuyu Before 1903.* Running some 1,400 pages, it is a paean to the social, cultural, and linguistic riches of the Kikuyu people. One section lists the traditional names of over four hundred species of plant; another gives the precise lexicon to describe the various colours and markings on goats, cattle, and sheep. In the person of the liberal Anglican, English devised Kikuyu's orthography, published Kikuyu dictionaries and grammars, built schools. For all its paternalism, the language doubled as the language of African aspiration. When, in the 1920s, educators proposed dropping English in favour of teaching in Kikuyu, Luhya, or Luo, many parents recoiled; English, they understood, was a potential leveller, a gateway to the wider world.

Ngũgĩ's own children, several of them writers, have this idea of English. I learn this towards the end of our conversation, when the young server comes for our plates. I am asking Ngũgĩ what his offspring think of his ideas; in place of speaking, he picks up my pen and notepad again and writes. Beside the names of his sons and daughters appear the titles of their respective books: *Nairobi Heat, The Fall of Saints, City Murders, Of Love and Despair.* All English. His son Mũkoma, the author of *Nairobi Heat,* is presently an assistant professor in the English department at Cornell. Is his father disappointed?

He isn't. Or so he says. He shrugs. Children will be children.

But even were his children to ever start writing in Kikuyu – as Francophone authors Pius Ngandu Nkashama of the Congo, and Boubacar Boris Diop of Senegal also write in Tshiluba, and Wolof, respectively

– other obstacles, cultural, technological, economic, remain. In indigenous languages no author can yet aspire to scrape much of a living. According to a report by the Kenya news agency, dated July 30, 2014,

In developing countries like Kenya, there is a serious lack of reading materials . . . In Kiambu County . . . there are no public libraries where people can quench their reading thirst, and this has seemingly led to poor reading culture within the area . . . In USA, a child is introduced to a library at as early age as five years, unlike in Africa where even university students are not acquainted to or are 'allergic' to libraries.

Another agency item, put out a few weeks before my meeting with Ngũgĩ, carried the stark headline 'Few Customers in Bookshops.'

Ngũgĩ remains cautiously optimistic about the future of Kikuyu and other African-language literature. He has reasons to hope. One is the annual Mabati-Cornell Kiswahili Prize for African Literature, co-founded in 2014 by Ngũgĩ's son, Mũkoma, which awards original writing in Kiswahili. The bigger one, by far, is translation. In February 2015, the Nigerian publisher Cassava Republic brought out *Valentine's Day Anthology,* a downloadable collection of short romance stories by leading African authors. Each story had been translated from English into an indigenous language such as Kpelle, Kiswahili, Yoruba, Igbo, or Hausa. And Cassava Republic is not alone. Ceytu, a new imprint in Senegal, recently translated *The African,* a novel by the French-Mauritian Nobel laureate J.M.G. Le Clézio, into Wolof.

Colleagues wait for Ngũgĩ at his hotel. Before taking my leave, I ask him if he could teach me one more Kikuyu word, the one word that everyone ought to know. In my notepad he writes, in the same firm, elegant hand as, forty years earlier, he wrote on prison toilet paper: *thayũ*.

Peace.

8

ICELANDIC NAMES

Jóhannes Bjarni Sigtryggsson has to decide whether or not *Cleopatra* is Icelandic. It is, he knows, a decision which cannot be taken lightly: in the balance lies the future self-image of a couple's newborn daughter. It is, furthermore, a decision which Jóhannes doesn't take alone. He shares the little room in the capital's three-storey National Registry building with a senior lecturer in law and another academic working in Icelandic studies. The three, meeting monthly, compose Iceland's *Mannanafnanefnd,* the Persons' Names Committee, charged with preserving the nation's ancient infant-naming traditions. Every month, sometimes six, sometimes eight, sometimes ten submissions from parents or prospective parents (along with cheques in the amount of 3,000 krónur, about £23) reach the committee. On average, between one half and two thirds of the names proposed will be approved by the members and duly entered in the register, which presently counts 1,888 boys' names and 1,991 girls'. The task is endless: parental creativity and the search for luckier names will always make the thousands already in circulation seem too few. In a country jostling with Jóns and Guðrúns and Helgas, a Bambi, a Marzibil, a Sónata, is an invitation to admire and remember. But occasionally parental creativity tries too hard. Then Jóhannes must send the submission back, telling the petitioner to pick another name and to be more careful about the rules that he and his colleagues are bound to enforce. It is with such a message that the committee replies, in May 2016, to the couple who wrote in with *Cleopatra*. The name is turned down on the grounds that the letter *C* has no place in the Icelandic alphabet.

If, politically, Iceland has long been one of the world's most liberal nations – the first European state to give women the vote, an early adopter of marriage between same-sex partners, the sole to date whose government has been led by a lesbian – in matters of language it is highly conservative. Personal names – many of which date back to the sagas and comprise nouns and verbs and adjectives that speak of the long bleak winters communally, ancestrally, endured (*Eldjárn*, 'fire iron'; *Glóbjört*, 'glow bright') – are thought of as an extension of the language, part of the national patrimony. So Icelanders preserve their names as the English preserve their castles: Icelandic names, like England's historic buildings, are listed; the committee, like a heritage commission, is appointed every four years to supervise, adjudicate, oversee.

The present committee was appointed in 2014. Jóhannes, then forty-one, was young. But his credentials are impeccable. He has a doctorate in Icelandic grammar. He has authored many learned papers (in one, he categorised fifteen different sorts of hyphens: the hyphens that link compound nouns or digits written in words, that denote hesitation, that show elided pronunciation, and so on). And, perhaps most importantly, he has the best of all names. Jóhannes Bjarni Sigtryggsson. A name in keeping with his family tree. His grandfather – and not any old grandfather, a celebrated poet of children's verse – had been a Jóhannes Bjarni; like many firstborn Icelanders, Jóhannes was named after his grandfather (who, to judge by the black-and-white photographs, he somewhat resembles – a resemblance the grandson, with his mousy hair worn short and his round glasses, cultivates). The poet's daughter Þóra married a Sigtryggur, Jóhannes's father. Jóhannes Bjarni Sigtryggsson: grandson of Jóhannes Bjarni, son of Sigtryggur.

Jóhannes is married and has three sons. When they became parents, Jóhannes and his wife stuck to tradition rigorously. They named their eldest son Guðmundur after the boy's maternal grandfather; their second they named Sigtryggur in memory of Jóhannes's father. Out of grandfathers to name their third son after, Jóhannes and his wife chose to call him Eysteinn in honour of Eysteinn Ásgrímsson, a fourteenth-century monk and poet known for the purity of his Icelandic.

Icelandic 'pure'? To the tourist for whom the country is all fairy-tale elves and unpronounceable volcanoes, the idea will be surprising. To a sympathetic outsider like myself, friendly with Icelanders from various walks of life, familiar with their language and landscape, it is more than surprising: it is absurd. Many in Iceland are far too playful with their words to handle them reverentially. Like the easygoing Australians – those inhabitants of a hotter, vaster island – who, more than

Britons or Americans, are fond of telescoping their commonest nouns (*relly* for *relative, ute* for *utility vehicle, ambo* for *ambulance*), Icelandic mouths are quick to dock lengthy words: *ammó* for *afmæli* ('birthday'), *fyrró* for *fyrramálið* ('tomorrow morning'). Readers of the popular broadsheet *Morgunblaðið* ask their newsagent for the 'Mogga.' The same goes for names. In Iceland everyone is on first-name terms. A Guðrún – the most popular girl's name in Iceland – is Guðrún even to five-minute acquaintances, even in the telephone directory (where *Jóhannes Bjarni Sigtryggsson, Icelandic specialist,* follows *Jóhannes Bjarni Eðvarðsson, mason,* and *Jóhannes Bjarni Jóhannesson, engineer*). But to her close friends and family, *Guðrún* ('Godly Mystery') is too formal: with a brother she answers instead to *Gurra;* to a childhood pal she will always be *Gunna; Rúna* is how an aunt on her father's side addresses her; another aunt, on the mother's side, calls her *Dunna*. And if this Gunna or Dunna comes from a village in the north, or from the Vestmannaeyjar, an archipelago off the south coast, she might be known as *Gunna lilla* ('little Gunna') if she is a short woman. Or if she is very thin, as *Dunna stoppnál* ('Darning needle Dunna'). If, working in a kitchen, she always has her sleeves rolled up, the villagers might call her *Gunna ermalausa* ('Sleeveless Gunna'). Our Jóhannes, the committee man, on the other hand, isn't anyone's *Jói;* and at the mere idea of *Hannes,* he frowns. *Jói orðabók* ('Dictionary Jói') or *Hannes nei nei* ('No, No Hannes') would make his thoughts crawl. To one and all, he is simply Jóhannes.

Jóhannes is a quiet man, a word puzzle fan, a vegetarian in a land of meat eaters. He wouldn't swat a fly.
So he is unhappy when a decision by the committee
leads to a parent's complaint in big and angry handwriting. He would like to tell the complainer, 'I'm sorry.
I didn't make the rules.' The rules decide when he has
to get out his red pencil. And Jóhannes is right. He
didn't make the rules; they were already in place, more
or less as they exist today, in his grandfather's time.
The history of language purism in Iceland is long and
complicated. Over the centuries, the form of this
purism has changed. When Icelandic readers in the
Middle Ages praised Eysteinn Ásgrímsson for the purity
of his work, they meant to praise its Lutheran austerity,
the spare lines shorn of affectation. Now when the
academics praise someone's speech or writing as 'pure,'
they mean it is an Icelandic unadulterated by foreign
words, foreign sounds, foreign letters (like the C in
Cleopatra). The shift requires a little explaining. To
understand Icelandic purism in its modern form, with
its strict rules and red pencils, it is necessary to look
back to events in the early nineteenth century.

The young authors writing in Icelandic in the early
1800s wrote under the sway of Romanticism: they had
a Romantic idea of their language's past. In their imaginations, Icelandic was a once beautiful woman who
had taken to her sickbed, laid low by an infection of
low German, Latin, and Danish loanwords. Danish, in
particular, the language of the island's colonisers, had
weakened Icelandic. The authors, in their writings,
sought to help their language convalesce. They mocked

certain Danish sounds, the way the men and women in Copenhagen chewed their words. They lauded the old, weather-beaten farmers in the island's faraway villages who, they said, couldn't comprehend the slightest Danish. Soon, *Danish* became more than a word of nationalistic displeasure: to act Danish was to dandify your speech with Parisian salon talk, turn your back on the homeland, put on airs. To speak Icelandic in the capital, Reykjavík, while avoiding the temptations of fashionably French-sounding Danish, was to speak in the manner of an honest, upright man, sincerely and unselfconsciously.

Iceland required the proud voice of a modern national poet, so the authors believed; the Germans had Goethe; the French had Molière; the English had Shakespeare. They promoted from their ranks a young naturalist by the name of Jónas Hallgrímsson. No one had written so beautifully about the island's fauna and flora before, nor so deftly sketched the ironies of everyday life for the little farmholder:

> *Hví svo þrúðgu þú*
> *þokuhlassi,*
> *súlda norn!*
> *um sveitir ekur?*
> *Þjér mun eg offra,*
> *til árbóta*
> *kú og konu*
> *og kristindómi.*
>
> *[Goddess of drizzle,*
> *driving your big*

cartloads of mist
across my fields!
Send me some sun
and I'll sacrifice
my cow – my wife –
my Christianity!]

Moreover, the naturalist showed a remarkable apti-
tude for inventing new words, words made out of
existing words, without having to borrow sounds or
notions from the obliging French, Greeks, or Danes.
Aðdráttarafl ('magnetic force', literally 'attraction
power'), *fjaðurmagnaður* ('supple', literally 'stretch-
mighty'), *hitabelti* ('the Tropics', literally 'heat-belt'),
and *sjónarhorn* ('perspective', literally 'sight corner')
are only a few of the hundreds with which he endowed
the language. Then, in 1845, he died at the age of thirty-
eight, and his naively bucolic vision of his countrymen as wise farmer-citizens became forever fixed.

Icelandic is self-sufficient: this became the cry of the
nationalists seeking independence from Denmark.
Poetry reduced to politics. In 1918, independence came.
But the aftereffects of Hallgrímsson's naive vision
persisted. The national obsession with rooting out
every trace of foreign influence in the language grew
fierce. So fierce that from time to time government
campaigns would lecture citizens on how to tell
Icelandic, Danish, and other foreign words apart. Listen
to *tónlist* (literally 'tone-art'), newspaper readers were
crisply informed, not *músík*. Take your shower in a
steypibað (literally 'pour-bath'), not in a *sturta*. *Smart*,

the Danish way of describing someone or something as 'tasteful,' was repudiated in favour of *smekklegt*. And as technology expanded and the world shrank, the purists' efforts only intensified. The minting of brand-new words became a full-time occupation. Since the 1960s, the country's universities have regularly put their eggheads together to rationalise the work. (The minutes of their meetings, like those of the Persons' Names Committee, are currently written by Jóhannes.) The work also includes keeping tabs on the media, lest a foreign word – these days, usually English – supersede any Icelandic coinages. Upbraiding fingers will be wagged at the television or radio presenter who forgets to say *jafningjaþrýstingur* instead of the current *peer-pressure*.

It is the paradox of the Romanticists' struggle that, though it was supposed to free the Icelander of all anxiety to perform, as well as erase any standards of speech beyond those words and sounds that come naturally to mind and tongue, the result has been quite the opposite. The poet and his pastoral vision, creating the ideal speaker, also created a new linguistic standard and divided the nation's consciousness. When those who live far from the countryside in the capital, as two thirds now do, catch themselves saying something from an American film, or a British pop song, they cringe. Some undergo a linguistic crisis and yearn for the best, the purest Icelandic, that Icelandic of Icelandics still spoken out in the sticks, to communicate authentic-ally. It is of this yearning, the narrator's, that the popular, prize-winning novel *Góðir Íslendingar (Fellow*

Countrymen), published in 1998, speaks: a forlorn young Reykjavík-dweller discovers a certain dignity – of speech, and thus of character – only among the isolated country folk (the translation is mine):

Up to this day the best Icelandic in the land was said to be spoken in Hali [a tiny farming community in the south-east of the country]. I was filled with admiration when I stepped inside this temple of Icelandic . . . the woman went over to the stove, stirred a large pot and made coffee . . . when I mentioned having heard that here was spoken the most beautiful Icelandic in the country, the woman called over from the kitchen, 'I don't know about that. Here we speak the East-Skaftafell dialect . . . some men once came here and said that the folk over in Hestgirði spoke the purest.' . . . I have become so self-conscious about my own speech that I turn every sentence over three times in my mind before I dare let it out.

'The best Icelandic in the land' – but the narrator's judgment isn't aesthetic, only notional. Not so much as a sentence of this 'purest' dialect does he hear, and it doesn't matter. It is in this Icelandic-as-idea that the narrator, his author, and many readers feel great pride; in their own everyday Icelandic, many feel unsure. Somehow, the two Icelandics coexist.

It is in a present reflecting this history that the decisions of Jóhannes and his colleagues on the Persons' Names Committee make news. Headlines like '*Manuel* and *Tobbi* get the green light, but not *Dyljá*' and '*Yngveldur* is allowed, but *Swanhildi* is banned' are common. The article under the former, published in May 2016, notes that the committee has adjusted the final names of four children of immigrant parents: 'the son of Petar is now Pétursson, the daughter of Joao becomes Jónsdóttir, the son of Szymonar, Símonarson, and Ryszard's daughter becomes Ríkharðsdóttir.' In the previous month, articles relayed the committee's curious decision to authorise the boy's name *Ugluspegill* (literally 'Owl Mirror') – an adaptation of a name for a prankster in medieval German folklore. 'Whilst indications exist that the name has negative connotations in Icelandic,' the committee conceded in its report, 'these are little known among the general public and aren't for that matter particularly negative or derogatory.' The committee concluded, 'A remote or uncertain risk that the name will cause its bearer embarrassment in the future is not sufficient reason to ban it. We therefore give the name *Ugluspegill* the benefit of the doubt.'

Reports like this have fostered anger toward the committee, which has been growing in some quarters of the country. Every now and then Jóhannes hears of a public figure who wants to see the committee closed down and Icelanders allowed to name their offspring however they like, but he only shakes his head doubtfully. He recalls what his colleague Ágústa says when she hears this kind of talk: wind us down and you'll end up with families that give their children only digits for names, and others that will want to go by an appellation seventeen names long. Above all else, Jóhannes and his colleagues worry for Icelandic grammar. In Icelandic, nouns have gender. Boys' names behave like masculine nouns, girls' names like feminine nouns. But how to determine the correct declensions of names like Tzvi, Qillaq, Çağrı? What if immigrant parents bestow on their son a name which, to Icelandic eyes, has the forms of a girl's? Jóhannes puts on a brave face, but he worries.

He has reason to worry. In 2012, an adolescent's parents took an old decision by the committee to court in an attempt to overturn it. Jóhannes's predecessors told the court they had simply followed the rules. They explained how the confusion in the case had all begun. One day, fifteen years ago, when they were in their little room at the registry building going through their post, a priest's baptismal paperwork caught their eyes. They assumed that the priest had messed up the forms. They telephoned the parsonage. No, the priest responded, he had not. He had written *Blær* in the space for the infant's name, and he had meant to write *Blær,* because it was thus that the baby girl had been christened. The committee members stopped him there. They were grammar-bound, they said, to invalidate the name: *Blær,* though the noun sounds sweetly, and though it means 'gentle breeze', is masculine. A masculine noun for a baby girl! What had he been thinking? The priest apologised: the name was so rare – apparently there are only five men named Blær in all of Iceland – that he hadn't given the matter proper thought. He tried to broker a compromise; the priest and the parents talked. But when he suggested fine-tuning the newborn's name to *Blædís,* a perfectly good girl's name, the couple would not hear of it.

The mother, Björk Eiðsdóttir, told the court her side of the story. She said she had found the name Blær in the much-loved 1957 novel *Brekkukotsannáll (The Fish Can Sing)*, by Halldór Laxness. (Mischievous, prickly, gifted Laxness is the country's only Nobel Prize in Literature laureate. As a young man, he changed his own name from Halldór Guðjónsson. As an author, he had no time for the language purists. The spelling in his novels had always been his own: he wrote *leingi* instead of *lengi* ('a long time') and *sosum* instead of *svo sem* ('about, roughly'), adhering closer to how the words are pronounced. And his novels frequently mocked the purists' naive Romanticism, tempering their pristine flower-coloured valleys with tales of rural destitution and sheep diarrhoea.) Laxness, in typical Laxness fashion, had made the Blær of his novel a female character; and Björk Eiðsdóttir decided that if she ever had a daughter she would call her Blær. In 1997 her daughter was born and baptised. When told by the priest that the committee was unhappy with her choice of name and had rejected it, she wrote pleading letters to the prime minister and the arch-bishop in vain. Five years later, when the family travelled to the United States, the girl's passport gave her legal identity as Stúlka ('Girl'). It was like a game. Whenever she stood before a uniform she was Stúlka, but with her parents, teachers, and classmates she was always Blær. Her mother added, pointedly, that her daughter had often been complimented on her name. Now the girl was fifteen. Before very long she would marry and pass the name down to her own children.

It was why mother and daughter had finally decided to sue.

When the solicitor general summed up on behalf of the committee, he allowed that *Blær* sounds less masculine than many other words. Even so, he said, it would represent a 'big step' for the court to determine that nothing prevented *Blær* from belonging to a child of either sex.

The lawyer representing the girl and her mother replied with a flourish. He said that the step the court was being asked to take was really rather small, since, as it turned out, there existed precedent in Iceland for calling a girl Blær: a Blær Guðmundsdóttir, born in 1973 (only a few months after Jóhannes), already appeared in the national registry. This Blær's mother had persuaded the committee there was a set of possible feminine declensions of the noun: if a woman called Blær gave you a gift, it would be from *Blævi* (whereas from a man called Blær it would be from *Blæ*); if you hadn't seen the same woman for some time you could say that you missed *Blævar* (whereas if your absent friend was a man you would be missing *Blæs*). Icelandic grammar, the lawyer concluded, was malleable. Societies change; their grammars change.

The court agreed. In January 2013 the judge, over the committee's misgivings, awarded the girl the right to go henceforth by the name of Blær.

Societies change. A hundred years ago children born out of wedlock were a rarity in Iceland. Today the opposite is increasingly true. No stigma attaches to the Icelandic mother who raises a child by herself. For this reason, among others, more and more sons incorporate the mother's name into their surname.

Grammars change. The obsolescence of the committee is only a question of time. Twenty years from now, or twenty-five, or thirty, a young man will walk among his country's fjords, gullies, and glaciers. He will be called Antónius Cleopötruson.

9

DEAD MAN TALKING

The man at the Manx language museum was enthusiastic, more than pleased to help; and within days of my electronic enquiry there arrived in the mail a copy of *Skeealyn Vannin* (*Isle of Man Stories*). The museum's pride these stories are, each told, in the words of the island's last native speakers, into recording machines seventy years ago and ever since preserved on acetate, then cassette, now digitally remastered, transcribed, and translated into English. Thick, big, and dense – hardly light reading material – the resulting collection, with its accompanying CDs, testifies to the compilers' tremendous dedication. The thought of this labour, that of a critically endangered language's tiny band of revivalists, was astonishing. How many unearning hours they had lavished on the collection's every voice, every page, every paragraph! Thanks to them, through death and decades of general indifference the stories had survived. They had come down to me, an Englishman in Paris, in a crumpled orange envelope postmarked at Douglas, the island's capital, intact. I handled the book reverentially. It promised revival, a reprieve from extinction; but can such a promise be kept? About it hung an unmistakable air of elegy. On the cover, black-and-white photographs of flat-capped, watery-eyed elders stared out at me.

Mann, or the Isle of Man, is a small island – 220 square miles, no bigger than Guam – in the middle of the Irish Sea. Ireland lies west; head due north and you'll make land in Scotland. To the south, at a farther distance than these other neighbours, is the other nation where a Celtic language is still widely spoken, Wales. The Celts settled Mann around the sixth century CE. The island was wind, rock, and ragwort; it was also, tradition tells us, fairies. The *mooinjer veggey* (literally, 'little people'), the isle's original possessors, had protected its shores from Rome by shrouds of a magical mist.

For centuries, history – Catholic saints, Norse berserkers, English nobles – came and went. History long ignored the Manxman's language. The island's stones were good only for runes, parchment only for Latin. Manx shouts and sweet nothings, calls and compliments, jokes and jeers, rose and vanished into the air like fairy mist. It wasn't until after the Reformation that local clergy began committing Manx prayers to paper, began easing their parishioners' sounds into a readable and reproducible script. In a bilingual 1707 tract entitled *The Principles and Duties of Christianity/ Coyrle Sodjeh*, the language attained at last the dignity of print. Years later, a Manx translation of the Gospels followed. But the church didn't publish Manx for the sake of elevating the language; the island's religious leader, the Englishman Thomas Wilson, thought it beneath his bishop's dignity to speak it with any fluency.

The real reason Manx got published was that so few of the islanders could read and comprehend English. An English vicar at the pulpit mouthing away like a fish plucked out of water, intelligible only by virtue of his familiar, dainty, slender-fingered gestures, the regularity of his droning, the occasional 'Jesus' and 'amen': this, in certain parishes, was the Sunday sermon. As late as 1842, in a letter to the isle's lieutenant governor, a group of clergy was compelled to write, 'In all the parishes of the island there are many persons who comprehend no other language than Manx . . . An acquaintance with this language is a qualification indispensably requisite in those who minister.'

But by then the islanders' language had already gone into decline and was retreating fast for the hills of villages. It was described as 'decaying' by its first lexicographer, Archibald Cregeen, in the introduction to his Manx–English dictionary:

I am well aware that the utility of the following work will be variously appreciated by my brother Manksmen. Some will be disposed to deride the endeavour to restore vigour to a decaying language. Those who reckon the extirpation of the Manks a necessary step toward the general extension of the English, which they deem essential to the interest of the Isle of Man, will condemn every effort which seems likely to retard its extinction.

The deriders were in the capital, the merchants and their customers: English speakers through and through, and eager importers of all things British. For generations, their small population held steady; their urban condescension toward the provincials remained limited. Then the nineteenth century brought with it the international craze for sea bathing. Suddenly, sand and salt breezes were commodities. Mann possessed both in abundance. The numbers of English speakers visiting the island swelled.

So that after history – its saints and berserkers and nobles – tourism (a new word) now came to Mann. The tourists outnumbered the natives two, then three, then five, then ten to one. They were families from Britain's northern mill towns for the most part. On holiday. Full of chat. Soon the island's towns grew noisy with them speaking English. Not the English of the islanders' vicars, mind you, all drone punctuated by the occasional *amen;* theirs was musical, easygoing, moneyed. The islanders in their woollen jumpers, on hearing the tourist talk, began to itch. They began to cross over to English whenever a vacationer walked into earshot. And so, little by little, the Manx people lost touch with their language.

Losing touch with the Manx language meant forfeiting its own peculiar treasure. Hundreds of years' worth of accumulated knowledge, each speaker's birthright, was jettisoned. The language had nurtured health. A great-grandmother, knowing the Manx for this plant or that flower, the traditional names transforming each into a potential remedy, might cure generations of her family of wheezy coughs, gammy legs, upset stomachs. It had promised boys adventure at sea and put food on the table. It had incited complicity – in endearments and turns of phrase – even between a farmer and his animals. 'In my youthful days it was in Manx we always spoke to our horses and our cows. Even the dogs themselves, unless you spoke to them in Manx, couldn't understand you,' wrote J. T. Clarke in 1872. All this, with every speaker's death, fell a little further out of everyday currency and into the realm of folklore.

At the last count, the 2011 census, barely one fiftieth of the island, which comes to a little over 1,800 inhabitants, claimed any knowledge of Manx. Of these, maybe several score – the most ardent revivalists – could speak the language more or less fluently. The numbers are catastrophic; it is a measure of the revivalists' ardour that they are yet able to read in them a sort of triumph. And it is true, the numbers are less catastrophic than they were a generation or two ago. They represent an improvement from the 1940s, when hardly one person in every hundred knew any Manx at all. It was during those bleak years that the revivalists recorded the stories of the island's last native speakers.

Yn Cheshaght Ghailckagh, the Manx Language Society, the revivalists called themselves. They were bookish, brilliantined young men who spent their weekends foraging for the last natives in the outlying farms and villages. Every rumour was pursued, every suggested name taken seriously. Doors of thatched cottages were knocked on; respondents were addressed in Manx, and, often, quickly apologised to in English. But there were also times – each a cause for relief and joy – when, after tramping across marshy fields or noticing some out-of-the-way turning in one of the isle's many nooks and crannies, a white-haired native Manxman or woman was stumbled upon. Then the young men would bring out their pencils and notebooks and prod the old speakers' memories with their questions: What could they remember about their schooldays? In what ways had the village changed in the past fifty or sixty years? Did they recall the Manx word for this bird or that vegetable? Any proverbs? Could they recite the Lord's Prayer? The young revivalists would visit the friendliest, or perhaps loneliest, speakers regularly. Some apprenticed themselves to the speakers' farms in return for several hours of conversational Manx per week. In this way the revivalists – future teachers at evening classes in the language – fine-tuned their accents, fleshed out their vocabularies, became more and more conversant.

Consternated too. Time was pressing; the speakers were frail and very old. Before long the revivalists would be out of native voices to listen to. So they set to recording them, committing the voices to a revolving disc, for the benefit of future learners. A recording van – a gracious loan from the Irish government – duly arrived on a cattle boat from Dublin on April 22, 1948. After hosing it clean of dung, the young men drove off on a tour that would take them up and down the island, from village to village, for the next two weeks.

Fear of failure must have been distracting. Bad roads awaited them; plentiful spring rain – drumming, voluble – threatened interruption after interruption; the 12-inch acetate discs (with space on each side for fifteen minutes of recording) were prone to blips. But the speakers' frailty was uppermost in the young men's anxieties. The speakers had got to an age at which their cottages possessed lives of their own. Water in a saucepan would occasionally boil by itself. Knitting needles would somehow find their way into a cutlery drawer. And in conversation the elders' attention frequently wandered. Dialogue was ever fraught. Words were left hanging in the air, suspended in mid-sentence. An answer proffered might return minutes afterward. Also, what would they make of the newfangled technology? They had never seen a turntable in their lives. Some of the cottages weren't even equipped with electricity. (In those cases, the men had to shepherd the occupants in their van to one that was.) What if they proved shy of microphones?

Shy as Mrs Kinvig. Pushing eighty. A smart, small-featured crofter who read the Manx Bible every day. She had raised ten predominantly English speakers in the tiny southern hamlet of Ronague. The surrounding countryside Mrs Kinvig knew like the back of her hand. She knew the spot near the hamlet where spilled water appeared to roll uphill. She knew the best place to catch sight and smell of the steam train bound for the capital. She knew the way to Castletown, an hour's journey on foot across sodden fields to the coast (a journey she had undertaken every morning as a young dressmaker). Her Manx, of the nineteenth century, she had acquired quite by chance; it was what her parents spoke whenever they wished to discuss something unintended for their daughter's ears. In her old age, between snipping turnips and tossing chicken feed, Mrs Kinvig kept up her Manx with Mr Kinvig, ten years her senior, and with the young revivalists who dropped in from time to time. But now when the men came in the van and asked her to repeat, from her store of anecdotes, their favourites, Mrs Kinvig grew tight-lipped. The microphone intimidated. No amount of encouragement from the men or her husband would coax the stories out of her. Finally, though, she relented – whether from fatigue or sheer embarrassment – and called up from memory several hymns:

> *Dy hirveish Jee dy jeean, [A charge to keep I have]*
> *Shoh'n raaue va currit dou, [A God to glorify]*
> *Dy yannoo ellan veen, [A never-dying soul to save]*
> *Dy chiartagh ee son niau. [And fit it for the sky]*

The recorders had better luck with the oldest Manxman, John Kneen of Ballaugh in the north, who was closing in on his hundredth birthday (and who, as it happened, had another ten years in him). *Yn Gaaue*, the smith, the villagers called him. Gnarled fingers resting on his walking stick, Kneen reminisced aloud about his years shoeing horses '*son daa skillin 's kiare pingyn son y kiare crouyn*' ('for two shillings and four pence for the four shoes'), when there were thirty smithies in the north and '*cha row treiney, ny boult, ny red erbee cheet voish Sostyn*' ('there wasn't a nail, nor bolt, nor anything coming from England'). To provide the smith with a conversation mate, the recorders fetched another Manx native, Harry Boyde, a former labourer; though lifelong neighbours – their cottages only five miles apart – the two men had never spoken before. Introductions must have gone well, for they spoke together in Manx for hours to the delight of the recorders who had them sit equidistant from the Mrs Kinvig-scaring microphone. With the young men, Kneen had often been loquacious; with Boyde now he turned suddenly attentive: peering out of his one good eye, a veiny hand cupping his working ear, he let his interlocutor do much of the talking, content to nudge one or another of Boyde's recollections along every now and then with a *'Dy jarroo, ghooinney?'* ('Really, man?') or a *'Nagh vel eh?'* ('Is it not?'). The men talked manure and rum and market life. They described an island on which booted feet had been splayed by riding horses, the riders little more than their titles: *yn saggyrt* ('the parson'), *yn cleragh* ('the clerk'). They told their stories without inhibition, with the great unselfconsciousness of the ancient. They thought back over their long lives to times when emotions ran highest. Kneen, loquacity returning to him, left Boyde to listen in amazement as he nonchalantly recalled having once seen a crowd of fairies, the little people, leap and frolic in the meadow at sundown: *'Va'n fer mooar gollrish mwaagh'* ('the big one was like a hare'). At the sound of the blacksmith, the fairies started and fled.

From Ballaugh the young men continued on the last leg of their tour of the island. They drove to the northern-most tip to record a brother and sister, John Tom Kaighin and Annie Kneale. Of Kaighin one of the recorders wrote in his notebook: 'Blind, aged 85, very lively and with an enormous voice.' Something of this liveliness is communicated in his story about the parson and the pig. One evening, at dinnertime, a parson came to the home of an old woman parishioner. He saw a pig eating out of a pot on the kitchen floor. The old woman invited the parson to sit, but the parson refused to sit unless she first '*cur y muc shen magh*' ('put that pig out'). The woman repeated her invitation, feigning calm and deafness; the parson refused to sit; the pig tucked into his pot. On and on this went, until at last, the parson snapped the old woman's patience. '*Cha jean mee cur y muc magh*' ('I won't put the pig out'), she blurted, since the pig at least brought money into the household, whereas the parson took it out.

Annie Kneale had a harder time than her brother in recollecting her Manx. Forgetfulness occasionally got the better of her. Her speaking was rusty. Sometimes she would lapse into English; a sentence begun in Manx might conclude in English. But she wasn't one to despair. And the young men were encouraging. Adroitly, they rephrased a question or overlooked a fumble; proffered words supplemented her sketchy responses. Vivid fragments of a former existence, with the men's help, gradually resurfaced. Girlhood memories: village folk, when the wind dropped, saying '*T'eh geaishtagh*' ('It's listening'); her mother, when porridge on the stove started to bubble, saying '*T'eh sonsheraght*' ('It's whispering'), then '*T'eh sonsheraght, gow jeh eh. T'eh jeant nish*' ('Take it off [the heat], it's done now'). Glimpses of a lost way of life, a vanished world in which even the wind and the porridge spoke Manx.

The year after the recording Annie Kneale died. She was survived by eight known native speakers.

* * *

To listen to the recordings at a lifetime's remove, in the warm and dry of a Parisian apartment, having fed the bright CDs into a modern player, was to complicate sentimentality. And yet the old men's and women's voices, as they filled my living room, moved me. The voices held frailty, but also a certain sly humour. In spite of the words lost in beards or expelled as whistles through missing teeth, they were strangely attractive. More than once, listening intently, I caught myself mouthing along.

I kept an eye on the transcripts as I listened. The words made me think of Welsh: the verb-first sentences, concatenations of consonants that intimidated the tongue, the sophisticated ballet of sounds hard and soft. The Welsh of my week-long vacations. Road signs in Wales – *ysgol* ('school'), *canol y dref* ('town centre'), *gyrrwch yn ddiogel* ('drive safely') – wouldn't have looked out of place on the island, I thought. I could almost envision the Eisteddfod, a large Welsh music and poetry festival, taking place on Mann.

When Brian Stowell was sixteen, he read in a local newspaper about the imminent demise of Manx. He got in touch with the article's author, a young revivalist by the name of Douglas Fargher, and joined the cause. Ten years later, in 1964, the two would drive south along gorse-lined roads into Glen Chass ('Sedge Valley') to tape the last native Manxman standing (and speaking).

'I got my Manx from the same men who had met and recorded the native speakers in 1948,' Brian tells me by telephone. The continuity is important to him; it confers authority. It is with an air of authority, briskly, that he speaks. 'Conversed every weekend with Doug. A live wire, he was. Had a fruit import business in the capital. Every weekend, long conversations with him and his workmates: six hours of spoken Manx on a Saturday; four to six hours on a Sunday. By the following Easter I was speaking the language fluently.'

Glen Chass was home to Edward 'Ned' Maddrell, a retired fisherman who had been the youngest of the 1948 speakers. In 1964 he was in his eighty-seventh year. Black-and-white photographs, and the odd colour Polaroid, of Ned Maddrell – all findable on the Internet – show a large man, broad-shouldered. His frame suggests a lifetime of manual labour, but he also seems to have been something of a snappy dresser – always in a shirt and tie – and approachable. A ruddy face handsome with composure. To his upbringing by a great-aunt, and the relative reclusion of a subsequent life at sea, Maddrell owed his Manx speaking. He could remember his childhood, when everyone in the village spoke Manx to each other. Not to speak it was to be an oddity, like a person deaf and mute. He could remember old Mrs Keggin, who lived in a thatched cottage at the foot of the road leading to Cronk ny Arrey Laa, and who had no English. (According to Professor Sir John Rhys of Celtic at Oxford, she was probably the last Manx monoglot.) Through Ned – whose schooling was in English – Mrs Keggin would barter her eggs for rolls every week with the travelling breadman.

Maddrell didn't let his being the language's safekeeper go to his head (though the attentions of the young revivalists, linguists, and the occasional journalist – some from afar – must have sweetened his final years). Brian recalls the courtesy with which he and his colleague were greeted and welcomed into the largish room that served as Maddrell's reception area. The two men came with cassette tapes – still a novelty in the 1960s – and eyed the turning spools as they and Maddrell, in facing armchairs, conversed. Perhaps to an outsider it might have seemed that they had come to take down Manx's dying words. But their conversation in the language, regularly punctured by light banter and gentle laughter, asserted optimism: Manx, in the shape of Brian and the other revivalists, would go on being heard and spoken long into the future; the succession was assured.

In spite of his great age and native credentials, Maddrell, with the young men, was modesty itself. He wouldn't speak in the manner of a master addressing his disciples. On the contrary, he deferred continuously to their education. 'If I make a mistake, don't hesitate in correcting me,' Maddrell told them in English at one point. 'You're scholars, and I'm not.' The deference posed its own risks.

'We wanted to double-check the correct pronunciation of *eayn,* meaning "lamb",' Brian tells me. 'We asked Ned to say the Manx for *lamb* aloud. Ned said *"eayn"*. My colleague, Doug, repeated what Ned had just said, but it came out a bit differently. Ned said, "That's it," and then imitated Doug's pronunciation. But we were aware of this sort of hazard.'

There were surprises. To the revivalists, who had thirty years of Manx learning between them, some of the old man's words and expressions were new. Maddrell was talking about his days and nights aboard rocking boats, amid nets and lazy crabs and low skies, when for 'star' he said *roltag*. But the dictionary's version was *rollage* – appearing right after its definition of *rollag*: 'the hollow an oar works in on the gunwale of a boat.' Had the old man made a slip of the tongue? Brian and his colleague knew otherwise. They knew that Maddrell's boat had brought him into regular contact with Irish sailors. For 'star' the sailors would have used their Gaelic word, *réalta*. Maddrell's Manx had simply adjusted to its environment, to the Irish sailors' comprehension; *roltag* was among the adjustments that had stuck. Equally unfamiliar was the expression of Maddrell's *'Va ny taareeyn er'* ('The terrors were upon him') – an inheritance, it turned out, from Old Irish.

For many years after Maddrell's death in 1974, Brian assumed custody of Manx. He worked by day as a physics lecturer 'across the water' in Merseyside, and then in the evening devised distance courses in the language and wrote promotional articles and sang traditional ballads for a vinyl LP. To remedy the lack of Manx literature – 'The Bible is no use, too difficult' – he translated popular children's tales. In 1990 his *Contoyryssyn Ealish ayns Cheer ny Yindyssyn* (*Alice's Adventures in Wonderland*) appeared. By that time attitudes toward the language had long softened to bemusement; it became possible for the island's schoolchildren to be taught a smattering of Manx.

Brian: 'Many pupils were learning French or Spanish already. To these, teachers began adding thirty minutes of Manx per week. A few words one lesson, the odd phrase the next. Nothing strenuous. The parents didn't object.'

So, having learned to say *Je m'appelle Jean* or *Me llamo Juan*, and to count *un, deux, trois, quatre, cinq*, or *uno, dos, tres, cuatro, cinco*, the pupils were taught to say *Ta'n ennym orrym John* and to count *nane, jees, tree, kiare, queig*. But not much else. Manx materials – books, exercises, trained teachers – were still in very short supply. The lessons were like the small change of the foreign language departments' resources.

The revivalists did not mind. The thought of children speaking any Manx at all delighted them. Brian explains:

Remember, not long before, most islanders had given up on the language. Manx had a reputation problem. It was associated with poor people and backwardness. Then it was associated with loners and eccentricity. We were embarrassed to speak it in front of others. I have a little story about that. I rang a pal on the island one day from my home in Liverpool. I spoke as usual in Manx, but he would only answer in English. If I said something like 'kys t'ou?' he'd reply, a little stiffly, 'Fine, thanks, how are you?' It was only afterward that I realised that my friend had been at work when I called. He had an office job. He didn't want his colleagues to hear him using Manx.

Brian gives me another story from around the same period: pub-goers who conversed over their beers in Manx sometimes found themselves on the receiving end of another drinker's anger, he said. Insults had been known to ignite into brawls.

Such incidents are now behind the revivalists, Brian says. He has seen for himself how much things have changed in the twenty-five years since he left England and physics and retired to Mann. People from all over the island show growing interest in the language; the number of learners rises census by census. The obstacles to revival no longer seem insuperable. Only, the workload has become too much for Brian. Some years ago, he handed the work over to Adrian Cain – an energetic forty-something. Among the revivalists, Brian and Adrian are household names.

Shortly after talking to Brian, I telephone Adrian to learn more about him and contemporary Manx. 'I'm originally from the south of the island, near Cregneash, Ned Maddrell's natal village. I had aunts who told of the scraps of Manx they heard around them growing up.' Much like Brian, he fell in with the revivalists as a teenager, and their zeal rubbed off on him; learning the language took up much of his early adulthood; years later, he wrote in it to Manx friends from the desk of his economics classroom in east London. In London, for a time, he got caught up in city politics – 'left-wing type stuff' – but quite soon he tired of the marches and slogans. His thoughts turned to home. 'I thought, defending a language is political. Not the signing of petitions and the like. The speaking and the teaching. Showing that Manx still matters. That's enough politics for me.' His return coincided with the growing media market for 'minority languages'; Manx, the fight to revive it, was suddenly newsworthy. Eager to talk up the cause, he gave interviews to reporter after reporter. 'Cats with no tails, motorbike races, offshore accounting – it was all that most people had heard about the island. The language offered another side.'

I ask him whether he finds the media tiring – that is, frustrating and disappointing. I can imagine that he is asked the same questions over and over.

'In a word, yes. Well, quite often. With some of the journalists' questions, you think, "Oh my god, where do you start?"'

Adrian's Manx fluency once earned him a spot on a British teatime television programme. He shared the sofa with a husky comedian of Caribbean descent. To the viewer, the contrast could hardly have been more startling. Just as noticeable was the language difference: the comedian's English boomed; Adrian's Manx sounded out of place. An hour before the programme, in the green room, he had agreed when the producer instructed him to perform a crash course in the language. But now, on air, it happened that the host and the comedian had almost no time – literally – for the course. For all Adrian tried to talk concretely, to enthuse, in the few minutes allotted to him, the men's involvement remained half-hearted, their responses glib.

The experience, alas, is not uncommon. Adrian, all

earnestness, accepts the invitation of a Belgian documentary maker, or a German press columnist, or a British radio host. He prepares for his bit part with care. He thinks up a number of good lines and simple examples. He irons his interview best. He arrives at the café or studio on the dot. He hopes only that the interviewer will do the language justice. But more often than not, the interviewer is late, heedless of the scene-setting that would engage a wide audience, and scantily prepped. Unsurprisingly, the result is a rushed, cobbled-together cliché: the language of the islanders is as hard to learn as it is quaintly exotic; its decline is sad; against all the odds it resists doughtily. It is enough to depress any revivalist. Worse, within a community as tiny as the islanders', jealousies talk. 'Some say of me, "There's Adrian again in the media bigging himself up."'

But Adrian prefers to remain upbeat. He is a professional optimist. All publicity, to his mind, is good publicity. Publicity attracts learners from overseas, a new type of visitor – the 'linguistic tourist' (a boon to the island's economy, which has suffered from the modern rise of cheap package holidays to Spain). A bilingual *No Smoking* sign – *Jaaghey Meelowit* – on a red-bricked street corner is to these tourists what a sea view was to Victorian vacationers.

'They are mostly young and come from all over Europe. The idea of Manx as a community anyone can choose to belong to is probably the principal draw. Our classes are very relaxed. In a way we're all learners. There's no one right accent to acquire: the German learners speak Manx with a German accent, the Swedes with a Swedish accent, the Czechs with a Czech accent. And that's fine.'

Backpacks, long blond hair, middle-class giggles: Adrian's conversation groups are a world away from the farmers and fishermen in their cottages. The new speakers' disparateness can occasionally strain a listener's comprehension. Adrian insists his learners have the freedom of the language.

'Some people have gotten quite into the old recordings of the native speakers and too involved in the small print of pronunciation. But ought we really to stick to 1940s fishing village pronunciations into the twenty-first century? Take *maynrey*, which means "happy".'

He says it 'man-ra'.

'Well, there are those who listen to the old recordings and hear "mehn-ra". They think everyone should say "mehn-ra". They say I'm saying and teaching it wrong. But the recordings are a tool, they're not gospel. "Man-ra" is my dialect of Manx. I say "man-ra", you say "mehn-ra". It's all good. I suppose I'm a bigger-picture person: as long as you make out what I'm saying, and I what you're saying, our Manx is correct.'

When he says this, I remember an argument among the revivalists over 'correctness' that Brian mentioned. Fifty years ago, many new words needed coining to match modern island life: the elders of the recordings had never ridden in a car or aeroplane, owned a computer, or gone to university. Several revivalists responded by putting forward their own ideas for this or that word. Adrian anticipates my question.

'It's like *parallelogram*. Is that a Manx word? I don't know what else to call it. I would just say "parallelogram". Same for *jungle*. I say "jungle". The geekier, into-authenticity types say *"doofyr"*. They take it from Irish. Both words ought to be acceptable.'

Grammar, too, can vary from speaker to speaker. Sentences are often simplified. Take a sentence like 'I saw her yesterday.' A fluent Manx speaker would say it *Honnick mee ee jea*. (*Honnick* is the irregular past tense form of *fakin* ['to see']). Most learners, however, prefer to drop the irregular form and say instead *Ren mee fakin ee jea* ('I did see her yesterday'). Likewise for *Hie mee* ('I went'), which they replace with *Ren mee goll* ('I did go'). And, being 'came' (*haink*) droppers, they say instead *Ren mee çheet* ('I did come').

'If Ned Maddrell or any of the other native speakers could be resuscitated and returned to the island today, they'd be astonished by the sight of these learners. The accents, some of the sentences, would probably strike them as rather odd. But they'd recognise the language as theirs. They'd understand and be understood when they spoke.'

Regularly, Adrian does the next best thing to resuscitating the dead: starting with Brian, he confers with the few elders, not native speakers but nonetheless fluent, who still remember them. Like the recorders before him, he preserves every anecdote he hears. Half a century ago, Brian tape-recorded his conversation with Ned Maddrell; these days, Adrian videotapes his conversations with Brian and uploads them to the Internet – complete with subtitles – for all to watch.

I watch Brian and the other men in Adrian's videos, watch their brows pucker with reminiscence as they talk. There is Davy Quillin of Port Erin, gray-haired, blue-jeaned, his posture sagging into his burgundy sofa, talking about going to sea at sixteen with his Manx lesson books. He gestures toward his books under a coffee table, just out of eyeshot. He speaks his Manx quickly, fluently, the way he once heard it spoken as a child. He remembers well the day when, on a walk with his father, he spotted two old men three-legged with walking sticks. When he and his father got closer, he heard his father whisper 'Listen.' The old men, deep in conversation, were *'loayrt cho aashagh ass yn Ghaelg . . . v'eh yindyssagh clashtyn ad'* ('speaking so easily in Manx . . . it was wonderful to hear them'). One of the men was Ned Maddrell. He doesn't know who the other man was. *'Garroo,'* Maddrell and the other man kept repeating, as the weather beat about them: 'Rough.'

In another of Adrian's videos, Derek Philips, bald and bespectacled, white-moustached, with the fleshy look of a retired butcher, reclines on his cream sofa and demonstrates his grasp of Manx. His Manx, he explains, dates back to the revivalists' evening classes he sat in half a century ago. One of the revivalists, a Port Erin bank manager, befriended him; thereafter, they would meet regularly in each other's homes and converse in nothing but the language. Philips's nearness to fluency made him gregarious. He was always on the lookout for other speakers. Once, from behind the shop counter, he ventured to ask an elderly customer whether he knew any Manx. He was hoping to at least cadge a new word or two. But the customer replied with indignation, 'Manx? Manx? That's nonsense, man. Manx is worthless!' He thinks the poor man must have had his Manx beaten out of him as a boy.

The videos – tens of minutes, all told, and mostly recorded in the intimacy of the elders' homes – offer a fair sample of modern Manx. For all their fluency, the men's lack of immersion in a community of speakers sometimes shows. Talking about gardening, Philips's lips pause before continuing with *poanraghyn* ('runner beans'). A moment later, he asks Adrian *'Cre t'an fockle son cauliflower?'* ('What's the word for cauliflower?'). *Laueanyn* – the Manx equivalent of 'gloves' – is something he will misremember. He is at his best, his smoothest, when delivering any of his striking anecdotes; by dint of repetition over the years he tells them at an impressive rate.

One anecdote stands out in particular. In the pub one evening, a darts player came over and asked him for a game. In a minute or two, said Philips. I'm just talking to the wife. The player nodded and returned to his darts; he was about to fling one when he collapsed before the board. Philips's wife was a nurse. She ran over and *'prowal dy yannoo yn stoo er y cleeu echey'* ('tried to do the stuff on his chest'). In vain. The player was dead. Hastily the pub was emptied of its drinkers; an ambulance was called. And as the Philipses waited for the sirens, thirty minutes or so after the darts player's collapse, his body briefly made as if to sit up. *'As dooyrt my ven dooys, va shen yn aer scapail woish'* ('And my wife said, that was the air escaping from him').

At the ring of the bell, the blue uniformed pupils of the Bunscoill Ghaelgagh primary school race each other out of their classrooms into the playground. Some kick a ball; some skip rope; some stand about and chat. The teachers, all ears, walk among them, always on their guard. If they catch any chatterers whispering in English, they stop and issue a gentle rebuke.

The Bunscoill, which was founded in 2001 by a handful of parents, is situated in the island's central valley and is the first Manx-language school. In fifteen years, its intake (ages ranging from five to eleven) has expanded from scanty to seventy. A fair number of the current pupils' parents aren't revivalists; they hope that primary education in a second language will make their offspring future wizzes at Spanish, German, or French. The revivalists, for their part, hope the school demonstrates that Manx has a future: if seventy children is not exactly a big number, it is at least seventy more than are studying in the recently extinct languages Klallam (of North America), Pazeh (of Asia), or Nyawaygi (of Australia).

Among the newcomers to the school last year was Adrian's son, Orry. Orry has a head start on most of his schoolmates; Manx has been instilled in him by his father from birth. At six, he is still small; he makes a small child's mistakes. He can mix up *eayst* ('moon') and *eeast* ('fish'). 'And one time we were on holiday in Norwich, and I said to my son *"Vel oo fakin yn shirragh shen ayns yn edd echey heose sy cheeill?"* ("Do you see that falcon in the nest up in the church?"), and he looked at me all confused and asked what the bird was doing in a hat. You see, *edd* can mean "nest" but also "hat".' Such mistakes are a young native speaker's rite of passage to fluency. Orry grows surer-tongued by the day.

Orry and his schoolmates are the first children to speak fluent Manx in over a hundred years. Visiting elders can hardly believe their ears when they sit in during lessons and hear the language in children's voices. Ditties and nursery rhymes unsung for generations have been given a second life; tales of modern life – of picnics, pop music, and computer games – fill sheets of paper before being read aloud.

The school employs four teachers and two class assistants. It isn't easy to find the staff. By some accounts, past class assistants have been far from word perfect. One of the newer teachers – a mother of pupils at the school – confided her pre-job interview nerves to a local reporter: she had had to brush up her grammar. For weeks prior to the interview, she had read nothing but Manx translations of children's books.

The headmistress gazes out of the windows at the passing traffic. Nearly everything beyond the school's weathered grey slate walls – tearoom sandwich boards, bus shelter graffiti, turned-on TVs, labels on milk bottles and cereal boxes and bags of sweets – is in English. English daily dwarfs the Manx that the children learn at school. English at the breakfast table, English outside the school gates, English during the bedtime story. At home, with his mother, even Orry 'defaults to English,' Adrian admits.

It is why, even in the heart of the revivalist community, anxieties persist. Only time will tell whether the playground's sounds signal a renaissance, or whether they are a 1,500-year-old language's last gasp.

10

AN ENGLISHMAN
AT L'ACADÉMIE FRANÇAISE

Sir Michael Edwards, the first and only Englishman at the French Academy, invited me to a teatime rendezvous last summer at his home near Saint-Germain-des-Prés. The tea, though, never arrived. It was one of those stiflingly hot Paris days when the thermometer temperature exceeds one hundred; perhaps the French part of my host's mind simply didn't consider it to be tea weather. Too polite – too English – I dared not suggest otherwise. I thought delicious thoughts about the milky cuppa I did not drink.

No matter. We talked, now in English, now in French, in the book-filled salon till late in the evening; I had so much to ask him about. At seventy-seven, Sir Michael's life and career had been long and rich. A scholar and author of verse who works in a French he learned as a British schoolboy, he had become one of the language's forty *Immortels,* its undying defenders and overseers, two years earlier. In the four hundred years since Cardinal Richelieu dreamed up the Academy, soldiers and clerics, chemists and numismatists, admirals and hotel-keepers, an African head of state, and – beginning in 1980 – even a few women, had all been admitted. But never, before Sir Michael, had an Englishman.

His election made headlines; the French papers were

particularly generous. The daily *Liberation* enthused, 'England has sent us a beautiful gift.' The reaction was different across the Channel, in London. The reporters there did not know quite what to make of the news. A fellow from Blighty refining French's official dictionary! They did little to conceal their surprise. 'The academy is famous for its tireless battles against "Anglo-Saxon" invasions of French, offering Gallic equivalents to Anglicisms, such as *courriel* instead of *email*,' noted the *Telegraph*. Explaining himself to his country's press, Sir Michael said, 'This is a moment of crisis for French, and it makes sense, I believe, for the academy to choose someone who comes from, as it were, the opposite camp but has become a champion of the special importance and beauty of the French language.' In one of his many interviews for the French media he was reported as saying that his favourite word was *France*.

'Actually, I don't know where that came from,' Sir Michael told me. 'I remember someone pushing a microphone to my mouth and asking me for my favourite word. My favourite French word, of course. The answer just fell out of me. Come to think of it I much prefer *rossignol* ("nightingale").'

The surprise at his election – and not all of it was British – surprised him. But he could entertain the surprise, up to a point. England and France, after all, had a history. Each, to the other, had once been an invader. Each, against the other, had gone to war. On the fields of Agincourt, thousands from both armies had been turned into arrow fodder. So the brows raised at Sir Michael's ascension to the Academy was, from a historian's point of view, understandable. From a linguist's, however, the surprise was unfounded. *Surprise* – the origin of the word is French. Like *election* and *history* and *armies* and *origin*. According to one estimate, one out of four English words was imported from France: to speak British English is to speak a quarter French. (United States English is another matter. Americans pump their trucks with gas, not with petrol (*pétrole*); cook zucchinis and eggplants, not *courgettes* and *aubergines;* shop at the drugstore, not at the pharmacy (*pharmacie*). Leaf-yellowing autumn (*automne*) in New York or San Diego is fall.)

For Sir Michael, the Queen, with whom he had an audience on her 2014 state visit to Paris, symbolises the close relationship between the two languages. A descendant of William the Conqueror, the monarch (whose motto is *Dieu et mon droit* – 'God and my right'), has a reputation for fluency in French. 'The Queen always speaks first. She complimented me on my living here: "A very beautiful city." That's about as much as I can tell you. Conversations with her Majesty must remain confidential.'

On a shelf stood a framed photograph of their meeting.

'Did you converse in English or in French?'

'In English. But at the state banquet later that evening, she spoke in French.'

The banquet took place in a chandeliered hall of the Élysée Palace. In the bright lights, the Queen, in white, sparkled with diamonds, and the red riband over her right shoulder told all that she belonged to the Legion of Honour. As she spoke, she turned the pages of her speech with white-gloved hands. Banquet etiquette – it is the royal custom to read from notes – but I was curious. What was Sir Michael's opinion of the Queen's linguistic performance? Suddenly, predictably, diplomacy came to him. He would only say, 'very good.' I remembered then that the speech had been televised live in France (the Queen is as popular as ever with foreign republicans), as had others she had given in the past, and recordings of them could be watched online. I found several, some days after the meeting with Sir Michael, and listened. No, Sir Michael did not speak the Queen's French. His pronunciation was even better, his accent even smoother. And later still, reading up, I learned that the Queen had picked up her second language not in France but within the walls of her father's palace; it turned out to be a byproduct of her cloistered upbringing, the work of a Belgian governess – proof, paradoxically, of her unworldliness.

(And, reading further, I came upon a contemporary account of the French spoken by the Queen's namesake, Elizabeth I: 'She spoke French with purity and elegance, but with a drawling, somewhat affected accent, saying "Paar maa foi; paar le Dieeu vivaant," and so forth, in a style that was ridiculed by Parisians, as she sometimes, to her extreme annoyance, discovered.')

Sir Michael – though he now moves in high circles, an easy converser with queens and presidents and prime ministers – is of humble beginnings and had to learn French the hard way. No Belgian governess for him! 'I was born and raised in Barnes, in southwest London. Very English, the pubs and parks and street names: Cromwell Road, Tudor Drive. Modest, too.' His father, Frank, a garage owner, worked in car parts. Sir Michael's aspirations were his mother's doing. 'As a girl she had written a play. Of course, nothing ever came of it, but it stayed with her, this dream of making a living out of words.' Her son was able to cultivate his ambition at an Elizabethan all-boys secondary school, Kingston Grammar. (*Grammar*, another French word, meaning 'book learning,' is a cousin of *glamour*.) 'I fell for French. I was eleven. I opened my textbook – *A Grammar of Present Day French with Exercises*, by J. E. Mansion – and there they were: *oui* and *non*. Such magic in so simple words. *Yes* and *no*. But how important! It was the same with other French words. They possessed an aura.' To the garage owner's boy, the words seemed bright and shocking with unfamiliarity. 'A new world opened to me: new ways of naming, of seeing, of imagining.'

And taking in Sir Michael now, the embroidered red lion on his navy-blue tie, his black leather slippers; looking into the ingratiating eyes behind the metal-rimmed spectacles; attending to his studied clubbability, his professorial way of speaking ('Bother', 'Vivat!') – so dapper in dress, so clerkish in manner – you can see how the British grammar school, at a remove of sixty-plus years, had laid the foundation for his present role. Had he always enjoyed his school lessons?

'No. I remember finding the textbooks dry, cold, unfriendly. Probably they put off many of my classmates.

I was never put off, but then somehow I knew that French was so much better than its textbooks.'

He recalled his French schoolmaster. 'Dr Reginald Nicholls. He had a spastic jaw. Quite the disadvantage for a language teacher.'

Dry textbooks, a teacher with a spastic jaw – and when, after seven years at Kingston Grammar, Sir Michael went to Cambridge, he was taught French 'as though it were dead'. All reading, no speaking. Taught only to decipher Montaigne and Voltaire and Racine. Once again, education might have thrown him off French forever. But Racine enchanted. Sir Michael went on to write a thesis in Paris on Racine. And flexing his memory, Sir Michael recited a little Racine to me:

> Moi-même, il m'enferma dans des cavernes
>   sombres,
> Lieux profonds et voisins de l'empire des ombres.

The citation is from *Phèdre*. And while we were on the subject of the play, Sir Michael said, 'French and English writers don't think and write alike. Racine taught me that. For instance, an English author would never write, as Racine does, *à l'ombre des forêts* ("in the shadow of forests"); he would write "in the shadow of a cedar" or "in the shadow of an oak tree".'

'Interesting. So the two languages configure reality differently. Could you tell me a little more about this difference? How would you characterise it?'

'I would say perceptions in French are more abstract – like hovering over an experience in a Montgolfier. It's the reason why French thoughts tend to be holistic, and the texts homogenous. Whereas English perceiving is earthier, detail-led, full of quirks.'

'What do you mean by texts that are "homogenous"?'

'Racine wrote many plays, but he wrote them with few words: two thousand. Shakespeare's contain ten times more. That gives you an idea of how few Racine needed. In French, the same word can be made to say several different things. Consider *attrait* – a typical Racine word. Applied to a woman it means "charms"; but use it to describe something vaguer, like the unknown, and you have to translate it as "lure": *l'attrait de l'inconnu*, the lure of the unknown. The "attractive-ness" of a town or city, an "interest" in some topic, feeling "drawn" to this or that – all meanings that merge in *attrait*. It's this quality in French that helps a text cohere.'

Out of modesty, perhaps, Sir Michael didn't mention Racine's links to the Academy. The subject of Sir Michael's thesis, the author of *Phèdre,* the model of the French language's 'purity' and 'eloquence' (the terms are Richelieu's), Racine, born less than four years after the Academy, joined its ranks at the age of thirty-three. So in Racine, Sir Michael also has an illustrious predecessor.

Racine took him to Paris. And in Paris he met and later married Danielle, a Frenchwoman. 'My children and grandchildren all have French citizenship. I would have become a French citizen myself much sooner had I known that it was possible to keep both nationalities. You see, I didn't want to lose my British passport.' With his French wife and British passport, Sir Michael taught French, English, and comparative literature for many years in Warwick, Essex, and Paris (with stints in Belfast, Budapest, and Johannesburg). In addition to his career in academia, he reviewed French and English poetry for the *Times Literary Supplement.*

'I was sent a book of poems by Yves Bonnefoy to review. Later we became firm friends. A quintessentially French author. It was he who gave me the idea to write in the language.'

Sir Michael has gone on to write many books, in both languages, including a study of Samuel Beckett, another favourer of French over his native English. In *Paris Aubaine,* a recent verse collection, Sir Michael even mixes the two together, sometimes within the same sentence: 'Inspecting her woodcuts, I thought the Seine too sinewy, turmoiled and yet, l'eau grise, sous la haute pierre, s'anime de guivres, se trouble là-bas, dans les remous de sa présence unearthly.'

Guivre, a heraldic term for 'serpent' (perhaps Sir Michael, composing his verse, had in mind the Serpentine River in London's Hyde Park), was, I thought, just the sort of classical and 'quintessentially French' word of which the Academy approved. But *guivre* had, in fact, been given no entry in the first edition of the Academy's dictionary, published in 1694. Lack of space – the work stretched to only 18,000 entries – was one explanation. A likelier explanation could be found in the early academicians' curious attitude toward words. Claude Favre de Vaugelas, a nobleman whose influence among his colleagues ran high, championed a vocabulary that avoided anything the tiniest bit provincial, vulgar, or technical. Keep French gentlemanly – such could have been his motto. Gentlemen in France had seemlier things to talk about than serpents.

The silliness in the dictionary's making was exceeded only by the slowness. Nitpickers expended weeks on the definition of a word like *bouche* ('mouth'). After fifteen years at the dictionary, Vaugelas and his fellow academicians had only got as far as the letter *I*; dying prevented him from reaching *je* or *jovial* or *jupe*. Years later, in a moment of crossness, Antoine Furetière complained about the energies wasted on pointless arguments:

Right is he who shouts loudest; everyone harangues over the merest trifle. The second man repeats like an echo what the first has just said, and most often they speak three or four at a time. When five or six are present, one will be reading, another opining, two chitchatting, one dozing. . . Definitions read aloud have to be repeated because someone wasn't listening . . . Not two lines' worth of progress is made without long digressions.

Furetière was also at work on his own dictionary. When the others at the Academy found out, they told him to stop. He refused; he had put the best years of his life – over thirty – into the project. For this refusal he was thrown out. The subsequent back-and-forth between the Academy and its former member came to be known as 'the quarrel of the dictionaries'. The Academy's, fifty years in the writing and still incomplete, was a mess. Furetière revealed that many of its entries weren't in alphabetical order; that a word as common as *girafe* ('giraffe'), was nowhere to be found; that the arguers could not decide whether to list *a* as a vowel or as a word. That the academicians reduced their dictionary to an exercise in dilettantism was bad enough; their dropping whole parts of the language for being less tasteful than others was to Furetière incredible. 'An architect speaks just as good French, talking plinths and stylobates, and a soldier, talking casemates, merlons, and Saracens, as a courtesan who talks alcoves, daises, and chandeliers.' Admirers of Furetière published his dictionary (containing an entry for *guivre*) in three volumes in Holland in 1690, two years after his death and four before the Academy at last presented its own, much sparser, to the court of Louis XIV. The Sun King preferred Furetière's.

Furetière's achievement threw into relief the Academy's shortcomings as a dictionary maker. One man had done what scores, over sixty years of fits-and-starts labour, had struggled – and failed – to do. As a parable of individual endeavour – one head better than a hundred – it is compelling. Even more compelling, and embarrassing for the academicians, was the publication in 1755 of the *Dictionary of the English Language,* by Samuel Johnson. 'Without any patronage of the great; not in the soft obscurities of retirement, or under the shelter of academic bowers, but amidst inconvenience and distraction, in sickness and in sorrow,' Johnson wrote of the seven (some accounts say eight, others nine) years it had taken him to define some forty-two thousand English words. Johnson had had to work fast to pay back the local booksellers whose commission money kept him in ink and paper. The speed had come at the expense of sleep and temper, but not – it was his pride – of quality. Critics, at home and abroad, were impressed. What quill power! Johnson, feigning modesty, called himself English's 'humble drudge'.

They were the remarks of a man looking back, separate from the past, now at ease in his accomplishment. At the start of his labour, though, Johnson's ambition had been much wilder. Having looked at a copy of the Academy's third edition (published in 1740), his thoughts, like the thoughts of the academicians in Paris, turned to fixing his language, making it exempt from the corruption of workers and foreigners. Johnson's initial aim wasn't, therefore, to record every word employed in England. His dictionary was to be selective; in compliance with his patriotism, the author would leave out many French words, or at least discourage the reader from using them. He didn't want his countrymen to 'babble a dialect of France'. Of *ruse*, he wrote, 'a French word neither elegant nor necessary.' *Finesse* he also deemed 'an unnecessary word which is creeping into the language'. *Spirit* used in the French sense, to mean 'a soul' or 'a person', was, Johnson added, 'happily growing obsolete'. Instead of *heroine* he recommended that Britons say and write *heroess*. But in the course of his long toil, Johnson's snobbery softened. Some words were simply too beautiful or too useful to care much about where they came from. Like *paramour,* which Johnson conceded is 'not inelegant or unmusical'.

Johnson, though he began his dictionary with an academician's cast of mind, closed it with quite another. The very idea of a language needing fixing struck him in the end as nonsense. Academies like the French one, he asserted, could never work: sounds are 'too volatile and subtle for legal restraints'; syllables cannot be put in chains, nor speech 'lashed' into obedience. 'The edicts of an English academy would, probably, be read by many,' he noted wryly, 'only that they might be sure to disobey them.'

The year after his election, Sir Michael was honoured with a ceremony welcoming him to the Academy. In keeping with an old tradition, the members attributed to him a word and definition from their dictionary (now on its ninth edition and counting some sixty thousand entries). They chose *universalité:*

> n. f. Ensemble, totalité, ce qui embrasse les différentes espèces. *L'universalité des êtres, des sciences, des arts.* En termes de Jurisprudence, *L'universalité des biens,* La totalité des biens.
>
> UNIVERSALITÉ signifie aussi Caractère de ce qui est universel, de ce qui s'étend à un très grand nombre de pays, d'hommes. *L'universalité de la langue française . . .*

(In my translation: 'a set, a totality, that which sub-sumes the different kinds. *The universality of beings, of sciences, of arts.* In jurisprudence, *the universality of goods [entirety of assets],* the totality of goods. *Universality* also refers to the condition of that which is universal, of that which extends to a very great number of countries, of people. *The universality of the French language . . .* ')

It is on such definitions that Sir Michael and eleven of his colleagues on the dictionary commission work. 'Thursday is dictionary day. The twelve of us sit around a long table for three hours in the morning, going through the latest revisions to be debated one by one. It might be a definition that needs tweaking. Or an entry's example that needs replacing. Or a neologism of some sort. We probably get through around twenty or thirty words a week. It sounds like a sinecure, but we take our duties very seriously. The atmosphere in the room is pretty solemn. You have to ask permission to speak.' He raised his arm, as if he were looking at the commission's chairman, and lowered it again.

'At the same time there's a camaraderie. All those hours together, we become real buddies, *on se tutoie* [meaning that they address each other with the familiar form of *you – tu*].'

The commission was currently scrutinising the possible meanings of *rude*. Unlike the English *rude,* the French has many uses: it can mean 'rough', 'unpolished', 'unsophisticated', 'harsh', and 'severe'.

'Not long to go before we're finally out of the *R*'s.' He smiled. *'Ars longa, vita brevis.'* An academician's joke.

It was not only the *R*'s that kept Sir Michael and his colleagues busy. Responsibility for answering usage queries from the public rests with the commission. In the old days, the queries arrived in envelopes and fussy handwriting. Now, they come typed in emails. *Courriel.* Not 'email'. The commission's answers appear on a dedicated page of the Academy's website. To Edwin S., who enquired whether one should say *Quand est-ce que tu viens?* ('What time do you arrive?') or *Quand viens-tu?*, a commission member replied that the latter was better, even if – it had to be admitted – the former was far more common. A certain Shiraga wrote in to query the French pronunciation of *bonzaï*; in its native Japanese, she remarked, the word is pronounced 'bonssai'. The French should always pronounce it with a *z* sound, the commission said, and added sternly, 'It is not a Japanese word, but a French word that has been borrowed from Japanese.'

Sometimes the responder is Sir Michael. He also contributes to the Academy's style guide (published, too, on the website): *Dire, Ne Pas Dire (Say, Do Not Say)*. According to the guide, for instance, 'we don't say' *il est sur la short list* ('he is on the shortlist'); *shortlist,* too English, is out. 'We say,' instead, *il est parmi les derniers candidats susceptibles d'obtenir tel prix* ('he is among the remaining candidates capable of obtaining such prize'). A roundabout way of putting it, to say the least. And on and on, in the same convoluted vein, the guide goes.

We don't say *une newsletter;* we say *une lettre d'informations.*

We don't say *une single;* we say *une chambre pour une personne.*

We don't say *éco-friendly;* we say *respectueux de l'environnement.*

So many proscriptions! Don't. Don't. Don't. It made for rather unpleasant reading.

But Sir Michael said I had got the guide wrong. The guide isn't anti-English. Many in the Academy he called Anglophiles, admirers of British and American novels, the words of Wordsworth and the like. No, he said, it is a matter of clarity. Many English words confuse. The 'artificial English', bizarre and stunted, that globalisation peddles. On posters in the Paris subway, on billboards near Notre Dame, in loud radio ads that resound along boulevards: *Just do it! Nespresso, What Else? Taste the feeling! This is Her! This is Him!* Sir Michael sees all around him landscapes disfigured by these meaningless slogans. 'It is something my colleagues have long and rightly bellyached about. The health of a nation,' he said, as if reciting, 'depends on the health of its language.'

Thus the aesthetic defence of French has given way to the ethical. No longer is the Academy protecting French for gentlemen; it is preserving French for the common man. But behind the ethical posture, as behind the aesthetic, the same anxieties, the same obsessions. A kind of language panic. In 1985 in a public speech to the Academy, Sir Michael's predecessor, the author Jean Dutourd, denounced 'the murder of syntax, the genocide of the dictionary' being committed by the boorish purveyors of 'Atlantic pidgin'. They had the 'rapacity of real estate developers,' Dutourd warned, who, given half a chance, would demolish the 'palace' of the French language to build a super-luxury high-rise over it. He wanted the French government to take action. He told his genteel audience he wanted to see a 'linguistic inquisition' in France. He urged the finance minister to create a grammar inspectorate whose task would be to comb the press, books, and ads for bastardised words. Anyone who published *nominer* ('to nominate') instead of *nommer* (the older French form), or who, under the influence of English, mixed up *sanctuaire* and *refuge*, would have to pay a fine of twenty francs. A tax on words!

Admittedly, Sir Michael is no Jean Dutourd, (whose books, before coming to the Academy, Sir Michael had never read.) He is a moderate. He does not believe in taxing words. Like the Academy's website, his moderation is a concession to the twenty-first century. 'French is changing, France is changing. I'm part of the changing face of France.' Behind his grey head, the frilly orange-red lampshade was distracting. 'Of course the language needs to be fixed to some extent, to remain readable one hundred years from now,' he continued, 'but we mustn't try to stop the future' ('We mustn't be Jean Dutourds,' he could have said). 'Can we really expect any human institution not to have its share of boring old farts?'

'We mustn't try to stop the future.' But I couldn't be sure Sir Michael meant that. I couldn't understand why the academy considers, for example, *jazzman* and *blackout* and *fair play* and *covergirl* all acceptable French words, but thinks *shortlist* too English, or degenerate English. *Blackout* turns up in several novels by Patrick Modiano, winner of the 2014 Nobel Prize in Literature. J. M. G. Le Clézio, France's other living Nobel laureate (he won in 2008), uses *covergirl* in his novel *Désert*. What is to stop either of them writing *shortlist* into a future work of French literature? Certainly not the academy's style guide. Neither writer – I note in passing – has shown any interest in joining the academy.

It was getting on – with all the talking, we had forgotten about the clock – but before leaving Sir Michael, I asked to see the famous books and rooms at the Institut de France that he and his fellow academicians use during their conclaves. He said that could be arranged. The academy would sit again in the autumn; I should send him a reminder then.

I did. And Sir Michael was as good as his word. After summering in Burgundy, he replied with a date, Thursday, November 12, and an address, 23 quai de Conti. (The assembly's address was superfluous. My apartment was little more than a stone's throw away: I walked past the pillared façade every time I crossed the Pont des Arts.) On the twelfth, at an hour when the afternoon *séance* was winding up, I presented myself to the young lady at the reception. I said I was there to see 'Michael Edwards.' The receptionist looked blank. I tried again. I said, pronouncing his name as if it were a Frenchman's, 'Michelle Edooar'.

'Ah, Monsieur Edooar!' She handed me a security badge, let me through the electronic turnstiles, told me where to go to sit and wait.

Across the cobbled courtyard was the door, closed, forbidding.

I tried the door. It creaked, then opened. Busts, tapestries, chandeliers. Presently Sir Michael came down a staircase. The setting's pomposity stiffened his gait. The academicians had been 'inside' for their weekly one-and-a-half-hour plenary session, he explained, during which the dictionary commission's suggestions were mooted. Thursday afternoons at the academy often went smoothly, but this one had been more difficult. He was about to tell me more, but apparently thought better of it. He changed the subject. 'Let me show you the library.'

The Mazarin library: the seat of French letters. Six hundred thousand volumes wall to wall. Here, amid the smell of vellum, musty leather, and millions of pages steeped in centuries, Sir Michael's 1965 thesis on Racine rubbed covers with the original plays; with a 1580 copy of John Baret's *Alvearie*, a multilingual dictionary – English, Latin, French, and a sprinkling of Greek – 'newlie enriched with varietie of wordes, phrases, proverbs, and divers lightsome observations of grammar,' which Shakespeare, in composing his own plays, is thought to have consulted; with *Notre Dame de Paris*, by Victor Hugo, who became an Academy member only after being turned down thrice; with the collected poems of Baudelaire, whose sole candidacy, in 1862, went nowhere.

Sir Michael said, 'I believe that's because he'd had a run-in with the law.' The Academy had also rejected

Molière (whose name has become forever associated with French: *la langue de Molière*), Pascal, and Zola at one time or another.

About his own candidacy, Sir Michael said, 'You hand-write a letter to each academician. And each letter has to be personalised. You're advised not to write too much. I may have gone over to a second page.' He admitted to having been unfamiliar with the names of several of the academicians; he'd had to mug up the titles of their books, their themes, and their styles beforehand.

Walking me through long corridors from the library to the Academy's assembly room, Sir Michael showed me a statue of Jean de La Fontaine, beside which, on his first day, as the 'new boy', he had been made to wait until called. Inside the stuffy assembly room, below the gilded portrait of Richelieu, forty plush red seats were laid out in an oval. Each academician's seat is numbered. Sir Michael's is Seat 31. Cocteau once sat there, as did the author of *Cyrano de Bergerac*, Edmond Rostand.

It was here, one Thursday afternoon, in seat 29, that the anthropologist Claude Lévi-Strauss gave his defin-ition of *boomerang*. And it was here, on another Thursday afternoon, that Lévi-Strauss persuaded his colleagues to modify the dictionary's entry for *rance* ('rancid'). According to him, the entry, which spoke of a 'disagreeable' taste and odour, betrayed a Western bias. For many cultures, he told the room, rancidity is part and parcel of their cuisine. His complaint was upheld: *disagreeable* was cut and replaced with *strong*.

I wondered whether Sir Michael would also leave a

mark on the Academy's dictionary. I wondered whether he would speak up, out of his Englishness, when the time came for the assembly to look again at *sandwich*, say, or *turf*.

He read my thoughts. Or perhaps his being back in the place where, only an hour or two before, emotions had flared and words had been crossed, moved him to recount.

'I suppose I bring to the Academy a way of speaking that's a bit different. I'm not above making a joke. At the same time I'm not afraid to speak my mind. I have opinions.' He stopped, looked around, lowered his voice and said in a confiding tone, 'I had a bit of a dingdong with Giscard.' Giscard is Valéry Giscard d'Estaing, nearly ninety, president of France between 1974 and 1981; he was elected to the Academy in 2003. The Englishman's and ex-president's dingdong was over how to sum up *vamp*. 'He started it. He claimed it means *une belle séductrice*. Pure and simple. I said "No, it means more than that." I said it came from *vampire*, and therefore suggests a dangerousness alongside the beauty.' Apparently much of the session was taken up with their dispute.

The French penchant for abstract, never-ending debate! After ten years living in France, I know it well. Frenchmen and Frenchwomen, it is widely said, pay little attention to the Academy, and in my experience they don't have to. A family lunch, lasting hours, is their academy. Sharing a wine with friends on a noisy bistro terrace is their academy. How the French love to talk! And to talk about talk!

Back at the reception the lady helped Sir Michael into his long black Burberry coat. He was off to a Schubert concert; he is an inveterate concertgoer. 'Our annual meeting is in three weeks' time. All the academicians will be there. Dressed up in our uniforms, you know, like a bunch of deposed South American junta generals. I'll see you're invited. I'll be presiding.'

The following evening, Friday, November 13, 2015, two miles from the Academy and my home, one hundred and thirty rockers, bistro diners, and café drinkers were killed in a coordinated attack by radical Islamist terrorists. Hundreds more were wounded. A state of emergency was declared.

Violence has a way of reducing the mundane to a pure ridiculousness. Already it had been hard enough to begin to write about the French Academy – its quaint rituals and pointless-sounding arguments – without making it sound absurd. Now 130 deaths. In comparison with them, what did it matter how a dictionary defined *vamp*?

But the more I thought about the violence – so ruthless, so flagrant, designed to wow and whet hatred – the less absurd the Academy, its work, its role, began to seem. Violence levels everything, turns bricks, bottles, and bodies indiscriminately into so much rubbish. The academicians, contrariwise, seek to discern, weigh, conserve. Violence silences; the Academy champions words. *La mort injuste* versus *le mot juste*.

President Obama, offering American condolences, reached for French words: 'The American people draw strength from the French people's commitment to life, liberty, the pursuit of happiness. We are reminded in this time of tragedy that the bonds of *liberté* and *égalité* and *fraternité* are not only values that the French people care so deeply about, but they are values that we share.'

And in the days after the attacks, throughout Paris, a book originally written in English, by an American, sold by the thousands: Hemingway's *A Moveable Feast*.

Once again, hearteningly, that close relationship between the two languages.

Three weeks later, with my invitation card, I waited in line at 23 quai de Conti before passing security. December's Academy was more austere in the chill air. Men in Swiss guard-like costume, all white gloves and red feathers, clanked to attention as we filed inside. We sat under the cupola – a building set aside for such grand occasions – concentrically. In the audience, French household names: a singer, a filmmaker, a telegenic author. And the capital's bourgeoisie.

Solemnly, the academicians came in and sat together. Each wore a green and black waistcoat, bicorn hat, cape, and épée. They were very white, very old, very male. (About the Academy's woman problem Sir Michael had told me, 'The Academy falls over backwards to elect women; we want women; not enough women apply.') Sir Michael asked for a minute's silence in memory of the victims.

After the silence, after the applause for the winners of the Academy's literary prizes, after a speech on the history of the French novel, Sir Michael cleared his throat and spoke. He spoke of violence. Not the violence of gunmen – the type, exercised by poets, (in my translation) 'against conventional perspectives on life and self, against tired clichés'. But the poet's violence, Sir Michael continued, was that of someone who 'cracks a nut, or bites into a fruit'. A 'violence of the poetic act', which is always accompanied by 'gentleness in the respect for the real'.

Reality, he concluded, responds to language. 'Reality,' he had told me, months before, in his apartment, 'is polyglot.'

11

OULIPO

Pity who, as typist, in 1969, had to turn a famous Parisian author's fourth manuscript into book form: *La Disparition* (*A Vanishing* or *A Void*). Our typist sits at his contraption and, palms cast customarily downward, starts jabbing away; but rhythm is hard to find. Fit digits, usually busy, now look to him lazy or numb. Tsk-tsking this or that digit won't do him any good, though. It's all about using his hands unorthodoxly. It's all about application, knuckling down. His right pinky works, his right thumb works. Clack, clack, clack . . . ding. Clack, clack, clack . . . ding. But application or no application, typos abound; a hand has its habits. So our struggling typist starts again, and again, until paragraphs, long and looping, catalogs of copious food and drink, slapstick situations, parodic imitations of Arthur Rimbaud and Victor Hugo, comic accounts of killings, sixty thousand-plus words total, finally all turn out without a slip – without so much as a whiff of *français*'s most common symbol twixt *d* and *f*.

A short illustration:

Anton Voyl n'arrivait pas à dormir. Il alluma. Son Jaz marquait minuit vingt. Il poussa un profond soupir, s'assit dans son lit, s'appuyant sur son polochon. Il prit un roman, il l'ouvrit, il lut; mais il n'y saisissait qu'un imbroglio confus, il butait à tout instant sur un mot dont il ignorait la signification.

(In my translation: 'Anton Vowl couldn't drift off. On had to go his lamp. His *Jaz* alarm clock told him it was past midnight. With much sighing Vowl sat up in his pyjamas, his pillow for a prop. Took a book to dip into; but saw in it only an imbroglio of ink, fumbling to grasp this or that word's signification.')

La Disparition's author was 'Gargas Parac'* (an alias of his making). Born in 1936, an orphan at six (his *papa* and *maman* both victims of Nazism), Parac was a sociologist by training – working in public opinion polls and as an archivist/information coordinator until his skilful and ambitious writing paid off. A first book, *Things,* was brought out in 1965 and won him instant standing and a major *prix*. But it was his falling in with OuLiPo (an acronym that roughly stands for 'workshop for writing within constraints'), a multinational group of *avant*-guard authors, that would prod his imagination to think about words and grammar in a surprisingly original fashion.

(Also in OuLiPo: Italo Calvino, writing in Italian, author of *Cosmicomics* (with, for narrator, Qfwfq), *Mr Palomar* and his *Città Invisibili,* in which Marco

---

* Editor's Note: With the fifth letter of the alphabet in play, the pseudonym is, of course, Georges Perec's.

Polo and Kublai Khan discuss Polo's distant and fantastic sojourns.)

Writing within constraints: for Parac, as for Calvino, it was an opportunity to pull back from linguistic norms and so avoid stock ways of using words. Parac's boundary was lipogrammatic: truncating his ABCs by a most popular symbol, and sifting through his vocabulary to pick out all words containing it. Up to two thirds of quotidian *français* was thus out of bounds for his forthcoming book. Just thinking of writing within such limits might throw many of us into a panic. But for Parac it was a stimulant to lift him out of his chronic author's block.

Lipograms had a long history, dating back to antiquity. Lasus of Argolis, writing around 500 BC, known primarily for his dithyrambs and for tutoring Pindar, was also an originator of hymns minus any sigma (*s*), its hissing sound grating on him. Lasus wasn't a 1-off. At Romanticism's high point, a Gottlob Burmann (1737–1805) of dainty constitution brought out an opus of soft-sounding lyrics that contain no harsh *r:* no *Frau*, no *dürr* ('thin'), no *rann* ('ran').

In Paris, Parac's lipogrammatic writing was a first. Its author had at his disposal a total of 25 ABC symbols, originating in Latin, not all individually so old. Copyist monks in Norman scriptoria had had no *j* or *v:* quills would put down *iour* for *jour* ('day'), *auait* for *avait* ('had'); it wasn't until 1762 that *j* was always told apart from *i*, and *u* from *v*. W, though known to Walloons, did not gain a national dictionary's sanction until 1964.

So Parac had *w* to put in writing (a joy to him, having a soft spot for it), *j* and *v*, *s*'s that Lasus did without, and *r*, which Burmann was always avoiding. But not what a schoolchild is normally taught to jot down – along with *a*, *i*, *o*, and *u* – whilst still small. Not what looks much as a mirror's portrait of a *3*. A most difficult taboo to work around: statistically it amounts to 1 in 7 ABC symbols in a Gallic book (by way of comparison, 1 in 8 in a Californian or British opus, 1 in 6 in Dutch). Our author had his work cut out for him.

As a backdrop to his writing, Parac had Paris's May 1968 riots: a capital in turmoil, young anarchists occupying public buildings, shouting slogans, scrawling graffiti against an out-of-touch administration. Orthographic acts of fury. No to *flics* (cops)! No to *intox* – short for *intoxication* (propaganda)! Plays on words, on symbols. An infamous placard said all it had to say succinctly, in just a handful of symbols: *CRS SS* (*CRS* is an acronym standing for *riot control units*). Thus, a community's *a*'s and *s*'s and *x*'s and so on could impart information in print, could signify, without always participating in a word. A word wasn't simply a unitary transcription of sounds; its constituting symbols (*f-l-i-c-s*, *i-n-t-o-x*, groupings as familiar as '*c-r-s* – *s-s*') also had things to say. This was Parac's crucial insight – his bolt of brain lightning.

Part artistic, part political, our wordsmith's goal was to show how important ABC layout is in how words in a book, a tract, a tabloid, a glossy, 'talk' to us and 'think' for us. Linguists call this *visual iconicity:*

symbols composing a word mimic, up to a point, what it talks about. Iconicity, at its most basic, says that short words using an ABC's most popular symbols – such as *rash* (*r-a-s-h*) – will, broadly, bring up day-to-day things; long words that consist of uncommon symbols or unusual combinations of symbols – such as *psoriasis* (*p-s-o-r-i-a-s-i-s*) – discuss knotty, atypical affairs. In addition, symbols' forms in union can act as a visual aid, portraying a word's topic. By way of illustration, think of *locomotion*. Its *l* is a pictorial fuming puff; its consonants, wagons; and its four *o*'s, what roll on rails. Similarly, Victor Hugo thought *lys* ('lily') a right orthographic form and its rival *lis* wrong. For *y* to his mind was akin to a sprig with consonants as blossoms surrounding it.

Browsing *La Disparition,* taking in its paragraphs, a striking thing is its fair lack of diacritic marks. So far as 'hats' on nouns go, only a small quantity show up: *août* ('August'), *trouvât* ('found'), *chaînon* ('link in a chain'). It turns out that as soon as you start amputating all words containing *français*'s fifth ABC symbol, you must also lock out much vocabulary that is put down diacritically. To purists, Parac's book, its words' physiognomy, is an optical shock. Almost an insult. A 'hat' on a noun – an offshoot of *Orthographia Gallica* (composition around 1300) – boasts symbolic status today. It's a bit posh. It's why folk who want to climb a social rung will occasionally add a 'hat' to nouns that don't don it, putting down, say, *ajoût* ('addition'); on all occasions it is right to put *ajout*. It's why a 1990 commission's proposal to simplify orthography by dropping *français*'s surplus 'hats' got swiftly told off by columnists and public outcry.

(Occasionally an ABC symbol can amount in authors' minds not to a plus but to a stigma: in Russia, around 1900, writings by anti-Tsar radicals would fashionably omit from nouns Cyrillic's 'hard sign', (*tvyordy znak*) – a ъ that marks a word's final consonant as non-palatal – associating it with backward-looking ritual.)

La Disparition is a zany whodunit, a curious story about abyss and oblivion, in which a man, Anton Voyl, fights awful insomnia. 'Why amn't I dozing?' Voyl asks. Doctors can only shrug. Psychopharmacologists' pills do not work. Gradually, Voyl and a fair many of his pals drop out of sight, go missing, vanish into thin air. Victims of booby traps? Shootings? Stabbings? Falling pianos? A mafia's hitlist? In hot pursuit, kith and kin toil to find a solution. Parac was a fan of Franz Kafka, and it shows. Mortality is a constant, as in *Richard III:*

> *I had a Richard too, and thou didst kill him;*
> *I had a Rutland too, thou holp'st to kill him.*

An odd mood draws you in, a combination of story and orthography. It's a story told in 25 parts (jumping straight from Part Four to Part Six); a taboo symbol throughout, conspicuous in its omission. Parac, handling his biro with brio, pulls it all off.

I watch an old black-and-gray TV program on which Parac, chubby, a dark billow of curly hair atop his brow, is stating that his story is auto-organising associations of ABC symbols. His book is its own author, a product of an 'automatisation of writing'. His collocutor and host, going off script, gasps quickly to his public watching, 'I should add that this isn't a *canular* ["hoax"].'

Automatisation of writing? But it is not only promotional tour talk. Think about it. By putting his mind primarily to symbols – not words or plot – Parac was following an original driving logic. Grammar is occasionally his own: a man's brain hurts with *'un fort migrain'*; folks talk with *'cordons vocaux'* ('vocal ribbons'), not vocal cords. If a word such as *lip* is out of bounds in *français* – damn taboo's fault! – our author puts down *'pli labial'* ('labial fold'). Lists push his story forward: *'son minois rubicond, mafflu, lippu, joufflu, bouffi'* ('his rubicund mug, ruddy, lippy, chubby, puffy'). Syntax also adjusts: a first paragraph, a following paragraph, a third, all start with, *'oui, mais'* (*'oui*, but').

'Portons dix bons whiskys à l'avocat goujat qui fumait au zoo.' This postscript in Anton Voyl's diary is also a pangram (a 'Parac' pangram, that is, containing 25 ABC symbols out of 26). A variant in *anglais* might go, 'a quick brainy fox jumps with a guava lizard.'

Parac also plays with words put in print by past virtuosos. Full lipogrammatic modifications of ballads, cantos, stanzas. To supply a notion of how this turns out, my adaptation of 'Daffodils', by William Wordsworth:

> *I, solitary as a cloud*
> *That floats on high past coombs and hills,*
> *Without a warning saw a crowd*

A host, of brilliant daffodils;
Along a pool, among hawthorns
Flapping and dancing in mid-morn.

Continuous as bright stars that glow
And glint-glint on our Milky Way,
Ranging in an undying row
Along a margin of a bay:
Six thousand saw I in a flash
Tossing gold crowns in sprightly thrash.

Vivid pool, but my daffodils
Outdid its sparkling in spirit:
A rhymist could not but turn gay
In such a jocund company:
I rapt – so rapt – but hardly thought
What fund this show to us had brought:

For oft, if on my couch I sigh
In vacant or in gray humour,
Gold will flash on that inward mind
Bliss of my on-my-own hours;
And so my soul with passion fills
And frolics with my daffodils.

From paragraph to paragraph throughout his book, always omitting what is twixt *d* and *f*, Parac brings in words from many lands. Thus crop up words in Latin: *oppidum civium romanorum* and *sic transit Gloria Mundi*; words in Italian: *Ah, Padron, siam tutti morti*; USA words: *It is not a gossipy yarn; nor is it a dry, monotonous account, full of such customary fill-ins as 'romantic moonlight casting murky shadows down a long, winding country road*'; and 'Saarland patois': *man sagt dir, komm doch mal ins Landhaus. Man sagt dir, Stadtvolk muss aufs Land, muss zurück zut Natur. Man sagt dir, komm bald, möglichst am Sonntag.* Truly, *La Disparition* is a multilingual work.

Today, translations of Parac's book sit on racks all around our world: in a city such as Hamburg (*Anton Voyls Forgang*), in Italy (*La Scomparsa*), in Croatia, Holland, and Romania. In Spain it contains no *a* (*a* is Spanish's most common ABC symbol); in Japan, no *i*; in Russia, no *o*. Translations in many lingos; who knows, a stalwart translator may soon put it into South India's Malayalam.

Today, too, computational analysis of *La Disparition*'s prolixity throws up fascinating linguistic data. Word clouds highlight its most-occurring nontrivial vocabulary: *sans* ('without'); *savoir* ('to know'); *grand* ('big') – most words for talking about small things contain Parac's taboo symbol. Also common: *mort* ('dying'); *mot* ('word'); *blanc* ('blank'); *noir* ('black'); *nuit* ('night'); *obscur* ('dim'). Statistical analysis also shows that, for all its innovation, Parac's book strictly follows Zipf's law (found by a Harvard corpus linguist, G. K. Zipf – it was said that if you bought him a Floribunda, G. K. would promptly count all its thorns). Zipf, probing and tallying a library's worth of words, found that in any book roughly half its total vocabulary will occur on only 1 occasion; a book's bulk consists of a handful of its most common words. According to my back-of-a-napkin calculations, *La Disparition* has a vocabulary of around

8,000. A tiny fraction, its 100 most occurring words, crop up so continually as to fill about half of its manuscript; just 400 of its most common words occupy four fifths. Thousands of its words occur only sporadically, turning up on just two occasions throughout Parac's publication; four thousand pop up on just 1.

Including:

Alunir ('to land on a moon'); *axolotl* (a Nahuatl word for a tropical amphibian); *finlandais* ('Finnish'); *hot-dog; infarcti* (plural of *infarction*), *opoponax* (a kind of myrrh – also, a 1964 story by 'Monica' Wittig, told with unusual childish narration); *pawlonias* ('Paulownias'); *roucoulant* ('cooing'); *taratata* (a cry of doubt); *uxorilocal* (a social anthropological word for a man living in proximity to his missus's clan).

(Warning: a quantity of Parac's 1-off words, such as *s'anudissant* – 'discarding all your clothing' – you won't track down in any standard dictionary.)

And looking again at *La Disparition*, as if with a magnifying glass, past its words, to its individual ABC symbols, you find Zipf's law is also at work. Most of its words contain *a*'s and *i*'s and *n*'s and *u*'s and *s*'s; not many contain *w*'s, *z*'s, *k*'s, *x*'s, or *j*'s. O, it turns out, gains most from Parac's bold plan: *o* round as a nought, a fitting symbol for omission.

If Parac (d. 1982, at 45, of a lung tumour) was still with us, h'd b a fn of txt msging. Not paying a thought to alarmist spoilsports who call it 'sad shorthand', 'dumbing down' or a 'digital virus'. No. To him, just gr8 2 play with ltrs in this way, making word drawings such as *bisouXXXXX* (a common nding to a Frnch txt msg, similar to *kixx* for 'kiss kiss') and wOw!, was^? ('what's up?'), and 'i'm off 2 bd zzzz.'

An ABC is a living, fluid, multivocal thing.

12

TALKING HANDS

Visiting a château outside Paris last Christmas, I stepped from a fire-warmed room to find the corridor in commotion. A cascade of kids, in frilly blue and green costumes, came rumbling and tumbling down the stone-slab stairs. On a festive club outing of some kind, the girls and boys in their fancy dress radiated excitement. There was the clack of their heels hurriedly meeting each step, the slap of little palms against thick walls, and the froufrou of flappy outfits as they raced and tussled good-naturedly. And intermingled with all of that, something like the burble of collective physical presence, of many simultaneous intakes and outtakes of breath. 'They can be *so* noisy,' muttered a middle-aged woman beside me. Yet among the group not a word was spoken. Not one shout or shriek. Then it dawned on me that they were deaf children; and, later, on my way out, passing through the grounds in the wintry gloom, I crossed them again, huddled now with their caregivers beneath a street-lamp, their busy fingers conversing.

La langue des signes française, the children's sign language – for it most definitely is a language, complete with syntax and morphology and slang – is one of many: every continent has its own distinct forms. But France can claim sign language as England claims English (that more famous global export). A Frenchman, Laurent Clerc, was instrumental in the creation of America's first school for the deaf, in Hartford, Connecticut, in 1817. The system of gestures and facial expressions Clerc brought to the United States had been learned as a boy in Paris, according to methods devised by one Abbot de l'Épée in the mid-1700s. In Yankee hands, the result would become ASL, American Sign Language, which in turn fed several of the forms used widely today across Africa and Asia. So it was that the French, they of the Gallic shrug, as famous as their Italian cousins for the expressivity of their bodies, taught the world to sign.

ASL's quirks, with their roots in the games and garments, the duels and ideas of eighteenth-century France, can mystify the observer. Why do the signs for *tomorrow* (literally, 'one [day] into the future') and *yesterday* (literally, 'one [day] into the past') enlist the thumb? Because the French count *one* on the thumb and not, as do Americans, on the index. The thumb that flicks out from under the chin to mean *not* in ASL is the French child's still-popular signal of defiance. *Cannot,* the right index striking down against the left, mimics the sort of swordfights of which Europe's noblemen were once so fond. And if you rightly see a reference to the head in *government* – in which the index lands on the signer's temple – it is only half an explanation: the hand carrying the index first rotates, perhaps drawing the revolutionaries' tricolour cockade.

I was in Montreal, then in Ottawa to see family, some months after that château visit; and, since my aunt spoke French and would have seen deaf patients during her many years at the local hospital's rehabilitation service, I shared my interest in ASL's origins with her. I had even prepared an anecdote (which Emily Shaw and Yves Delaporte recount in one of their *Sign Language Studies* papers), of American commentators interpreting *stupid* – the peace or 'V' sign raised to the forehead – all wrong; they claim it represents prison bars that curtail the mind. The gesture, in fact, emulates horns and for a very good French reason: *bête* can mean 'stupid' but also 'beast'.

I never did get round to telling my aunt, Margo Flah, the anecdote. I did not need to. In her youth, she said, she had learned to sign ASL conversations while working at US centres for the deaf. On moving to Canada, she could be seen signing the evening news on television. Car crashes and parliamentary speeches and lottery millions passed through her fingers. People would come up to her in the street. It was a whole side of her I had never known.

Margo offered to teach me the signs of the alphabet. She sat me opposite her at one end of the long table in the living room; and, following her lead, I closed the fingers of my left hand flat over the palm. '*A*,' she said. Then my hand, mirroring hers, contorted into a *B*, a *C*, a *D*, an *E*, an *F*. 'Wait!' She broke off. 'Are you left-handed?' ASL users, she told me, normally sign with their dominant hand. I write right-handedly, but for some reason, have always thrown with my left. So the choice of hand seemed to depend on the nature of signs: were they something like objects that you pitch through the air, or more like calligraphy, with the space around your body for paper and your hand and arm for brush? I decided to switch to my writing hand. Margo looked relieved. She continued, letter by letter, on through the Zorro-like flourish of *Z*. Her kindness, her patience, made an enthusiastic student of me. Several times, we ran through the gamut and lingered over those letters my hand did not get quite right. How sharp the practised signer's eye is! Fingers a fraction too low over the bent thumb formed an inadequate *E*. The fist of *S* needed to be a little tighter. Too much bending of the index spoiled the *X*. Once or twice, leaning over the table, Margo took my hand in hers and gently adjusted my fingers as though setting the time on a slow watch.

'I have two friends you should really meet,' Margo said after we had finished signing the alphabet. She invited them round to the house, some mornings later, for coffee and cake.

Notwithstanding his French name, Michel David was raised in an English-speaking Ottawa household. He was a bright-eyed man in his early sixties. He wore Russian green trousers, a chequered shirt beneath a beige sweater, a trim white beard, and, in his left ear, a cochlear implant device.

Monica Elaine Campbell, a few years younger than Michel, was tanned and stylish in a blue-and-white silk scarf, matching blouse, and denim skirt. Her wrists jangled with bracelets, and on the fingers of both hands she had several gold rings. Unlike Michel with his device, she could not hear. She read my lips when I spoke.

We sat around the plate-laden table, in a warm smell of brewing coffee – Michel and Monica Elaine side by side, facing me, with their backs to the window and the bright April sun. Such preliminaries assured Monica Elaine adequate light to see every movement my face and mouth made. Carefully, without hurry, I articulated my words, first of introduction, then of questions for them; in response, Michel and Monica Elaine used ASL (the prevailing sign language in Anglophone Canada) and spoken English.

Michel said, 'I'm small-*d* deaf,' meaning he did not consider himself culturally deaf; he did not have ASL as his first language. Hearing loss ran in the family, though. As a small boy sitting in the kitchen, he saw his grandmother's cousin communicate in gestures. Around the same time, he noticed that he had a toddler's sense of balance; he had to gaze at a fixed point in the distance as he approached it in order to keep himself plumb. (At night, the lights out, if he ventures to the bathroom without his device he says he stumbles along the landing 'like a drunk'.) He lay in bed for hours and felt his ears ache. At nine, it seemed to him his parents were always stepping out of his earshot. Forty-decibel loss, the audiologist told them. And that was only the beginning. The son's earshot grew narrower and narrower by the year. Voices, even the loudest, even the most familiar, became patchworks of guesses and memory. In his teens, he turned

to ASL; he attended evening classes. Then, one afternoon, while cutting his parents' grass, he heard his ears
pop and then nothing. 'I thought the mower had died
on me.' He was twenty. It was the beginning of thirty
years in total deafness.

Michel's face was turned toward Monica Elaine as
he spoke. She watched his brisk, fluent signs; read the
accompanying expressions on his face; and, when he
had finished, nodded at the gesture for her to begin.
To take a turn. The pouring of pepping coffee, the
slicing of aromatic cake were understandably distracting;
and it is considered rude, Michel added, to sign with
your hands full.

Monica Elaine was prelingually deaf: her deafness
had been discovered at fifteen months. But her parents
did not want their daughter to acquire sign language
– that would have meant either having to learn ASL as
a family or else packing the infant off to a distant
school for signing children, and neither option seemed
acceptable. So Monica Elaine stayed put with her
hearing siblings – three brothers and one sister (a future
teacher of the deaf in the United States) – amid the
picturesque quays and harbours of Charlottetown,
Prince Edward Island.

Presently a new school opened on the island, a school
for the lip-reading and speaking deaf. Monica Elaine's
parents took the John Tracy Clinic correspondence
course to prepare to send their daughter there. They
acted on advice from Los Angeles (Spencer Tracy,
John's father, was the clinic's main benefactor). The
advice was straightforward enough: speak to your deaf

child in slow, complete sentences; include her in the everyday life of the family; treat her like any other inquisitive little girl. They followed these principles to the letter, making Monica Elaine fully 'speech ready' for when school started. When she was four, it did. In front of a mirror she was taught to form letters in her mouth: the *p* on the lips, the *t* by raising the tip of her tongue to the palate, the *h* by misting the glass with her exhaled breath. Touch – a nervous hand raised to the teacher's cheek and throat – communicated good pronunciation. Little by little the pupil acquired her soft, measured voice.

Monica Elaine's voice is pleasant, easy to understand. Rare are the occasions when she drops a sound from a word ('migle' instead of *mingle*), or replaces one for another ('cazier' for *cashier*). And yet behind the ease lay much personal pain. Growing up, she felt herself 'teetering between two worlds', unsure of her identity. Her childhood confusion turned to an adult's anger at not being on signing terms with so many of her peers. Deeply she regretted that her brain had been made to resemble that of a hearing person's. One day, she resolved to study ASL. Never too late to learn. She was thirty-seven.

What courage and perseverance! And it became clear as the morning went on how hard Michel and Monica Elaine had had to fight to lead full lives. They grew up at a time when the attention given to deaf adults' talents and aptitudes was small. Both had nonetheless found in themselves sufficient reserves of determination and self-belief to earn degrees and launch careers.

Michel: 'People saw me as a gardener. I had a diploma in horticulture. But then I thought, wait, I can do more. I have a mind. I remembered reading a James Michener novel, *The Source*. The story opens on an archaeological dig in Israel. I got the dollars together and flew to Tel Aviv. I was twenty-one. I spent three months on a kibbutz. Then I took a boat to Cyprus and Greece. I hitchhiked everywhere: through what was then Yugoslavia, Austria, Germany, France, Belgium, Holland. When I was done with Europe, I flew with a Soviet carrier to Japan. I was there several months.'

'Did you use sign language with your hosts?'

'Yes. That's the great thing about signing, wherever you are you can usually make yourself understood. And I found foreigners in general to be much more patient with a deaf person.'

The experience emboldened him to pursue academic studies on his return home. 'The hardest course for me was French. I was forever mixing up the meanings of *entendre* and *comprendre*.' After a bachelor's degree in psychology, he moved to Toronto to obtain a master's in social work, started a support group for deaf adults there in 1986, and later became a mental health counsellor with the Canadian Hearing Society.

Monica Elaine: 'The professors at my university would sometimes walk up and down as they gave their lectures, which made it impossible to lip-read what they were saying.' During one of these, the professor spoke for two hours, pacing up and down before garish striped wallpaper: the 'visual noise' was uncomfortably loud.

And yet she persevered. Mathematics and biology, 'very visible subjects', allowed her to play to her strengths. She went on to a long career in human resources.

'At around the time I learned to sign I became involved in palliative care for the deaf.' Too ill to write, many deaf patients were deprived of the comfort of communicating with family and friends. She went from hospital to hospital, from deathbed to deathbed, interpreting patients' last signs, reading bloodless lips. For this, in February 2016, Monica Elaine was invested into the Order of Ontario, the province's highest civilian honour.

Michel and Monica Elaine had something else in common, I learned. Their respective partners are hearing. Communication within the couples consists of a mix of speech and signs. Michel is a father of five: all are native signers. 'Their first sign, at around ten months, is "milk".' He clenched his fist. 'Like this. Like pulling an udder. They use it to mean "food". They point to milk, crackers, juice and use the same sign. It goes to show how early signing babies can generalise and develop concepts.'

Michel had his cochlear implant operation twelve years ago, at 49. His life changed. 'My daughter Jessica was three. What a joy to hear her crying!'

He could jump at a clap of thunder, thrill at classical music, or savour the quiet of deaf meetings. But the implant, he acknowledged, is considered a threat in many quarters of the culturally deaf – the big-D Deaf – community. Some fret that it encroaches on their way of life, threatens Deaf pride, undermines the use of

sign. Michel, though, did not think in terms of either/
or. He considered himself multilingual. He knew the
Deaf culture (as did Monica Elaine), for which he had
respect. 'They are very honest. Honest to the point of
bluntness. If you have put on weight since they last saw
you, they will sign that you look fatter. At the same
time, they won't come right out with their problems.
They will tell them within a story.'

And Michel and Monica Elaine could understand
the Deaf community's vigilance, its desire to cherish
all that made sign language so distinctive.

Monica Elaine said, 'For instance, to say something
like, "Have you ever been to New York?" you use the
signs for *New York, touch, finish,* and *question.*'

She signed them for me. To express *question,* the
signer's index bends and points in the direction of the
person being asked. The upper body also leans forward.
ASL grammar, Monica Elaine continued, expanding,
is spatial. For instance, leaning forward can also mean
the signer is describing something that will happen;
likewise, leaning back can indicate that the talk is of
what is already in the past. If the signer leans a little
to the right (assuming the signer is right-handed) the
other person knows that whatever is being signed about
occurred only a matter of minutes ago.

One hour conversing with Monica Elaine and Michel
grew into two, then three. We were surprised when our
host, my aunt, told us. We had not seen the time pass.

I raised the fingertips of my open right palm to my
chin and lowered them in the direction of Monica
Elaine and Michel, recalling a sign Margo had taught

me at the same table several days before. I was thanking them: for the morning in their company, for all that they had taught me about ASL, and for so much more besides.

13

TRANSLATING FAITHFULLY

The Neapolitan writer Erri De Luca, a nonbeliever, is nonetheless a devout student of the Old Testament. Every morning he rises at five o'clock and begins his day by turning to a verse, a psalm, a story, in the original Hebrew. De Luca is not a scholar; he is wakened by a lean body that, for many years, routinely rose before dawn to labour in cesspits, to lug baggage across airport tarmacs, to work aloft on scaffolding, coffee-smelling and sunburned. Through those years of hard labour, his morning minutes with the Bible, each day nudging his self-taught Hebrew a few words nearer to fluency, fortified him. 'During the period when I was exposed to the Bible for the first time, I was in the desert of my life, and I needed a desert book,' De Luca told a reporter for the Hebrew-language newspaper *Haaretz* in 2003. He reads its stories for the self-scrutinising distance they give him, for the sober elegance of their sentences, for the company of their characters. He reads them as literature ('the God of Israel,' he once wrote, 'is the greatest literary character of all time'). Other books – he thinks – suffer in comparison to the Bible. To his mind, only the Bible makes no attempt to flatter the reader.

The habit De Luca, now sixty-six, acquired half a

lifetime ago has become the foundation of his prize-winning career as a writer: alongside biblically-cadenced novellas, plays, and poems are his translations into Italian (each an exercise in idiosyncrasy) of Exodus, Leviticus, the tales of Samson and of Ruth, the stories of Noah's Ark and of Jonah gobbled down by a whale, commentaries on the Psalms, the Tower of Babel, and Ecclesiastes.

'The sun also rises'; 'There is nothing new under the sun'; 'To every thing there is a season, and a time to every purpose under heaven.' Ecclesiastes is perhaps the best known of the Old Testament books. 'Vanity of vanities; all is vanity.' This is perhaps the best-known line in all Ecclesiastes, from the Latin *vanitas vanitatum, omnia vanitas*. De Luca, in his translation (I am working from a French version in *Noyau d'olive,* by the excellent Danièle Valin), manages to make it sound intriguingly unfamiliar. The original Hebrew, he tells us, is *havel havalim*, Havel being the figure in Genesis whom English readers know as Abel, brother of Cain – mankind's first murder victim. In De Luca's interpretation, he is the first waste of life. *Havel havalim:* 'waste of the waste'.

It is the rare Bible reader who learns Hebrew in order to peer beneath the layers of centuries of translation. For most believers – let alone nonbelievers – Hebrew remains *lingua incognita*. I am impressed by De Luca; I want to learn more; I decide to get in touch. We frequent the same literary circles; and, a few days after sending him an email (in French, a language he learned while working in Paris as a day labourer), he replies graciously.

Ancient Hebrew has around five thousand root words. Like other Semitic languages, its written sentences are composed only of consonants; vowels have to be supplied by the reader's imagination. Imagination, De Luca tells me, is something the Bible's translators often had in excess; they saw things in the words that were not there, as sky gazers see hunters instead of stars and fortune tellers see death instead of tea leaves. *'Etzev,'* he writes, when I ask him to give an example. In Genesis 3:16, God declares that women will give birth with *etzev*. The translators write it as 'sorrow' or 'pangs' or 'pain.' But De Luca says the 'pain' is illusory. 'The divinity does not condemn women to suffer.' Where *etzev*'s letters, *ayin, tsade,* and *bet,* occur together elsewhere in the Old Testament, they are understood to mean 'toil' or 'effort'.

De Luca answers no when I ask him whether he considers translation to be an art. 'For me, it's an exercise in admiration. He who admires knows his place, doesn't want to usurp the author. He who admires retains the distance necessary for admiration. It's a bad translator who thinks he knows best.'

Sober as De Luca is in his translating, a stickler for simplicity, he revels in punctiliously restoring the Hebrew's original poetry. He confides to me his joy at opening the pages of his chestnut 1984 edition of the *Biblia Hebraica Stuttgartensia* – bought long ago in a Milanese bookstore – and 'brushing away their dust with my eyelashes'. Psalm 105:39 shows De Luca's eye to advantage. It tells of the Israelites in the desert, where God at one point spreads (in other translators' vague words) 'a cloud for their protection' or 'a cloud for a covering'. Only in De Luca's suave Italian: *'una nuvola come tappeto'* ('a cloud for a carpet') do the words shine.

'We must pass over the little streams of opinion and rush back to the very source from which the Gospel writers drew . . . the Hebrew words themselves.' It could be De Luca writing (in recent years he has also turned his attention to the New Testament). But this is Saint Jerome, writing more than sixteen centuries ago, in a long Latin letter about the origin of *hosanna*. Jerome, then a young Roman priest with a training in the classics, was already at work on his Vulgate translation of the Bible, already figuring out how Matthew, Mark, Luke, and John should sound in Latin.

The Gospels were composed in Hellenistic Greek; the Septuagint, a translation of the Old Testament completed two centuries before the Christian era, was made in Hellenistic Greek. Hebrew had long retreated to the synagogues (*synagogue* itself a Greek New Testament word for *beth k'neset,* a 'place of gathering'). But by the fourth century CE, Greek, supplanted by Latin, had become as incomprehensible to many Christians as Hebrew. A good Latin translation of the Bible was required. The first attempts, from the Septuagint, were bad. It took Jerome, a scholar who knew his Greek, his Hebrew, and his Latin, to put them right. 'Why not go back to the original . . . and correct the mistakes introduced by inaccurate translators, and the blundering alterations of confident but ignorant critics, and, further, all that has been inserted or changed by copyists more asleep than awake?' he wrote in the preface to his translation of the Gospels in 383. Years later, in about 391, in a little cave in Bethlehem where he had settled, he began translating the many books of the Old Testament *iuxta Hebraeos* ('according to the Hebrews') by candlelight. *Bereshit bara Elohim et hashamayim ve'et ha'aretz. In principio creavit Deus caelum et terram.* In the beginning God created the heaven and the earth.

De Luca learned Hebrew while 'in the desert of [his] life'; Jerome learned Hebrew while living in the desert. The desert was east of Antioch, and it was during his ascetic time there in his twenties, struggling to 'endure against the promptings of sin' and tame an 'ardent nature', that the young monk decided to distract himself by learning the language from a Jewish Christian. He later reminisced:

Thus, after having studied the pointed style of Quintilian, the fluency of Cicero, the weightiness of Fronto, and the gentleness of Pliny, I now began to learn the alphabet again and practise harsh and guttural sounds. What efforts I spent on that task, what difficulties I had to face, how often I despaired, how often I gave up and then in my eagerness to learn began again.

Eventually, through perseverance and the encouragement of his teacher, Jerome's Hebrew grew to the point where he could find a certain elegance in its succinctness and the multiple meanings of its words. He could relish, for example, the wordplay in Jeremiah 1:11–12, in which God asks Jeremiah, 'What do you see?' Jeremiah replies, 'I see a branch of an almond tree (*shaqed*).' God responds, 'You have seen well, for I am watching (*shoqed*) over my word to perform it.' He could smile, pages later, on reading in Jeremiah 6:2–4 that shepherds (*roim*) shall flock to Zion, since the city had just been compared to a comely woman and the same Hebrew letters spelled out *lovers* (*reim*). And by assimilating Hebrew grammar, he could place this new knowledge side by side with that of Greek and Latin to contrive original and theologically pleasing arguments: *ruah* (Hebrew for *breath* or *wind* or *spirit*) is, he noted, a feminine noun; its Greek counterpart, *pneuma*, is neuter; in Latin, *spiritus* is a masculine noun. Ergo, the Holy Spirit is genderless.

In his high regard for Hebrew, Jerome differed from many of his contemporaries including Augustine, who thought translating the Bible in this way a waste of a Christian scholar's time and intelligence. It was enough accuracy, Augustine believed, to put into Latin the Old Testament books from their versions in the Greek Septuagint, which were widely held to have been divinely inspired. God had first spoken in Hebrew; but He had then gone to the trouble of repeating Himself in Greek. For the Greek reader, the Old Testament held no secrets. Jerome disagreed. There was the matter of mistakes. Working from the Septuagint amounted to translating a translation: always a risky endeavour. An earlier Latin scribe, sticking to the Septuagint, translated Psalm 128:2 as '*labores fructuum tuorum manducabi*', an ambiguous phrase whose meaning became an object of dispute: did understanding it as 'the fruit of labours' make any more sense than as 'the labours of fruits'? Jerome could only shake his beard in dismay. A simple blunder; the scribe hadn't realised that the Greek *karpoi* means 'fruits' but also 'hands' (literally, 'wrists'). The original Hebrew text actually reads (as in the King James translation): 'For thou shalt eat the labour of thine hands.'

Jerome and Augustine corresponded. Each attempted to persuade the other of the respective merits of translating from the Hebrew and translating from the Greek. But each man was stubborn; their arguments well rehearsed. Augustine's dubiety was not to be assuaged. With an eye to shaking his correspondent's resolution, he wrote with an alarming report from a North African bishop. If we believe Augustine, the bishop read aloud from Jerome's translation of Jonah – and prompted a riot in the pews. God, the bishop read, had given Jonah the leaves of a *hedera* ('ivy') for a parasol. While the bishop was speaking, he heard a sudden commotion, angry voices. Ivy? What do you mean, ivy? Frescoes showed Jonah recumbent in the cool shade of a gourd. Gourd was the plant named in the Septuagint. (Modern scholars identify the castor-oil plant as the likeliest candidate for the Hebrew *kikayon*.) 'Gourd! Gourd!' the congregants shouted in unison. The bishop was silenced. He read no more for fear of losing all control over his flock.

But Jerome was unrepentant. And eventually his bible superseded its competitors. For the next thousand years it was to his Vulgate translation that clergy the world over referred.

In early sixteenth-century Germany one of these clergy members was stout, sharp-tongued Martin Luther. Luther admired Jerome, a scholar after his own heart, and he resolved to do for the Saxons what Jerome had done for the Latin-literate monks. He studied the Septuagint in the Greek, Jerome's Vulgate in the Latin; he consulted the Old Testament's ancient Hebrew. Translating from the Hebrew he found particularly hard going. 'Oh, God! What a vast and thankless task it is to force the Old Testament authors to talk German,' he complained in a letter to a friend. 'They resist, unwilling to abandon their fine Hebrew for coarse German just as if a nightingale were asked to abandon its sweet melody for the cuckoo's song.' Elsewhere he compared the translator to a ploughman smoothing out his lines until a reader might run his eyes over three or four pages without ever imagining they had once been ridden with clods and stones.

The clods and stones were any words or turns of phrase that might get in the way of the ordinary German reader's appreciation. More than Jerome, or any other translator of the Bible before him, Luther thought of his reader: he wanted his work to be read and understood by the men and women in the street. So shekels and denarii, funny money, became *silberlinge* (literally 'silver-pieces') and *Groschens;* centurion became *Hauptmann* (literally 'captain'). When Paul, in his First Letter to the Corinthians, cautions those who speak in tongues to speak sense, lest they turn 'barbarian' to listeners, Luther has Paul say not *barbarian* but *undeutsch* ('un-German').

If one part of Luther's intention was to bring the Bible to the hausfrau and the bratwurst-seller, another was to remove what he considered was the papist gloss Jerome and others had put on certain words. The angel Gabriel who hails Mary (*Ave Maria*) as *'plena gratiae'* ('full of grace') in Luke 1:28 had been mistranslated, Luther argued. Gabriel would have greeted the young woman in Hebrew, with the same sort of warm words that he'd addressed to the prophet Daniel: *ish chamudoth* ('dear man'). Luther translated the angel's greeting to Mary thus: *'Gegrüsset sei du, Holdselige'* ('Greetings to you, dear, happy one'). No mention of grace at all. Heresy! But a printing press knows nothing of heresy. Within a generation of Luther's death in 1546, one hundred thousand copies of his Bible were being read, scribbled in, and argued over.

The Bible is the world's most read book – and the most translated. According to the Wycliffe Bible Translators (named after the fourteenth-century Bible translator John Wycliffe), the complete text is presently available in 554 languages; the New Testament alone exists in a further 1,333. Nearly two thousand different translations, yet they represent only a fraction of the world's languages: another four, possibly five, thousand still await an alphabet, an orthography, literacy. So every month, Wycliffe continues to dispatch its linguist-missionaries to the farthest corners of the globe. Every month, following their faith, men and women swap their suburban house and youth for years in a village hut, embowered in rainforest, in which to listen, repeat, write, and teach.

Someone at Wycliffe gives me the name and email of a returned missionary whom I could contact. I look up Andy Minch online and watch a video of a man in his late fifties, thin, bald, and quiet-spoken. In the mid-1980s, Andy and his wife, Audrey, flew from Illinois to the remote West Sepik province of Papua New Guinea, where they spent an engrossing, exhausting twenty years as Bible translators into the Amanab language. Living like the four thousand villagers, the Minches scuba dived in rivers, fed their dengue fever bananas and possum meat, and raised three children.

Andy's faith has been in the Minch family for generations: before Americanising the surname, his Prussian great-grandfather called himself 'Muench' (Monk). His parents often had missionaries over for dinner. As a boy in drab Chicago, Andy sat mesmerised by one grey-haired visitor who regaled him with a description of Australian foliage and aborigines. The memory of the old man's adventures stayed with him, and, after contemplating careers as a magician and locksmith, then graduating from the Moody Bible Institute, then marriage, the memory became something else, something insistent. Something like a calling. 'It was in the fall of 1982 that my wife, a nurse, and I, a teacher, pretty much out of the blue looked at each other and asked, "Isn't there more we could do with our life?"'

Wycliffe gave them a training in linguistics. But neither was fully prepared – how could they have been? – for their new jungle life. Barely fifty years before, the Amanab people had never seen a wheel or ice; they cut with stone axes and built their houses out of vines and brushwood, without nails. 'We arrived,' Andy writes to me, matter-of-factly, 'before the introduction of mirrors.'

'Early on, I took pictures of my village friends standing by their houses. That way I could learn their names and where they lived. I had a friend, Nimai, help me identify people. I showed one picture and after some examination of this two-dimensional representation, he declared, "That is Somangi." I showed another. "That is Wahlai." I showed another and he just could not figure out who was in the photo. Finally, I said, "That is you." "Oh," he laughed, "I wondered why he had my shirt on."'

How had he and Audrey learned the villagers' language? There had been no grammars, no classes.

We began by pointing to things like a rock until someone said *foon,* the word for rock, which I promptly wrote, using a phonetic alphabet. I'd point to a tree, and someone would say *li.* I'd point to a spider's web, and they would say *ambwamuhlaunalala,* at which point I would walk over to a rock and weakly say 'foon.' That was the beginning of our language analysis. We were like helpless babies, not knowing the language or culture. The people were so kind and gracious in caring for us as we lived among them learning their worldview and their wisdom.

Amanab is a language unlike any the Minches would have heard or read about in America. The distinction between horizontality and verticality is essential. To say, 'the book is in the house' in correct Amanab requires you to recall whether the book is lying on the floor or table, or sitting on a makeshift shelf. If the former, you say *buk rara gi* (literally, 'book house is'); if the latter, you say *buk rara go*. Similarly, you can't just say 'put it here'; you have to say something like *wanayi faka* ('put it here horizontally') or *wanayi foful* ('put it here vertically').

And you have to be as precise when talking about the past in Amanab. To tell someone you had drunk *bu* ('water') was to tell your listener either that your thirst had been quenched recently, in which case it was correct to say *'ka bu neg'*, or some time ago, in which case you said *'ka bu nena'*. So there were two pasts in the Amanab imagination, not one: the near-to-present and the more distant.

The villagers, Andy and Audrey discovered, ordered days with their bodies. The Minches' translation of Genesis has the Creator separating the light – day – from the darkness – night – on the 'left pinkie day' (the first). Mankind He makes in His image on the 'left wrist day' (the sixth). On the next, the 'left elbow day', God, basking in his accomplishment, deservedly rests.

Penei, a youngish male convert, helped the Minches with their translation. The story (in Genesis 43) of Joseph, whose brothers bring a gift of honey with them to Egypt, Penei greeted with an uncomprehending look. The only wild honey known to the Amanab wasn't sweet-tasting, Andy explains. (The honey's lack of sweetness, together with the absence of any village cows or goats, is why Exodus's famous 'land of milk and honey' in Amanab reads as a 'lush and productive place'.) When, verses later, the pharaoh's minister discloses that he is Joseph – 'I am your brother whom you sold into Egypt' – another Amanab fine distinction is needed. No single word for *brother* exists in Amanab. The Minches and Penei chose *sumieg*, meaning 'younger brother'. *Kaba negerni sumieg* ('I am your younger brother') is what the Amanab-speaking Joseph says.

As the years passed, more and more villagers observed and learned the ways of literacy: how to hold a pencil; how to decipher the missionaries' squiggles – like sago beetle tracks – on the pages, and transmute them, via the corresponding motions of the tongue and lips, to sound. A second Amanab man, then a third, joined Penei and made lighter work of the Minches' translating. And their children, who spoke accentless Amanab with the villagers' girls and boys, also helped them sharpen up their understanding.

Andy: 'I had struggled for months to figure out the meaning or purpose of a little word – *me*. One day I overheard my five-year-old son use it as he played on the front porch. I rushed out and asked him what he had just said. He quickly answered, "I didn't do it; it wasn't my fault!" "No, no," I said, "you just said a sentence with *me* in it." He nodded. I eagerly asked him, "what does *me* mean?" My son thought about it a moment and replied, "I don't know, sometimes you use it and sometimes you don't."'

The five-year-old's words put his father onto a new train of thought, leading him to the solution.

'It turned out *me* was a polite form; it softened a command, like the English *please*. We may say, "close the door" or "close the door, please." Sometimes we use it and sometimes we don't.'

Occasionally a Biblical passage contained a word that had no meaning whatsoever for the villagers. Then the Minches and their assistants would have to think up a substitute. Such a passage turned up in Luke 9:62, in which Jesus warns, 'No one who puts a hand to the plough and looks back is fit for service in the kingdom of God.' But *plough* was a word written for Greek and Hebrew eyes; the Amanab, jungle people, had never handled seeds, had never sown. The translation stalled. Andy and Audrey tried out various ideas on Penei, without success. Finally, the comparison with working the land, they threw out altogether. Instead, the Minches drew on the hunting life they had come to know. Luke's plough they turned into arrows. The verse became 'No one who shoots an arrow and looks back is fit for service in the kingdom of God.'

I ask Andy if translating a verse into Amanab had altered how he saw it. I wanted to know whether the experience of putting the Bible in another people's words had given him a new way of looking at his faith. It had. Andy points to the moment in John 21 when the resurrected Jesus questions Peter. In English, it is the same question thrice asked: 'Do you love me?' But in the original Greek, Andy tells me, John employs two words for love. Some scholars say the two – *agape* and *phileo* – are synonymous, but Andy, when he came to translate the passage for the Amanab, didn't read it that way. *Agape* here, he thinks, is the stronger word, signifying unconditional love; *phileo,* is something milder, fuzzier, like 'to have affection for'. So when Jesus asks and asks again if Peter loves (*agape*) him, and Peter replies, 'You know that I love (*phileo*) you,' the disparity between the two men's words pushes Jesus to finally demand, as if in resignation, 'Do you [even] love (*phileo*) me?' – 'Do you even have affection for me?'

Andy: 'In Amanab, *membeg* is "to have affection for." It is the standard term for expressing love. But there is another, *oningig lugwa*, which is more dynamic. It means "to continually hang your thoughts on", similar to when in English we say, "He is really hung up on her."' Jesus, Andy thinks, was telling Peter that affection wasn't enough; the believer was being asked 'to continually hang his thoughts on Jesus'. The Amanab term – so vivid, so picturesque – made the message clear. With it, Andy saw his faith in a new light.

Luther, five centuries earlier, wrote, 'We shall not long preserve the Gospel without languages. Languages are the sheath in which this sword of the Spirit is contained. They are the case in which we carry this jewel. They are the vessel in which we hold this wine. They are the larder in which this food is stored. And, as the Gospel itself says, they are the baskets in which we bear these loaves and fishes and fragments.'

14

A GRAMMAR
OF THE TELEPHONE

On the afternoon of April 4, 1877, Caroline Cole Williams of Somerville, Massachusetts, paced her salon, waiting for a smallish black walnut box to talk. The box had been left on a low shelf that morning by her husband, Charles. Charles worked in Boston, some three miles from the house, as a telegraph maker; between his shop and the house he had strung a wire. Caroline knew nothing about wires. She knew nothing about galvanic batteries, rheostats, resistance coils, magnets, or blue vitriol, either. To her mind, the box looked like nothing so much as an item in which a woman might keep her jewellery. Charles, always busy, had not thought to offer her any word of explanation; he had simply told her to listen out for a signal and to respond if and when the box came to life. But that had been hours ago, and Caroline, with one eye on the grandfather clock, was understandably beside herself with impatience. Perhaps it was all one big joke, this business about talking telegraphs. She was on the verge of giving up listening when a faint tapping made her all ears. 'Caroline?' the box said suddenly and quite distinctly. 'Caroline, can you hear me?'

Paper – cheap, light, pliable – had been the voice's emissary for centuries. Letters were the next best thing to meeting distant family and friends face to face. So it is easy to imagine how astonished the housewife must have felt at hearing, not reading, her absent husband's words, at recognising his voice conveyed neither by paper nor in person, at understanding him over a distance of three miles. Small, the voice in the box, and yet it was unmistakably Charles's. She had heard it a thousand times before: coming in a shout from another room of the house, behind her back, whispered tenderly into her ear. But never till now had she heard it like that, sounding neither near nor far, assertive yet uncertain, as if groping along the line.

'Charles,' she cried back. 'I can hear you. Can you hear me?' He could. And so too, having dropped in on him at the shop, could Charles's acquaintance – the device's inventor – Alexander Graham Bell.

The newspapers were soon full of Bell's invention. 'Oral Telegraphy' announced a headline in the *Times* of London dated September 19, 1877. 'The only difficulty which presents itself here,' the article remarked, 'is whether the Telephone can be made to convey any loud preliminary signal before beginning to "speak" . . . for we are hardly advanced enough as yet to adopt the suggestion of one enthusiast, and to go about with a sort of telephonic helmet or skull cap on our heads, ready for action at any moment.' Two months later, a long editorial noted, 'A great change has come over the conditions of humanity. Suddenly and quietly the whole human race is brought within speaking and hearing distance.' It went on, 'Already 500 houses in New York converse with one another; 3,000 Telephones are in use in the United States.' If the telephone had not yet taken off in Britain, it was because of 'that indistinctness of utterance, that slurring over of important consonants, and that dropping of the voice at the end of a sentence, which all foreigners observe in us. The Telephone will prove a severe test of both our speaking and our listening powers.'

The Telephone: so new, so exotic, so extraordinary that the word was printed with a capital *T*. The first callers took pointers from the press on how to talk into their contraptions. A *New York Times* article published in November 1882 advised readers not to yell 'at the top of [their] lungs' when responding to a call. Nor did they have to shut their eyes. Between the mouth and the transmitter, the article continued, a gap of not less than three and not more than eight inches ought to be allowed.

Language, too, had to catch up with the new technology. What was the appropriate response to a *ring-ring-ring?* To open the conversation with something like 'Good morning, sir' or 'Good evening, ma'am' was to run into problems: morning in Boston is afternoon in London; a 'ma'am' replying to the greeting might turn out to be a 'sir'. Shorter, blander words were required. Bell, a boating man, thought the answerer should always start by exclaiming 'Ahoy'; it was not one of his better ideas. His rival, Edison, proposed the alternative: 'Hello.' A new word, newer than *ahoy* (*hello*'s first attestation as greeting went back only twenty years), its novelty matched the technology.

Hello also held the idea of surprise – 'Why, hello!' – the creeping up of the caller's voice on its interlocutor. Perhaps that was the reason Edison's word chimed in with the telephoning experience. Or perhaps *hello* thrived because it was at once brand new and very old, because it had deep roots. It had grown out of the fourteenth-century *hallow* – to urge hounds on with shouts – the same word that gave us *holler*. And *hail*, going back further still, had used similar sounds to address the invisible presence of God.

Quickly, *hello*s spread like wires through the United States and far beyond, to every corner of the world. By the end of the century, there were more than a million telephones in the Union alone. *Hello*s by the million, in every language: *Allô, Hallo, Allo, Halo*. And out of these parallel wires and curt new words emerged the beginnings of a grammar of the telephone.

But this grammar – telephone English – went largely unnoticed and untaught for a long time. There were few fictional depictions or guidelines in textbooks; to believe the textbooks, nobody rang anyone at all. The *Second Book in English for Foreigners in Evening Schools,* published in 1917 by Frederick Houghton, was an exception:

Mrs Smith: Central, please give me North 3-5-8-9. Is this North 3-5-8-9? May I speak to Mr Miller?

Mr Miller: This is Mr Miller. What do you wish?

Mrs Smith: This is Mrs Smith on Flag Street. My little girl has diphtheria and the inspector from the Department of Health has just now quarantined us. Please send me to-day three loaves of bread, a pound of butter, a quart of milk, and a can of corn.

But it would take linguists another fifty years to start paying attention to how talk really works.

Wayne A. Beach, professor in the School of Communication at San Diego State University, is a world authority on phone calls. He has spent years of his career studying the ins and outs of over-the-line conversation, of language in action: emergency calls, romantic calls, nagging calls, chitchat calls. I wanted to learn more about his research. He replied to my email with his mobile phone number, then to another message with his landline number, and I dialled at the time we had arranged to talk. I waited. I counted the rings that I was making from Paris. Five. Six. Seven. Eight. Nine. Finally, an answering machine told me, in a child's voice, that they were not there and to leave a message after the beep. I left no message. I replaced my handset on the base station, waited several minutes, and redialled. I got the answering machine again. Maybe the professor wasn't in. Or was in the shower. Maybe he had completely forgotten about my call. It was enough to bring out the grumbler in me. Telephones! I only rarely adopt the caller's posture and have no mobile phone of my own. So when, with a wary hand, I called for the third time, and a man's 'Hello?' cut the ringing short, he heard my sigh of relief. The professor apologised; he had just got up; it was breakfast time where he was. 'Let me tell you, I can't remember the last time I had a conversation on this phone,' he said. 'I prefer mobile phones. Politicians here know our landline numbers; we get lots of calls come election time. I never normally pick up when it rings. You listen to your messages on the machine, delete them, call whoever it was back on the mobile phone. The mobile phone has become personal; the landline's a fallback for emergencies only.'

I asked him about research dissecting telephone talk. What was it that had drawn him to the telephone in particular? He said, 'I was raised in rural Iowa, in a town of four hundred. After the Second World War, my father worked for the local phone company. Climbed the poles, ladders, too; ended up becoming the manager of the district office. The technology side of things always fascinated me as I was growing up. I remember the old black dial phones, and the terrible pink and brown and green ones. My parents' was mounted on a wall in the kitchen; in those days, the cord dictated the circumference of our conversations: my mum talked while cooking and washing the dishes.'

Talking on the telephone, the professor said, was 'the purest form of conversation' – that was what made it so fascinating to him and his colleagues in the language and communication field. 'All you have is the voice: its prosody, its accent, its intonations. It's the voice that has to do all the work. Now, if we were having coffee or wine together, the talk would be different. The voice would be accompanied by our gazes and our gestures and the expressions on our faces. But the telephone does away with body language, does away with all those props. It's the elevation of speech.'

Conversation Analysis (CA), Professor Beach's speciality, began in the 1960s as a response to Chomsky's abstract approach to language. 'It was pointless – according to Chomsky – to want to study the features of natural conversation. Pointless because talk was thought to be random, chaotic, *degenerative* – full of mistakes, lisps, unintended puns. You were supposed, as linguists, to sit in your armchair and theorise sentences. All headwork, no legwork. No listening to and recording speakers, no getting your theories messy with how folk actually spoke.'

Then along came Harvey Sacks and Emanuel Schegloff. 'It was they who said, "There's order at all points; nothing's a coincidence in natural speech."' Words were not mere abstractions; they were 'speech objects', things that talkers used purposefully to create and shape every kind of social activity. The illumination had come to Sacks during a stint on a helpline. Beach explains:

He was a volunteer listening to all these callers in distress – some were tight-lipped, but others poured out their hearts to him. All the calls were recorded, so one day he asked his supervisor if he could get permission to write them down verbatim. Hundreds of conversations. He spent many weeks and months listening to them, making notes, performing analysis.

He found patterns, rules. One rule – an elementary one – was that the answerer always speaks first. Another was that certain words and units of speech like *hello* or *how are you?* come in pairs, helping to give the conversation its rhythm and flow: 'How are you?' 'Fine, thanks. How are you?' In their separate studies, Sacks and Schegloff showed that telephone openings are highly tidy and predictable. In their tidiness and predictability, the recordings studied didn't conform at all to Chomsky's idea of words that bumped and stumbled one into another. A meeting voice-to-voice begins with a ring and hellos, followed immediately by words used for the purpose of putting a face to the voice: 'Hello, John?' 'Yeah.' 'It's Denise.' Only then do the greetings appear: 'Hi.' 'Hi. How's things?' And after the greetings comes the reason for the call: 'I was wondering whether you might like to catch a film.'

Different openings, Sacks observed, produce different kinds of talk. One *hello,* more often than not, simply produces another. But if a call's recipient opens with something like 'This is Mr Smith; may I help you?' it elicits a response along the lines of 'Yes, this is Mr Brown'; it is a way of drawing a name out of the caller without having to ask for it in so many words.

It turned out that many conversations run on questions – or on units of speech resembling questions. Often, a question is intended as a signal, a preemption of something else: 'What are you doing tonight?' is an invitation disguised as a question; a child's 'You know what, Mummy?' is a way of obliging the parent to listen to a monologue.

Conversation, in Sacks and Schegloff's view, amounts to an exercise in collaboration; every time an unsuccessful joke, an obscure reference, or an ambiguous remark threatens misunderstanding, the voices on either end of the line go back and straighten out the kink before moving on. It's this instinctive collaboration – the ability of talk to right itself – even between strangers, even speaking at a great remove, that Chomsky, from his armchair, had not seen.

Professor Beach said, 'Sacks died young, in a car crash, but his ideas survived him. We use them today in everything from court cases to psychology. And many linguists decided to become conversation analytic.' But when the professor and his colleague and friend Robert Hopper entered the field, it was still new. 'We were both PhDs in communication, but we were very much feeling our way around. I drew some of my ideas from people in sociology; Robert travelled to England to meet conversation analysts there. We wanted to break out of the Chomsky mould, focus on people, study how conversation is produced moment by moment by its participants, together, since it can't be done individually. The problem was to know how to go about it. There was no one in the discipline to teach us how to do this; no one mentored us; it requires so much time, so much money. We had to work things out by ourselves.'

'We were teaching at American universities and Robert asked his students – back in the days when such things were possible – to record their conversations over the telephone. It was a sort of assignment he would give them. Of course, the recordings were submitted anonymously. And Robert would transcribe each only after a year had passed and any news in them long grown stale.' In this way, inciting students to donate their conversations, he collected many recordings; made many transcripts, noting every *mm* and *uh huh;* and built up his very own conversation library. From family, friends, and colleagues, Professor Beach did the same. But it was Hopper who wrote up the findings in a book, *Telephone Conversation,* published in 1992.

'The telephone summons [ringing bell] renders us vulnerable to a form of spoken intrusion,' Hopper writes. 'To begin every telephone speech event with a summons is to swing speaking's ecology toward purpose and to make dialogue asymmetrical in favour of the caller.' This asymmetry qualifies Sacks and Schegloff's insight about telephone talk as collaboration. The telephone, Hopper reminds us, distributes the labour – the 'talk work' – unequally between caller and answerer. 'The caller acts, the answerer must react.' The time and the day, but also the opening topic of conversation, are all of the caller's choosing. Resisting the topic can be hard. I know this from experience: over-cheery call-centre voices glumly and obediently listened to; a friend's inconvenient wish to carp or gossip indulged; a radio journalist (how did he get my number?) who would not take no for an answer. One price most of us pay for being poor, Hopper argues. The rich have always someone – the secretary, the personal assistant, the receptionist – to pick up their receiver. Most of these professional answerers have been, and still are, women.

Telephone Conversation details how female speech, from the 'hello girls' hired for their 'clear voices', instantly became corseted by Bell's technology; women's talk relegated to a cottage service industry. The book cites research by a feminist scholar and ethnographer of the telephone, Lana Rakow.

Like many women up and down the country, Rakow's smalltown Midwestern subjects were raised to be frequent and docile phoners. They called for the repairman, took down messages for the husband, enquired about an out-of-state in-law's health. They gained the operator's brisk, ladylike manners. They picked up at all hours, in their aprons, nightgowns, or summer dresses. They spent their lives at society's beck and call.

'Robert was right to feel gloomy about the telephone's role as a social leveller, as an equaliser,' Professor Beach said. Voices could be just as poor, as female, as black, as foreign, as a face. Old prejudices could transfer to new technologies. 'And today's mobile phones haven't improved things,' the professor continued. 'They go off everywhere, at the worst moments, heckle us into responding. They become addictive. Just the presence of a mobile phone at the bedside permits deep sleep to be intruded upon. They're the tail that wags the dog.' The professor's animosity, I discovered, was the animosity of a father, and directed in particular against the excesses of texting and online forums.

My son is twenty-one, my daughter is fifteen: between them, they send and receive eight thousand text messages every month. They're constantly connected. More contacts, fewer friends. It is depersonalising: it takes away from the intimacy and the immediacy that only face-to-face and telephone conversations can produce. Social media is a kind of shell game.

I had not expected my conversation with the professor to deepen and darken. I had expected something lighter, more technical (and indeed, in an aside, Professor Beach had spoken about the joy of transcribing telephone conversations, a skill for which he had developed an extraordinary knack over the years; he could, he said, close his eyes at any point in our conversation and visualise the utterances as an analyst's transcript, complete with stops, brackets, asterisks: all the transcriber's punctuation). I had expected something less personal. But like Sacks and Schegloff before them, Beach and Hopper had focused their studies on people and the ways in which they converse; the professor could not imagine separating the talker from the talk. 'One day I learned that Robert was dying. He had been diagnosed with cancer. We had known one another personally and professionally for twenty years. It was hard. The last months we talked a lot on the telephone. He was angry: he felt the doctors were withholding things from him. Communication between him and the doctors broke down. He told me, "I'm managing optimism."' Robert Hopper died in 1998 at the age of fifty-three. 'It was incredible. One out of every three Western families was being touched by cancer, and at the same time I realised we still knew very little about how people talk about and through the illness. It was high time we found out. I was lucky enough to be able to do something about that gap.'

Among the professor's call collection was a shoeboxful of cassette tapes. The tapes were like those his former colleague had once received: a donation to Professor

Beach's research from a graduate student who asked only that the names of the participants remain confidential. In his study, not long after Robert's funeral, the professor played the tapes and discovered that the student had recorded himself, his parents, and other relatives, talking on the phone about the mother's cancer, from the time of her diagnosis to just hours before her death: sixty-one conversations over a period of thirteen months. It was the first natural history of a family's cancer conversations. 'The transcripts of the tapes took me a long time to finish. Their contents were close to me. Naturally, Robert was on my mind. So was my mother. Not so long before Robert's death, cancer had also taken her. Listening to the student's words, I felt they could have been mine: I had once been that son.'

SDCL: MALIGNANCY #2:1

Mom: <u>He</u>llo.

Son: Hi?

Mom: Hi.

Son: How ya doin'.
    (0.2-second pause)

Mom: O:h I:'m doin' okay. = I gotta-
    (1.0-second pause)

Mom: I think I'm radioactive. Ha ha.

Son: $He- uh$ Why's <u>that</u>.

Mom: Well you know when y'get that <u>bo:</u>ne scan so they-

Son: Oh did [they do it already?

Mom: [Gave me that.]
Yeah they give you a <u>shot</u>. Then ya have ta (.).hhh
drink <u>wa</u>ter or <u>co</u>ffee (.) tea, >whatever the hell
you <u>want</u>< (0.4) in <u>vo:</u>lumes of it.

Son: Mmm hmm.

Mom: hh A::nd I'll go down at about ten thir:ty.

Son: Mmkay.
(0.4)

Mom: So (.) anyway.

Son: Hm:m. =

The transcripts' grammar had much in common with
that found in other kinds of telephone call: there was
the same *hello,* the same more or less orderly taking
of turns, the same sorts of repeating words and sounds.
But the professor noticed subtle departures from the
rules. Closeness between the talkers did away with any
need for names: a young man's *hi* sufficed for the
mother to recognise her son. Another departure came
when the son asked, 'How ya doin'?' and not 'How are
you?' A speech object like *how ya doin'?* is no question.
It commiserates. The person on the other end needn't
answer 'fine.' The mother didn't. She recycled the son's
doin', and added to it *okay* and a long pause. With
each part of this answer, the son would have heard the
badness of the impending news.

Viewed one way, the situation is quite simple: the

mother has news; the son wants her news. But, of course, it is not so simple: the two talkers being close, the mother's bad news is also the son's. And so the mother was tactful. She did not say, as in a book or a letter, 'I got a bone scan,' 'They gave me a shot,' 'I had to drink water . . . in volumes.' She said, 'Y'get that bone scan,' 'They give you a shot,' 'Ya have ta drink water . . . in volumes.' *You* or *y'* or *ya* are inclusive – also distancing, and therefore softening.

SDCL: MALIGNANCY #2:2

Mom: So. (0.4) It's <u>r(h)e::al</u> °<u>b(h)a:d</u>°. ((voice breaks))
(0.8)

Mom: ((sneezes))

Son: pt .hhhh I guess.
(0.4)

Mom: And uh: >I don't know what else to <u>tell</u> you.<
(1.0)

Son: hh hhh Yeah. (0.2) um- ((hhhh)). Yeah, I don't know what to say <u>ei</u>ther.

Mom: No there's nothing to say. >You just-< .hh I'll- I'll wait to talk to Dr Leedon today = he's the cancer man and =

Son: = Um hmm.

Mom: See what he has to say, and (0.4) just keep goin' forward. I mean (.) I might be real <u>lu</u>cky in five years. It might just be six months.

(0.4)

Son: Yeah.

Mom: °Who knows.°

Son: pt .hhh Phew::.

Mom: Yeah.

Son.   .hh hhh (.) Whadda you <u>do:</u> with this kind of thing. I mean- (.)

Mom: >Radiation <u>chem</u>otherapy.<
      (1.4)

Son: Oh <u>bo:y</u>?

Mom: Yeah.
      (0.5)

Mom: My only <u>hope</u>- I mean- (.) my only <u>cho</u>ice.

Son: Yeah.

Mom: It's either that or just lay here and let it <u>ki</u>ll me.
      (1.0)

Mom: And that's not the human con<u>di</u>tion.

Son: No. (1.0) I guess [not.]

Mom: [No.] (.) So that's all I can tell you (°sweetie°).

As the professor transcribed the tapes and then analysed the transcripts, he was struck by the mum and son's deft use of talk to switch between despair and hope. 'There's nothing to say,' the mother said.

But she kept up the conversation, drawing on talk's resources together with her son, turn after turn: 'You just . . . just keep goin' forward . . . my only hope . . . my only choice.'

'I came away from the tapes and the transcripts full of that feeling of hope,' the professor told me. 'It was the lesson I learned.' Hope, in the midst of life-shattering news, because there was always something more to say. Just a few words could comfort, reassure, inspire. Words and semi-words like *mmkay, phew,* and *oh, boy* all did valuable social work. 'I wish more doctors, nurses, and caregivers understood this. Medical explanations are not enough. Conversation is so important. Listen to what patients say and respond. But the problem is, how to respond. Many doctors lack the know-how.'

A lack Professor Beach is also addressing. Some days after we hung up – 'tuned out', in the professor's words – he sent me a video. He had condensed the many hours of telephone recordings into an eighty-minute play, put the mum's and son's words into the mouths of a pair of professional stage actors, and obtained funding to have *When Cancer Calls* performed in venues before thousands of medical professionals and patients and their families. When I watched the video, it occurred to me that the professor's choice of medium for his project had been perfect: the telephone and the theatre go back a long way. Bell's first public demonstrations of his instrument, in early 1877, took place in music halls. Days after she had paced her salon in Somerville, waiting for her husband's voice to come,

Caroline Cole Williams sang over the brand-new line to rapturous audiences in New York.

To judge from the video, the professor's audiences were equally enthusiastic. They cried and clapped and chuckled as the tall young actor in shirt and slacks, and the older, heavier actress in a dress and with an IV drip, delivered their lines.

Such complicity is understandable: the dialogue performed was extraordinary in its ordinariness. We hear these words every day, in cafés, in offices, in homes. They are the freight of every family's telephone line.

15

CONVERSATIONAL HUMAN

The dumb presence of objects is a silence that prods human beings into speech. It seems never to matter that the exchange is naturally one-sided. To dolls and teddy bears little children jabber incessantly. Gardeners can sometimes be overheard urging their budding tulips to flourish. 'Geez, you're beautiful,' a Chevrolet-lover says, under his breath, at the sight of his future possession. And, in our era of high technology, more and more among us mutter at computer screens.

But unlike the toys, plants, and cars, the wizziest computers have begun to talk back. Ask yours for the date of your next dentist appointment, the height of Mount Everest, the number of miles that separates New York from Sacramento, and an electronic voice – if you can call it a voice – with an unplaceable accent responds. Some tinker with the voice until it sounds plausibly male or female; turns flirty or sniffy; brags or dodders. You can also customise it to answer to a name.

Fifty years ago, 'Eliza' was one of the first programs to simulate human conversation. When the user at MIT hit return after typing 'Nice to meet you,' so prompt was the 'Oh . . . ?' from Eliza, which flashed up in chunky letters on the screen, so disarming, that it could feel as though a woman waiting on the other side of the computer's connection had just sent it. In the space of a few stock words, a few canned phrases, the character of an inquisitive psychoanalyst would take shape. To such an extent that the user mightn't be sure if what he sensed was empathy or his imagination.

User: Nice to meet you.

Eliza: Oh . . . ?

User: My name is John.

Eliza: I don't care about names, please go on.

User: I have a fear of computers.

Eliza: I see.

User: What should I do?

Eliza: What do you think?

User: Should I trust you?

Eliza: We were discussing you, not me.

'The fact that individuals bind themselves with strong emotional ties to machines ought not in itself to be surprising,' the program's Pygmalion, Joseph Weizenbaum, wrote in *Computer Power and Human Reason,* which was published in 1976. 'The instruments man uses become, after all, extensions of his body.' What pens had been for our hands and bicycles had been for our legs, computers were becoming for the brain: a computer could do the remembering and the calculating of a thousand men. But would a computer ever learn to converse? To employ human language with wit and feeling and creativity? Weizenbaum himself was sceptical. 'Eliza was such a small and simple step. Its contribution was, if any at all, only to vividly underline what many others had long ago discovered, namely, the importance of context to language understanding.' Without the user's indulgent fancy to flesh out the bones of the program's palaver, no dialogue could ever get going: Eliza was little more than a speak-your-imagination machine. Weizenbaum was sceptical, and none of the later generations of 'chatbots' gave him any grounds to reconsider.

One of the latest and most talked-about is 'Evie', a youngish bot with blinking green eyes, smiling pink lips, and flowing brown hair (it seems that bots are almost always made to look and sound like women). According to its makers, Evie comes out with statements that have all been acquired at some point in the past ten years from the things people type to 'her'. For this reason, its database of possible answers is vastly bigger than anything Eliza had to draw on. Even so, there are some strange moments when I attempt a chat with the pixelated face on my computer screen. A remark I type to Evie about Buster Keaton leads it to reply – actually to spit out, Spock-like – that I am 'making sense'. 'Am I?' I ask. 'Yes, you are the love of my life.' A clumsy play for empathy, if ever there was one. I change the subject. I try books. I wonder about her reading habits. Is she in the middle of a novel? Incorrectly, Evie replies, 'You already asked me that.'

'It's already uncanny how good some of these robots coming out of Japan look and sound,' Professor Naomi Susan Baron tells me over the phone from her American University office in Washington DC. Perhaps the linguist is attempting to dampen my scepticism about the value of chatbots. Baron is the author of a recent academic paper, 'Shall We Talk? Conversing with Humans and Robots', which is how I discovered her work. I ask her outright whether computers will ever master conversational Human and she says, 'That's the $64,000 question. I don't have a firm answer to give you, but I'll say this. Take syntax. Very complicated, all the ways in which people build sentences out of words and phrases. Very complicated, and yet computers can manage that nowadays. They passed that hurdle. Conversation is the next hurdle. Maybe insurmountable, maybe not.'

Professor Baron has been doing a lot of thinking about the likely features of authentic 'computer talk', comparing it with the various kinds of talk linguists usually dissect. Talking with a computer would be much like talking with a foreigner or a pet or a child, she thinks. 'Like child-directed speech. We used to call it motherese until we realised there were also stay-at-home dads. The characteristics are fairly consistent across cultures, across classes: the parent speaks to the child with higher pitch, greater articulation, slower delivery.'

Foreigners, pets, children: all have lower status. To believe Baron, machines may never be permitted to become our conversational equals. 'They'll be designed to do our bidding. To be informative. Or entertaining. Or both. But we likely won't accept any backchat from a computer. We won't want them to tell us things we don't want to hear. Conversation is about control, about power. Raising your voice, for example, or switching topics. We won't want to relinquish that power.'

Computers that grovel, pander, flatter, cheerlead. Computers as feel-good coaches, coaxing surplus calories out of their owners: 'Keep up the good work!' 'You've done just swell!' Baron envisages them as sponges, possessing magpie memories: not only would they ask you what you want for your birthday and anticipate the very moment when to dial up this or that friend, but also recall – via data-collecting bracelets and questionnaires – your every action and concoct menus based on how much of what and when you ate. Vigilant and obsequious. 'It might ask you whether you enjoyed the particular brand of spaghetti Bolognese you had three days ago on Wednesday.'

'But wouldn't that sound a little too pernickety? Too much like a computer?'

Baron laughs. She knows people who talk like that, she says. But she wants to make a bigger point, and to do so she describes toy robots presently manufactured in Japan for kids – toy robots in the form of seals. In Tokyo, these seals sell by the thousand. They squeak and look cute and cuddly. They are, in other words, only vaguely like a real seal: no sliminess, no sharp teeth (to shred and devour fish), no pinniped odour. No person in his or her right mind would buy a toy that resembles a seal too closely. The same would be true of any machine that prattles, argues, blunders as human beings do. 'You'd take such a machine back for an immediate refund. Too wordy. Always veering off-topic. Interrupting, mixing up meanings or forgetting whatever it was it wanted to say next. At the very least, you'd trade it in for a less human, more computer-like model.'

And here Baron returns to where she began our conversation: the Japanese generation of humanoid robots. 'There's a tipping point where human-like becomes too human,' she says. She brushed up against that point on a recent trip to Asia. 'I was at the airport. I went up to a lady at the counter. Only, the lady wasn't a lady. It was a robot. Complete with eyelashes, uniform, and good manners. Very polite. When I approached, it gave a little Japanese bow.' Baron's response was pure stupefaction. 'I bowed back. Then the robot spoke a greeting. It enquired how it could be of assistance to me. To see the robotic lips move was eerie. Watching the gestures, hearing the words coming out of those lips, I felt my flesh begin to crawl. I couldn't help that.' An automaton, all wax and wires, hidden in the convincing guise of a demure customer service employee, but the linguist found the encounter disquieting, disorienting. Even so, she thinks the squeamishness we feel now needn't always constitute an obstacle. She can recall the period, in the seventies, when the first home-answering machines were similarly off-putting. Older folk in particular were frequently at a loss for words when confronted with the shock of the beep. 'But there's nobody there,' they complained to Baron, then a young researcher.

The appropriate phrases, short, crisp, unruffled: the 'Hi, it's Grandma,' and the 'Just wanted to hear how your day went' and the 'Can you give me a quick call when you get this?' all took time to acquire. But acquire them, in the end, they did.

I think it's fair to say that Baron is a techno-optimist. But she is quick to raise her own reservations. What if the naive, the vulnerable, are taken in by online talk sharks? By programmers without all their scruples, whose chatbots preach, browbeat, coquet? As Baron says this, tales of email scams – a heart broken here, a thousand dollars lost there – come to mind; if you think that a swindler's typed-out text can achieve so much, it is easy to imagine how much more persuasive a sweet-talking program might be. Indefatigable, unpunishable, they would roam freely along the Internet's electronic byways, ready at all hours to snare, to trick, to fleece their next victim.

'And what if a robot won't accept no for an answer?' Baron wonders. There is unease in her voice. She asks me to picture an old woman living in a nursing home. The home doesn't have enough care-takers to go around. To save time, the staff assigns the old woman a talking robot. The robot is strict: three times a day – morning, noon, and night – it must see that its patient takes her tablets. But say the old woman is headstrong. She wasn't always sick and old. Say she was once a bigwig, her career filled with clash and ego, and now cannot stand to take orders from a jumped-up cash register. 'Ms. Henderson, it's time for your medication.' The robot repeats itself when the old woman pretends not to hear. 'Ms. Henderson, you must take your medication,' it intones. 'Your pulse is currently five beats below the normal level.' It utters something about blood sugars too, but the old woman still doesn't budge. With her child – assuming she had a child – she might relent and throw the little blue pill into her mouth; with a nurse – after a respectable amount of fuss – she would finally sluice it down with a large tumbler of water. But with a robot? Never!

'If the old woman is still sprightly, if her faculties are still intact, then she can always go for the off switch, I suppose. That's the big difference between robots and humans: having an off switch,' Baron says. What, though, she worries aloud, if the rules of human–robot conduct prevent this, forbid patients to unplug their caregiver? Robots talking patients into obedience on one side, and, on the other, patients unwilling to hear the machine out: textbook conditions for a shouting match. 'When you overhear a row between neighbours in an apartment block, or between a customer and a member of staff in a shop, that's unpleasant enough as it is. How then would we react to a war of words in which one of the protagonists is mechanical?'

A program that sweet-talked or squabbled persuasively would have a good chance of disproving Descartes. Three hundred years before the digital computer was invented, he wrote this in his *Discourse on Method and Meditations on First Philosophy:*

If there were machines which bore a resemblance to our bodies and imitated our actions as closely as possible for all practical purposes, we should still have . . . very certain means of recognising that they were not real men . . . they could never use words, or put together signs, as we do in order to declare our thoughts to others. For we can certainly conceive of a machine so constructed that it utters words, and even utters words that correspond to bodily actions causing a change in its organs (for example, if one touches it in some spot, the machine asks what it is that one wants to say to it; if in another spot, it cries that one has hurt it, and so on), but it is inconceivable that such a machine should produce different arrangements of words so as to give an appropriately meaningful answer to whatever is said in its presence, as the dullest of men can do.

Descartes' language test was a thought experiment, a seventeenth-century defence of the specialness of human reasoning; it wasn't intended to be operational. But the Turing test (named for the British pioneer of computer science, Alan Turing), first advanced in 1950, proposes a simple means of putting through its paces a program's ability to talk.

It goes like this. An 'interrogator' sits alone in a room before a computer screen. He is wielding a keyboard, and the messages he sends in quick succession go out to a pair of respondents in separate rooms. One is a man or woman who replies to the interrogator's questions and comments as any man or woman might. The other is a program, built to interact just like a confirmed talker. The interrogator has five minutes to tell the two apart. Puns, jokes, idiosyncratic turns of conversation: all forms of talk are permitted. If, after the dialogues, the two remain indistinguishable, the program passes, and we can discard Descartes' objection: a machine will be said to have conversed.

Yet Eliza, Evie, and the other chatbots remain very far indeed from passing. So far, in fact, that you wonder whether Descartes' *inconceivable* means not only unthinkable – as a flying machine might once have been unthinkable – but also impossible: impossible as a pig that flies. Talk as somehow fundamentally robot-proof. The clumsy exclamations, the non sequiturs, the wise-cracks that time and again fall flat as card castles: the computer's failure to say the right things does indeed seem telling. Its prowess in other disciplines, improving by leaps and bounds, makes the failure all the more remarkable. Programs have for years outperformed even the strongest chess masters and played draughts to perfection – literally. At the time of this writing, a program has for the first time bested a human champion at the ancient strategic game of go. (The human in question, a twenty-three-year-old South Korean, boasted in a pre-match press conference that he would wallop the machine five to nothing; he lost four to one). And then there are the face-recognising programs, the knee-bending robots, the quiz-show-question– answering machines. Only the computer's language smarts still leave much to be desired. Only in the area of language is a Cartesian disdain toward the machine still tenable. The computer stands tongue-tied; and its silence grows newsworthier by the year.

It isn't for any want of trying. For twenty-five years, an American businessman has reportedly offered $100,000 to the designer of the first chatbot crafty enough to fool a majority of its human interrogators. Every year programmers enter their latest creations – complete with first names, family names, and biographies – into the competition; and, in fairness to the programmers, every now and then one of the more gullible or unimaginative judges mistakes a bot's 'quirkiness' for a foreign adolescent's snark. That is the exception, though. The most fluent, thoughtful, engaging texts the programs never manufacture. They never 'produce different arrangements of words so as to give an appropriately meaningful answer to whatever is said in its presence, as the dullest of men can do'. Not once has the businessman been at any risk of being separated from his money. And the annual publicity the media gives his contest certainly does his business no harm.

Some days after my discussion with the linguist, it occurs to me that I might need the advice of someone who spends time with these programs. Someone who knows bytes from RAMs. I don't have that kind of knowledge. A computer's innards are a total mystery to me. I talk it over with a friend, an information technologist. He goes quiet. Then he tells me not to delve into the technical side of things, that it isn't necessary, and he suggests a name: Harry Collins. Not a computer specialist as such, it turns out, but a sociologist doing interesting work on the Turing test.

I email Collins and make an appointment to call his

office at the University of Cardiff. He sounds like a man between meetings when I telephone. For an instant, I dread having to tell my friend that the discussion with his sociologist came to nothing; but, very quickly, my anxieties evaporate. Collins's tight schedule (in the course of our conversation he mentions having three academic textbooks in the works) has made him curt, but also focused. He explains everything briskly and precisely, and I feel grateful for that succinct precision: a question saver.

On the present chatbots and their makers:

'The businessman's contest is nonsense. It's only a measure of *doing best* rather than of actually *doing*. But even the least worst bot can't talk. It can't converse. It can't use human language appropriately. Some of those in the field – the most optimistic – say, "Wait another twenty years. You'll see." Frankly, I don't believe them. They're hypesters; they'll say anything.'

On the Turing test:

I've performed my own experiments that are variations on the Turing test. The same setup: the computer screens and the keyboards and the separate rooms and so on, but with a second person on the receiving end rather than a program. Human to human, as in real life. The idea is to better understand how we humans communicate, how we make ourselves understood, how we employ language to pass for one of us and for a particular kind of person.

In one experiment, we had a group of colourblind subjects. We said to them, 'Reply to the interrogator's messages as though you can see colours just fine. So, for example, if an interrogator asks for your favourite colour, you type back, *blue* or *yellow* or *fire-engine red,* you name it, even though you've never seen anything blue or yellow or red in your life. The idea was to explore how they performed in the language of those who see the world in colour.

Collins reports that his subjects performed flawlessly. Without difficulty they discussed flower arranging, spun tales about playing snooker, described their impatience at waiting in their car for a traffic light to turn to green. Their interrogators couldn't tell whether they truly saw colours or not. The reason, he explains, is that the subjects had all been immersed from birth in a colour-seers' society; in it, they had acquired the language down even to familiar expressions about 'seeing red' or 'feeling blue'.

Collins took Turing's imitation game a stage further in a second experiment. He asked a group of blind subjects to converse at distance with their seeing inter-rogators. The subjects used screen-reader software to hear the keys pressed and the words formed as they typed their answers. 'They had all lost their sight when very small, by the age of two or three, so they had no memory of the visual world.' Even so, the interrogators were unable to determine from what their correspond-ents wrote that they were blind. The visually impaired, raised in a society in which vision predominates, had 'sighted language'. To every question they could provide a 'right type of answer'.

I ask Collins for an example of the questions posed

to these subjects. 'Well, for instance: "Around how many millimetres must a tennis ball drop from the line to be considered out?"'

They had never held a racket or swivelled their head left and right, left and right, to follow a tennis match. But they knew family or friends who had. And some of them had listened to sports commentaries on the radio.

'Then we turned things around. We asked a group of seeing subjects to type-talk as though they couldn't see and had no memory of having ever seen. In a word, to use "blind language." They couldn't. The interrogators, who were all blind, were able to tell right from the opening question that the subjects were only pretending.'

The interrogators' first question was the simple-seeming 'How old were you when you went blind?'

Collins: 'The subjects would say things like "two years old" or "I lost my sight when I was three", whereas a blind person will reply something like "It began when I was two, and I was registered blind at three and a half."'

Because they hadn't been raised by blind parents or mixed with blind friends, the seeing subjects had never learned how the nonseeing speak among themselves. They had never learned that to speak of going blind was to speak of a gradual process.

'What the subjects in these tests, the colourblind and the blind, were doing wasn't guesswork. They weren't attempting to "talk the talk". Something else, something far more interesting, was going on. The subjects displayed a sort of language know-how. They knew what *green* or *tennis* meant, but more importantly they knew precisely how talkers in green-seeing and tennis-playing societies use these words in everyday conversation. They could reproduce the appropriate conversational behaviour at any given moment. They could "walk the talk."'

Language as a stand-in for the body, as a substitute for direct experience: conversational Human is the outcome of our 'talk walking'. Often that talk is small. But, Collins adds, there are occasions when it has to turn more complicated, when we must address a lawyer, for example, or a doctor, and conversation suddenly comes less easily. Even so, most defendants and patients manage. An average person's language know-how can be surprisingly broad and deep.

To test just how deep, in 2006 Collins performed his most impressive experiment. On himself. 'I'm a sociologist of scientific knowledge. I've spent my career studying the men and women who do gravitational wave physics. I've hung out with them. Talked for hours on end with them. Immersed myself in their community. Now, I can't perform any of their calculations, no one would ever let me loose on a soldering iron, but I can talk just as they talk. So one day I decided to put my money where my mouth was.'

Collins asked a panel of gravitational wave physicists to send him and a gravitational wave physicist a list of questions to answer separately. One of the questions went as follows:

A theorist tells you that she has come up with a theory in which a circular ring of particles is displaced by gravitational waves so that the circular shape remains the same but the size oscillates about a mean size. Would it be possible to measure this effect using a laser interferometer?

The physicist wrote back:

Yes, but you should analyse the sum of the strains in the two arms, rather than the difference. In fact, you don't even need two arms of an interferometer to detect gravitational waves, provided you can measure the roundtrip light travel time along a single arm accurately enough to detect small changes in its length.

Collins, simulating a physicist, replied:

It depends on the direction of the source. There will be no detectable signal if the source lies anywhere on the plane that passes through the centre station and bisects the angle of the two arms. Otherwise there will be a signal, maximised when the source lies along one or other of the two arms.

Out of the nine judges on the panel, seven considered the quality of the answers to their questions to be identical. Only two dared to identify the nonphysicist. Neither chose Collins.

'Apparently, in one of his responses the genuine physicist had drawn on ideas from a published paper. I hadn't come across that paper. I had to come up with my own answer. The two judges thought, "Only a genuine physicist could write something like this."'

Collins passed this version of the Turing test because, like his colourblind and blind subjects, he had gained enough 'interactional expertise'.

'Reading papers and books and newspapers alone

won't do. You have to spend time, lots of time, in conversation with people who know from experience what they are talking about.'

I agree with Collins. I tell him his theory matches my experience as a writer. In preparation for my novel *Mishenka,* the story of a Soviet chess grandmaster's intuitive search for meaning, I spoke in person and at length with several grandmasters. I visited the home of the former world champion Vladimir Kramnik, the feller of Garry Kasparov, and drew anecdote after anecdote out of him. In Paris I sat backstage amid the analysts and reporters at a tournament in which the world's strongest players were competing. All in order to put myself in my character's thoughts.

'Exactly. *You* can do that, whereas no one can think up a way computers might ever socialise. They haven't a body. Maybe they don't require a lot of body. Maybe a tongue and larynx, a pair of ears and eyes, the mechanical equivalents, would be sufficient. But then, how do we go about embedding the machine in a human speech community? Making it a participant in the circulation of meanings? The very notion seems to me to be a nonstarter.'

A nonstarter. Why then did Turing foresee fluent machines in the course of the twentieth century? (He wrote that machines would likely converse with ease by the year 2000.) Probably, with his head for data, he assumed that conversation could eventually be boiled down to a science. Many other intellectuals of the postwar period believed that the brain was a squishy computer, that human language was nothing more than a digital code. The metaphor, even its critics admit, has the benefit of being seductive. Clingy. Through the disappointment of failed predictions, its popularity has survived.

'I blame Chomsky,' Mark Bickhard says. Bickhard, a philosopher of language, is speaking to me from his home in Pennsylvania via Skype. He says,

Back in the fifties, his work was all the rage, and of course it still has clout. Essentially, Chomsky claims that you and I understand sentences because of their structure – the order of the words, rules of grammar, and so on. I have two things to say about that. One, yes, of course language is in part structural, but then so too are any number of skills. Fire-making, for example, has its own syntax: you have to perform all the different subtasks – collecting the tinder and kindling, striking a match, blowing onto the logs – in a particular order for the fire to burn. That doesn't make fire-making a language. Two, learning theory has come a long way since the fifties; we now know the huge role played by situational context, and the many intricate semantic relationships between words, in how we communicate.

Bickhard is seventy, with a bald dome and intense eyes. As befits a resident of Bethlehem (Pennsylvania), he has a prophet's long white whiskers. Behind his desk, on a side table, sit thick books atop even thicker books. He came to language by accident, he tells me. Forty years ago, after writing a dissertation on psychotherapy, he was told it contained too much maths and was asked to include a chapter on language. Bickhard spent 'a whole bunch of years' studying and rejecting every linguistic model then available. But his fascination with what makes language language remained.

Language, according to Bickhard, is dynamic. Like Rorschach blots, words need to be constantly interpreted, and always require us to do some filling in. 'You walk down some old wooden stairs, and one of the steps creaks. Instantly, you know what that creak means: "Gee, I'd better get off – it's about to break." Well, the same sort of thing happens all the time with words. A father who hears his little boy say something like "I buttoned the calculator" understands him perfectly, knows exactly how to respond to him, even if, strictly speaking, the words themselves are nonsense.

'Or, imagine you hear someone shout, "Roast beef at table three needs water." Nonsense, too. Unless, that is, you're sitting in a restaurant. Waiter talk.'

In any given situation, the meaning of a word, a phrase, unfolds dynamically. It cannot be second-guessed. 'You're in a restaurant. You take a menu and you order. "Roast beef," you say. The waiter returns a while later with your dish.' For Bickhard, there is always much more going on than meets the ear. 'What we have to ask ourselves is this: how does uttering "roast beef" change the social reality in which the speaker participates?'

Does uttering *roast beef* carry a higher social value than, say, *pork chop?* Does the waiter, who took you for a vegetarian, see you henceforth with different eyes? Does the utterer's friend at the same table conceal a grin as his memory sings 'This little piggy had roast beef, this little piggy had none'?

'Words transform the world around us. Learning a

language is learning how *roast beef* transforms a situation compared to *roast chicken* or indeed *I'm tired, the one over there,* or *See you around.*'

I'm listening carefully to Bickhard, my pen running fast with notes, when all of a sudden our connection gives out: the philosopher vanishes in mid-sentence. Minutes pass. Finally, he calls me back and the screen fills again with his white beard and navy-blue sweater and the side table pile of books.

Humans in conversation, he concludes, update and modify social reality from moment to moment. Meanings are broached, negotiated, tussled over. Big things are at stake. Computers, on the other hand, inert and indifferent, 'can't care less' about meaning. It is this can't-care-less-ness that will forever keep them imitating people's words.

I care about the philosopher's words. They can change me, and I let them. When I turn off my laptop it feels warm. I notice that. Not the warm of a friend's hug or handshake; only of electricity, I think. But without it, how much less of the world's meaning would our brains transform, convert?

ACKNOWLEDGMENTS

I am grateful to my first reader, Jérôme Tabet, and to our respective families – in particular my mother, Jennifer, and my siblings, Catherine Tabet, Nicole Thibault, Raymonde Tabet, and Patrick Tabet – for their abiding encouragement. Also to my friends Ian and Ana Williams, Oliver and Ash Jeffery, Sigriður Kristinsdóttir and Hallgrímur Helgi Helgason, Valgerður Benediktsdóttir and Grímur Björnsson, Laufey Bjarnadóttir and Torfi Magnússon, Linda Flah, Claire Bertrand and her family, Valérie Leclerc and Arnaud Salembier, Jérémie Giles, Helen and Rick Zipes, Aurélia Chapelain and Didier Delgado, Agnès and Nicolas Ciaravola, Emilie and Jérôme Jude, Sonia Velli, Caroline Ravel, Yoann Milin and Marianne Cruciani, Guy and Nadine Landais, Martin Johnson and Kristina, and Leandro Jofré for the many hours of stimulating and multilingual conversation.

The essay 'Talking Hands' would never have been possible without the hospitality and unstinting support of Margo Flah and Jean-Philippe Tabet. Immense thanks to Monica Elaine Campbell and Michel David.

For their precious contributions to this book, my gratitude also goes to Erin McKean; Les Murray and Margaret Connolly; Eszter Besenyei, Peter Weide,

Ulrich Lins, Renato Corsetti, and Ken Miner, W. H. Jansen; Ngũgĩ wa Thiong'o, Wangui wa Goro, Gichingiri Ndigirigi, and Evan Mwangi; Richard Ringler; Brian Stowell, Adrian Cain, and Paul Weatherall of Manx National Heritage; Sir Michael Edwards; David Bellos and Gilles Esposito-Farese; Erri De Luca and Andy Minch; Wayne A. Beach and Lana Rakow; and Naomi Susan Baron, Harry Collins, and Mark Bickhard.

Thank you to my English and American editors, Rowena Webb and Tracy Behar, and to their diligent teams, particularly Ian Straus, Pamela Marshall, and Kathryn Rogers – and also to Andrew Lownie, my agent.